W9-ADD-488

The Genetics of the Skeleton

The Genetics of the Skeleton
Animal Models of Skeletal Development

D. R. JOHNSON

Senior Lecturer in Anatomy,
Department of Anatomy,
University of Leeds

CLARENDON PRESS · OXFORD
1986

Oxford University Press, Walton Street, Oxford OX2 6DP

Oxford New York Toronto
Delhi Bombay Calcutta Madras Karachi
Kuala Lumpur Singapore Hong Kong Tokyo
Nairobi Dar es Salaam Cape Town
Melbourne Auckland

and associated companies in
Beirut Berlin Ibadan Nicosia

Oxford is a trade mark of Oxford University Press

Published in the United States
by Oxford University Press, New York

© *D. R. Johnson, 1986*

All rights reserved. No part of this publication may be reproduced,
stored in a retrieval system, or transmitted, in any form or by any means,
electronic, mechanical, photocopying, recording, or otherwise, without
the prior permission of Oxford University Press

British Library Cataloguing in Publication Data
Johnson, D.R.
The genetics of the skeleton: animal models of
skeletal development.
1. Skeleton   2. Mammals—Genetics
I. Title
591.1'852   QP301
ISBN 0–19–854163–5

Library of Congress Cataloging in Publication Data
Johnson, D.R. (David Roderick)
The genetics of the skeleton.
Includes bibliographies and index.
1. Bones—Diseases—Genetic aspects—Animal models.
2. Bones—Abnormalities—Genetic aspects—Animal models.
3. Bones—Growth—Animal models.   I. Title. [DNLM:
1. Bone Diseases—familial & genetic. 2. Cartilage
Diseases—familial & genetic.   3. Disease Models,
Animal.   WE 140 J66g]
RC930.4.J64   1986   616.7'1042   86–683
ISBN 0–19–854163–5

Set by Cambrian Typesetters, Frimley, Surrey
Printed in Great Britain by
Butler and Tanner Ltd,
Frome, Somerset

Preface

In 1963 Hans Grüneberg published *The pathology of development*; coincidentally I entered his Department as a postgraduate student in the same year.

Grüneberg's book integrated the genetics of skeletal disorders. Instead of merely listing genes and their effects he classified them into categories. His book was thus at the same time a catalogue of skeletal disorders and an attempt to arrange them in a philosophical framework.

Twenty-two years later *The pathology of development* is inevitably dated and long out of print. Many new genes have been described in the interim and much new work done on the genes Grüneberg included which has increased our knowledge and changed our thinking. Even more importantly the tools of the trade have changed. *The pathology of development* makes only passing reference to biochemistry and none at all to ultrastructure and electron microscopy. A combination of new techniques, and the routine use of cell and organ culture has made the subject very different today from what it was 20 years ago.

Concepts have also changed. Grüneberg's section headings are, in the main, useful and I have retained them. However, it becomes increasingly apparent that there is now no real distinction, for example, between membranous and cartilaginous skeleton—they both form parts of a single continuum. The new ideas on segmentation and the abandonment of the idea of resegmentation of the vertebral column have also necessarily changed our views on the action of some mutants on the axial skeleton just as developments in the fast-moving study of limb differentiation lead us to reinterpret others. In some areas there has been notable growth; to cover this I have added sections on the face and teeth.

Each section of this book is planned as follows: first, a brief description of the development of the area which is our immediate concern. Second, a description of relevant mutants, arranged as far as possible in a logical order or at least grouped by similarities—I cannot impose order where none exists—and lastly an overview of the area discussed and the possible relevance of the conditions described. I hope that each fairly self-contained chapter will thus provide a guide to the material available for study and an attempt to see what processes in normal development they affect. The centre of the stage is occupied by the mouse, but other laboratory animals are included where points of interest arise, or where an organism (like the chick) provides a better experimental system for current techniques.

The question of the relevance to man of the conditions described in these pages inevitably raises its head. In 1947 when Grüneberg wrote *Animal genetics and medicine* he was clearly of the opinion that there was much of clinical importance to be learned from animal models. By 1975 he was disenchanted to find that what he considered to be 'skeletal genes' were usually something else entirely, affecting the skeleton only in a secondary way. In general the use of animal models seems to be helpful only when a simple chemical defect is involved; this relationship is examined more fully in the concluding section of the book.

When the proposal for this book was being evaluated by OUP readers I was given two sorts of advice. One reader said expand the clinical part; another said omit it—the clinician will want pure science. In fact I have compromised and included clinical references where they seem pertinent, a decision which will no doubt bring howls of anger from clinicians and non-clinicians alike!

Leeds   ————————————   D. R. J.
December, 1985

REFERENCES

Grüneberg, H. (1947). *Animal genetics and medicine*. Hamish Hamilton, London.

—— (1963). *The pathology of development*. Blackwell, Oxford.

—— (1975). How do genes affect the skeleton? In *New approaches to the evaluation of abnormal embryonic development* (ed. D. Neuberg and H. J. Merker) pp. 354–9. Georg Thieme, Stuttgart.

Acknowledgements

I should like to thank all the publishers, editors of journals, and individuals too numerous to mention who have given me permission to reproduce copyright material, especially illustrations.

I am also grateful to many others: to the libraries and Medical Illustration services of the University of Leeds for their co-operation in assembling and photographing material, to Steven Paxton for making innumerable photographic prints, to Jean Lowe and Sarah White for help with office skills, and, last but not least, to my wife Ann who drew or redrew many of the line illustrations.

Contents

1. Models of disease processes

In 1947 Grüneberg published a book pointing out links between animal genetics and medicine which 35 years later are taken for granted; this itself is a testimony to the inherent rightness of the ideas expressed and the utility of animal models. The essence of Grüneberg's arguments still stand today, but we can update them. Animal modellers today are more at home with the electron microscope, the amino acid analyser, and the computer than with the microtome, the camera lucida, and the planimeter. Yet, essentially, the arguments stand.

Laboratory rodents (usually mice and sometimes rats) are, on the whole, remarkably similar to man as regards their genetic constitution, although their physiology may vary in important respects. Genes seem to be ultra-conservative and some combinations of genes whose effects are easily recognizable are found in the same chromosome and separated by approximately the same distance in most unlikely bedfellows whose family trees must have diverged a very long time ago (Searle 1968). The laboratory rodents accordingly may be supposed to display inherited diseases closely similar to those of man. This book concentrates on the skeleton for the very good reasons that the author is (a) most at home there; and (b) somewhat daunted by the massive task that would fall to anyone attempting to review the whole of the field of inherited disease.

Inherited diseases in man tend to present, as do all disease processes, as a set of symptoms. This means inevitably that some of the assemblages of symptoms which we see as a disease may be heterogeneous. This is a problem in man where genetical data are often scanty and limited to a few individuals. In mouse large numbers of individuals can be bred readily and fairly cheaply and heterogeneity noted; assemblies of symptoms having similar facies may be checked against each other and identity or dissimilarity proven, usually in the space of one mouse generation, a matter of weeks.

Environmental causes are often a factor in genetic disease: two patients suffering from the same condition may differ because of their genome (of which more later) or because of differences in social background, upbringing, diet, or housing. Our laboratory models can be caged, aged, and fed identically, and, if necessary, be raised in contact with an identical range of pathogens.

Many inherited diseases in man are, fortunately, rare and consequently thinly spread. A specialist in such a disease may therefore see only a few cases in a lifetime. Animal models are not rare in this sense; once a useful model has been obtained a large number of affected individuals may be rapidly amassed.

Human diseases are often not seen until they make the patient uncomfortable enough to present himself to his doctor. By this time the condition may be advanced, especially in some diseases which are troublesome only in middle or old age. In the laboratory the onset of the disease may be studied. Furthermore fetal material, often unobtainable in man, is available for study. In more than one instance (e.g. Hurler's syndrome) such a study has led to the development of a scanning technique allowing mothers at risk to be tested; if they are carrying a genetically abnormal fetus they can be warned in time to take such action as they may consider necessary.

Human pathology is limited in scope: it is, of course, based on the study of the cadaver post mortem and on the study of biopsy specimens and blood samples. Otherwise techniques must be largely non-invasive. In animals fresh material from any tissue or organ is easily obtained and may even be fixed intravenously, a method producing excellent preservation of such notoriously difficult areas as brain and pituitary gland.

With the advantages outlined above the animal model should be more widespread than it is. Why is it not more widely used? The difficulties in using animals for research can be classified into two groups, intrinsic and extrinsic. Extrinsic difficulties are more apparent than real. First, communication. Clinicians are busy people and often cannot read as widely as they would like. There is almost certainly a problem in the right people not knowing what is available. This book is an attempt to right that particular wrong. Second, material. Many clinicians do not know where to obtain the necessary stocks. This is an avoidable problem. Appendix 1 contains a list of some of the major genetic stockholders who will be pleased to deal with enquiries. Technology has now progressed to a stage where many mouse strains are stored as frozen embryos at the two- or four-cell stage of development and can be thawed out as required and implanted into a foster mother. Third, the maintenance of stocks requires a little skill; without this stocks can be lost and operators discouraged. This is a minor problem; laboratory technicians can easily be taught the necesary skills.

The intrinsic difficulties are more subtle. The problem is this: the skeleton is essentially a secondary structure. Skeletal genes often alter the skeleton *en passant* as a result of activity elsewhere. This is, of course, not unique to the skeleton—it is a property of any morphological 'system' whose shape and size may be affected by very distant rumblings. Formally we could argue thus. If a mutant gene acts directly on the skeleton then its normal counterpart would also act on the skeleton and the mutant might be considered to be capable of telling us something about skeletogenesis. If the action of the gene is at several removes from the skeleton, for instance involving the vertebral column secondarily in acting on the neural tube, then the gene is a 'neural tube gene' and tells us little about skeletogenesis. The fallacy of this argument should be obvious. *Sensu strictu* there are very few skeletal genes, perhaps just those specifying the make-up of the proteins and polysaccharides forming the skeleton and the enzymes specifying them and governing their metabolism. The characteristics of the skeleton are determined by interactions during development and are thus based upon secondary, tertiary, or quaternary effects.

According to some authorities this makes the whole business of skeletal genetics something of a lottery, a lottery furthermore in which 'people without tickets often win' (Grüneberg 1975). This rather pessimistic view, however, underrates the case for animal models. The genes which affect the skeleton indirectly do so in man and in mouse; the multiple nature of the conditions we study is seen in both systems: chondrodystrophy and osteopetrosis in man, for example, are not now regarded as single entities. They affect the skeleton indirectly, demonstrating that epigenetic information is essential for normal development and that a structure must be in the right place, and of the right size, at the right time in order to develop normally. Less than 2 per cent of our DNA comprises structural genes; the rest includes regulatory sequences which work in a way not yet understood. The first chemical morphogen, if such substances exist, has yet to be isolated. The indirect effects of skeletal genes may in fact be mirrors of imperfections of the regulatory genome which controls gene products.

Whether we subscribe to the 'spanner in the works' theory of Grüneberg (1975) who suggests that a non-skeletal gene tells us no more of the structure of the skeleton than does the study of broken glass about the structure of a window, or to the epigenetic theory, the ultimate result is a congenital abnormality, an integral part of pathology. It is in the study of developmental pathology that the immediate future of animal models lies.

GENETICS AND SKELETAL ABNORMALITY

This book is not directed at the geneticist. We assume that the reader is medically or biologically trained and is thus familiar with basic genetics. However, basic genetics is not necessarily a great help in discussing the mutants described in the following chapters. We shall be dealing with phenotypes as well as genotypes and therefore need to understand the relationship between genes and the environment. In practice this can be split into three categories dealing with the major gene, its modifiers, and the genetic background. Consideration of the genetic background leads to the concept of the inbred strain, and work on inbred strains leads us to consider the non-genetic variables which contribute to the phenotype.

The major skeletal gene

This is a unit of convenience. In the history of small animal genetics many genes have been isolated which (a) have discernible effects on the skeleton and (b) behave in a regular Mendelian manner producing relevant ratios and phenotypes which allow us to classify that particular gene as a Mendelian dominant, semidominant, or recessive. These are the 'major skeletal genes' and form a major part of the armoury of the mammalian geneticist. In some cases the effect of the gene may be so great as to be incompatible with life. If death occurs in the prenatal period, such a major gene is said to be a prenatal lethal. Many major skeletal genes have some deleterious effect on the lifespan of the individual which carries them.

Modifier genes

The next rank in the hierarchy of genes is the modifier gene. Suppose we chose the two mice in a litter segregating for a major tail-shortening gene which had the shortest tails and bred from them. At the same time we could also choose the two mutant mice with the longest tails and breed from them. If we continued this process for several generations we would expect to select a line of mice with long tails and one with shorter tails, although both carrying the same major tail gene. Such experiments have been performed (Dunn 1942) and the predicted results achieved. The difference in tail length between our two lines exists because we have collected together, by selection, *specific modifiers* of the gene. In one line they will be positive modifiers, enhancing the effect of the gene, and in the other negative modifiers, reducing its effect. They are *specific* modifiers because they affect only the tail length of mutant individuals; normal littermates are unaffected.

Genetic background

Specific modifiers are present in all mouse stocks; when a small number of them is present genetic experiments can be performed and their number estimated. In this case they can be catalogued and even mapped on chromosomes. But the number of genes present in each cell of our bodies is very large, and they are all potential specific modifiers. When a large number of modifiers is present and individual modifiers cannot be distinguished, we speak of their global effect as being due to the genetic background.

Inbred strains

To a certain extent we can minimize the effects of the genetic background by inbreeding. The mouse is tolerant of a large degree of inbreeding, so much so that at the time of writing many inbred strains (Staats 1981) have passed 100 generations of brother/sister mating and some have reached 175.

The logic of brother/sister mating is as follows: let us suppose that the genotype of a pair of parents is Aa. Their offspring will comprise three types of mice AA, Aa, aa in the ratio of 1:2:1. If we mate these at random, two lines (AA x AA, aa x aa) will produce only homozygous offspring when they are brother/sister mated and the other 14 possible combinations will produce a mixture of heterozygotes and homozygotes. The heterosis of the stock should decrease by 19 per cent per generation (the mathematics need not concern us here) so that after 20 generations we can be certain of almost complete homozygosity. Extensive tests on inbred strains of mice (Deol *et al.*, 1960) have shown that this prediction is born out in fact.

We have, of course, only considered one gene pair. The same selection procedure will also produce homozygosity for all other gene pairs, but we cannot guarantee which allele of a pair will be fixed. One inbred strain may end up AA, BB, cc, DD, ee, FF . . . . , a second aa, bb, CC, dd, EE, ff . . . , and a third AA, bb, CC, DD, ee, FF . . . . . We have eliminated genetic variation in each strain (if there is no mutation) but we still have considerable variation between strains. We might therefore expect the tail length of our mutant to be different in each strain, once we had introduced it and removed the heterogeneity it brought with it. This is usually so.

Variation within an inbred strain

Once we have removed the genetic variation from an inbred strain, we have the nearest possible equivalent to a clone of animals, of a population of identical twins. The differences between inbred strains are clearly due to the different genes fortuitously fixed between strains. But variation also exists within strains. Apart from differences between the sexes, which might be due to genes carried on the sex chromosomes, any such variation must be due to 'noise' in the developmental process or to other non-genetic factors. Maternal physiology, maternal age (Searle 1954*a*), and diet (Searle 1954*b*; Deol and Truslove 1957) are known to be important in this context.

Pleiotropism

When dealing with skeletal genes one is a long way from the primary effect. Skeletal genes are pleiotropic, or, to put it another way, skeletal genes produce syndromes. The origin of these syndromes is, of course, of interest to skeletal geneticists.

Grüneberg (1938) distinguished between genuine and spurious pleio-tropism. Genuine pleiotropism was held to be the result of multiple primary gene effects, spurious pleiotropism to be due to a single primary gene effect triggering off a series of 'symptoms' or developmental events in different systems of the developing embryo. Such an assemblage, known as a 'pedigree of causes', is a useful tool in sorting out the complicated systems of developmental action which are frequently encountered.

Genuine pleiotropism is now held to be unlikely; the doctrine of one gene, one enzyme came with the flowering of cell biology in the 1960s. In some cases apparently baffling syndromes have been shown to be the result of 'spurious pleiotropism' at the enzyme level. The achondroplastic rabbit (described in Chapter 3) has been shown to have an underlying defect in a mitochondrial enzyme. For reasons as yet unknown the level of the enzyme is critical in cartilage, producing chondrodystrophy but not in liver where the deficit was measured. The mottled mouse (Hunt and Johnson 1972; Hunt 1974) has a characteristic and puzzling syndrome which was eventually traced to a series of enzymes requiring copper as a cofactor; the primary defect here seems to be concerned with a defect of copper transport across the gut.

In other cases we may have to rethink our concept of the gene. Careful genetic linkage experiments and new techniques of chromosome staining have shown that the se gene, for instance, is in reality a small deletion. Now a small deletion, except under rigorous and expensive scrutiny, will behave exactly like a single gene. Lethal alleles at the albino locus and the T^{hp} mutation are also small deletions. It may well be that the pleiotropisms which puzzled Grüneberg, such as the grey coat and osteopetrosis of the grey lethal mouse, may in fact be due to genuine pleiotropisms, the 'gene' in this case being a short deletion comprising a length of DNA perhaps including a number of adjacent genes which have been removed from the genome. Genes are often shuffled in the history of a species; there is little evidence that genes with a similar function are adjacent in mammalian chromosomes. A random deletion may therefore give us a syndrome with two roots which it may be very difficult to reconcile.

REFERENCES

Deol, M. S., Grüneberg, H., Searle, A. G., and Truslove, G. M. (1960). How pure are our inbred strains of mice? *Genetical Research* **1**, 50–8.
—— and Truslove, G. M. (1957). Genetical studies on the skeleton of the mouse. XX. Maternal physiology and variation in the skeleton of C57BL mice. *Journal of Genetics* **55**, 288–312.
Dunn, L. C. (1942). Changes in the degree of dominance of factors affecting tail length in the house mouse. *American Naturalist* **76**, 552–69.
Grüneberg, H. (1938). An analysis of the 'pleiotropic' effects of a new lethal mutation in the rat. *Proceedings of the Royal Society* **B125**, 123–44
—— (1947). *Animal genetics and medicine.* Hamish Hamilton, London.
—— (1975). How do genes affect the skeleton? In *New approaches to the evaluation of abnormal embryonic development* (ed. D. Neuberg and H. J. Merker), pp. 354–9. Georg Thieme, Stuttgart.
Hunt, D. M. (1974). Primary defect in copper transport underlies mottled mutants in the mouse. *Nature* **249**, 852–4.
—— and Johnson, D. R. (1972). An inherited deficiency in noradrenaline biosynthesis in the brindled mouse. *Journal of Neurochemistry* **19**, 2811–19.

Searle, A. G. (1954*a*). Genetical studies on the skeleton of the mouse. IX. Causes of skeletal variation within pure lines. *Journal of Genetics* **52**, 68–102.

—— (1954*b*). Genetical studies on the skeleton of the mouse. XI. The influence of diet on variation within pure lines. *Journal of Genetics* **52**, 413–24.

—— (1968). *Comparative genetics of coat colour in mammals*. Logos Press, New York.

Staats, J. (1981). List of inbred strains. In *Genetic variants and strains of the laboratory mouse* (ed. M. C. Green), pp. 373–6. Fischer Verlag, Stuttgart.

2. The membranous skeleton

MESODERMAL CONDENSATIONS

Skeletal tissue can mean many things to many men. A broad definition of the skeleton would include the hard tissues, those which can be mineralized, as well as ligaments, tendons, fibrous and other connective tissues along with vascular, nervous, and haemopoietic components. We shall take a narrower view and confine ourselves primarily to the mineralized or mineralizable tissues.

Following Orvig (1967) we can list these as follows:

1. Bone—intramembranous and endochondral;
2. Cartilage—calcified or not;
3. Enamel;
4. Dentine;
5. An inevitable rag-bag, an insoluble residue of intermediate forms.

Moss (1964) defined the basic cell type which produces these tissues as the *scleroblast*, implying that the tissues are closely related. Such a classification, however, cuts across other preconceived ideas: the scleroblast may be mesodermal, ectodermal, or of neural crest origin (osteoblast—mesodermal/ectomesenchymal, makes bone; chondroblast—mesodermal/ectomesenchymal/ectodermal, makes cartilage; odontoblast—mesodermal, makes dentine; ameloblast—ectodermal, makes enamel).

The question of which of these tissues, or more accurately whether bone or cartilage, is primitive has been debated. Cartilage appears first in the mammalian embryo, cartilage precedes bone in endochondral ossification, and the cartilaginous fishes (Chondrichthyes) were considered primitive so cartilage was assumed to have evolved before bone. In fact this assumption is not verified by the fossil record: enamel, dentine, calcified cartilage, and membrane bone are all of equal antiquity, appearing in the earliest Ordovician vertebrates. Endochondral bone is first seen in the middle Ordovician Osteostrachi, then in the Silurian acanthodians and some placoderms, and is prominent in those groups (rhipidistian sarcopterygians) which gave rise to the land vertebrates and in the actinopterygian (bony) fishes. Bone is seen as serving a supporting role as it does today and as an outer defensive shield (Berrill 1955; Schaeffer 1961; Denison 1963; Romer 1963). The dermal (membranous) skeleton may also have served as a reservoir of, and as a permeability barrier to conserve calcium and phosphorus. The ultimobranchial gland (later the parathyroid) helped in this function (Berrill 1955; McLean and Urist 1968; Smith 1961; Tarlo 1964; Urist 1962, 1963, 1964; although Moss 1963; Denison 1963 disagree). Cartilage also served as mechanical support, and allowed rapid growth especially in embryonic and larval forms (Romer 1942, 1963, 1964; Berrill 1955; Denison 1963).

Initiation of skeletogenesis

The first sign of imminent skeletogenesis is the appearance of mesenchymal condensations. These arise in areas where a bone is to form by intramembranous ossification and where a cartilage is to appear (whether or not it is later to be ossified); they are not unique to the skeletal system, being also found as the precursors of muscle masses. The position of a condensation defines the position of the skeletal element it represents, and the shape of the condensation defines the basic, although not the detailed shape of the element. The cells which form the condensation may be of local origin, as in the limb buds or the skull (Hall 1971) or may arise at a distant point and migrate to the place where the skeletal element is destined to appear (mesodermal cells migrate into the lower jaw of the chick (Jacobson and Fell 1941); ectomesenchymal cells migrate from the neural crest to the skull, mandible, and pharynx (Johnson and Listgaren 1972); muscle cells migrate from the somites to the limb buds (Christ *et al.* 1977). Clearly we must ask some questions about condensations: what is the stimulus for condensations to appear, why are some cells affected specifically, and what controls the condensation process?

The stimulus

Holtfreter (1968) showed that the cartilage of the chick head was formed by condensations which had interacted with pharyngeal ectoderm. Since this fundamental finding the refinements of the system have been discovered, largely by work on the normal chick. Consider the region round the chick eye. The eye is surrounded by scleral cartilages, scleral ossicles (neither seen as such in mammals), and membrane bone. All these skeletal structures are derived from the neural crest ectomesenchyme. During their migration the environment of these cells (from which the stimuli for condensation and differentiation may come) includes epithelial cells, mesenchyme derived from mesoderm, and extracellular spaces containing cellular products such as chondroitin sulphate, collagen, fibronectin, hyaluronic acid, and many more. Interaction with some or all of these elements will determine the where and when of condensation and possibly the type of element to be formed.

We now know some of the parameters governing these changes. Cartilage is formed only if the cells come in contact with epithelial cell products during migration (Bee and Thorogood 1980). The bones of the lower jaw form after contact with epithelia of mandibular arch after migration (Tyler and Hall 1977). Scleral cartilages are formed only after neural crest derived cells have interacted, after migration, with extracellular products formed by pigmented retinal epithelia (Reinbold 1968; Stewart and McCallion 1975; Newsome 1976). Scleral ossicles can only be formed if the cells react with scleral papillae within the scleral epithelium, each papilla forming the basis of a single ossicle (Coulombre *et al.* 1962; Palmoski and Goetinck 1970; Johnson 1973).

What is the specificity of this message? The epithelia concerned may be in sheets in the mandible or highly organized in the papillae (Fyfe and Hall 1981); heterotypic ectoderm will stimulate the formation of mandibular membrane bone (Hall 1978*a*, 1981) although the age of the epithelium is important. Such heterotrophic interactions allow us to experiment with the age and type of the ectoderm, the age at which the ectomesenchyme becomes responsive to the ectodermal signal, and whether a particular piece of mesenchyme will respond in a particular way to a particular signal, i.e. is the 'bone' signal a specific signal or a 'condense' signal which is interpreted as 'form cartilage', 'form bone' by different groups of ectomesenchymal cells? Hall (1981) looked at the induction of the scleral ossicles in the chick with the following points in mind: (1) the timing of the induction; (2) whether scleral ectomesenchyme would respond to mandibular epithelia by forming bone and vice versa; (3) whether the specificity of the scleral ossicles lies in the ectomesenchyme or the epithelium.

He isolated the relevant tissues from developing chicks, combined them appropriately and either grafted them on to the chick chorio-allantoic membrane or grew them in organ culture on millipore filters. His results may be summarized as follows. The induction of scleral ossicles was time dependent. Ectomesenchyme from stage 36 (Hamburger and Hamilton 1951), 10-day chicks could form bone in the absence of scleral papillae; earlier stages could not. Separation and recombination of the elements before stage 35 inhibited bone formation and heterochronic recombinations uniting 'young' epithelia and 'old' ectomesenchyme and vice versa failed to induce bone. This is in contrast to the interaction between the epithelia and ectomesenchyme of the mandible (Hall 1978*a,b*) where separation and recombination is followed by bone formation. Scleral ectomesenchyme, however, responded to foreign mandibular epithelia at all stages by forming scleral ossicles. Mandibular ectomesenchyme was also acted upon by scleral ectoderm. The interaction is thus permissive (Wessels 1977; Saxen 1977) providing a suitable environment for the initiation of differentiation but not specifying it. The osteogenic ability of scleral papillae therefore probably does not depend on their specialized morphology. Collagen has been seen beneath scleral papillae when induction is in progress (Van de Kamp 1968) and has been implicated as active agent in mandibular epithelium (Bradamante and Hall 1980) and may be a common factor in permissive osteogenesis.

The morphology of the bone in Hall's transplants was always characteristic of the ectomesenchyme rather than the inducing ectoderm. The morphology and histology of scleral and mandibular bone is distinctive yet they are both derived, as is cartilage, from similar ectomesenchyme. Two alternatives are possible: either clones of neural crest cells are specific for certain tissues or the cells are conditioned in some way by their environment. Jacobson and Fell (1941) reported that cells destined to become cartilage arose in separate mesenchymal centres outside the mandible (each induced by, or associated with, a transitory epidermal thickening), then migrated into the mandible where they differentiated. Again are these parcels of cells influenced by the ectoderm, or something else, or are they qualitatively dissimilar and do they migrate to different rallying points?

Hall (1982*a*) found that within the mandibular arch of stage-22 chick embryos the cephalic half contained more cartilage-forming ectomesenchyme than the caudal half. The caudal and proximal parts contained rather more osteogenic tissue. The most active osteogenic centre corresponded to one which Jacobson and Fell noted as forming the angular, splenial, and surangular bones. This throws little light on the clonal versus environmental determination of cell type, but indicates that the division has already occurred by stage 22. Mesenchymal control of morphogenesis seems to be the norm, and occurs elsewhere, for example, in the limb (Hinchcliffe and Johnson 1980) and in the teeth (Kollar and Baird 1969). Goedbloed (1964) showed that the development of the mouth cavity, the middle ear cavity, and the external auditory meatus also involve ectodermal–mesenchymal interactions.

In the limb bud, mesenchyme will eventually come to belong to one of two basic types, broadly fibroblastic or chondroblastic. Are these lines mutually exclusive or is fate determined by position? Small blocks of limb mesenchyme can be transplanted or grown in culture (Searls 1973). In the chick wing, condensations can be seen at stage 23, and, at stage 25, cartilage matrix can be distinguished by staining. Blocks of labelled chondrogenic mesoderm from stage-24 wing buds regulate according to site up to donor stage 24, but if grafted into older limb buds will form ectopic cartilage. Since there is no significant migration into or out of such blocks of tissue, Searls (1967) suggested that he was not dealing with an initially random mix of pre-cartilage and pre-fibroblasts.

Newman (1977) concluded that the difference between fibroblast and chondroblast seemed to depend on cell density. He disaggregated stage-25 mesoderm and grew the cells at less than confluent densities (i.e. so that the cells were not in contact with each other): they became fibroblasts. The same tissue, cultured as a single mass of tissue, produced cartilage.

The mechanism

Once the message (if there is a message) has been received by competent cells how do they change their behaviour so as to form a condensation? Embryonic mesenchyme is a rather diffuse spongy tissue with large intercellular spaces and it seems that condensations arise by the local abolition of this space rather than by mitosis. An increased rate of cell division has been reported in some blastemata (Hale 1956; Jacobson and Fell 1941) but autoradiographic experiments on chick limb buds (Janners and Searls 1970; Thorogood 1972) have failed to demonstrate it; in fact the cell division rate in the blastema tends to fall with time as more specialized cells which divide less frequently are formed (Abbott and Holtzer 1966; Flickinger 1974).

Prechondrogenic condensations of 'closely packed cells in which the nuclei are almost in contact with each other' (Grüneberg 1963) have been described on many occasions in tissue prepared for light microscopy (Zwilling 1961; Montagna 1945; Ham 1969). In electron micrographs some studies (Gould *et al.* 1972*a*,*b*; Searls *et al.* 1972) have failed to demonstrate them, although others (Godman and Porter 1960; Schmidt 1968; Minor 1973; Thorogood and Hinchliffe 1975) have described their appearance under the electron microscope. Thorogood and Hinchliffe (1975) made a quantitative estimation of cell packing (nuclei/unit area) and found an increase of 60 per cent in regions where blastemata became visible in light microscope preparations. Uncondensed areas, studied by both TEM and SEM showed a loosely constructed cellular meshwork with large intercellular spaces and with an occasional bunch of collagen fibres. Cells were stellate with many filopodia and joined by tight junctions or zona occludens. The earliest condensations were similar, but with a decrease in the size of the intercellular spaces. The cells here tended to have similar filopodia. The next stage involved the appearance of matrix components, banded collagen fibres, and granular or amorphous mucopolysaccharides. Characteristic whorling of blastemal cells was reported in amphibia (Anikin 1929) and, to a lesser extent in the chick (Ede 1976) when the blastema is cut in cross-section.

What properties of these mesenchymal cells cause them to behave in this way? If we mix together a relatively small number of cells (say 1 million) of different types in the laboratory and place them in a suitable medium, we find that they aggregate together, first in twos and threes, then in larger numbers. Eventually (Moscona 1960), they aggregate into regions containing only one cell type, or different cell types arranged in a particular way. What cells recognize as 'like' is an open question. In a mixture of chondrogenic and kidney cells from mouse and chick, aggregates consisted of just chondrogenic cells or just kidney cells, albeit of both species. In other experiments aggregation was according to species. The aggregation pattern also varies with time. Curtis (1961) found that *Xenopus* cells disaggregated in midblastula normally reaggregate with the ectoderm outside, mesoderm in the centre, and endoderm inside. If the ectoderm was allowed to reaggregate for four hours before the other tissues were added, the resulting mass had a mesodermal core and a coating made up of ectoderm and endoderm. After six hours reaggregation the endoderm was outside and the mesoderm inside with the ectoderm sandwiched between them.

Cell adhesion seems to be specific. We do not know if 'like' cells have a fit analogous to an antigen–antibody bond or if stickiness is just differential. Either mechanism would lead to a sorting of cell types from a mixture. Embryonic cells that stay together tend to reinforce the initial bond by interdigitation of membranes or the formation of desmosomes.

Increasing the adhesiveness of a group of cells would lead to progressively larger areas of cell membrane being in contact with each other and the cells being drawn closer together; this in fact is a formal description of the process of condensation. In *Dictyostelium*, a slime mould, aggregating cells secrete cyclic AMP and form spiral aggregates similar to those seen in transverse sections of blastemata (Durston 1973). Because of the similarity of the pattern generated it has been suggested that a similar state of affairs, in three dimensions, exists within a blastemal condensation. Ede *et al.* (1977) showed that re-aggregating chick cells grown in culture formed chondrogenic foci and that cells around these foci moved towards them in a zigzag pattern. Cells nearer the centre of the condensation moved at the same speed but in a more nearly circular path. Toole (1972) considers that condensation involves cell immobilization (a cell moving in a circle is effectively immobilized) following removal of hyaluronidate, whose presence is associated with cell mobility.

THE MUTANTS

The talpid (ta) chick*

We may well expect a mutant affecting condensation to have a widespread effect on cartilage, cartilage replacement bone, and muscle. The talpid series of mutations in the chick (talpid[1] Cole 1942; talpid[2] Abbott, *et al.* 1959; talpid[3] Hunton 1960) are so called because of their short, spade-like wings which resemble somewhat the forelimbs of a mole (Fig. 2.1). All are recessive lethals, none has been tested for allelism, and Cole's talpid[1] is now extinct. Abnormalities typical of the group were described in talpid[3]

---

* Mutants are of the mouse unless otherwise stated. The symbol (if assigned) follows in brackets with the chromosome number if known. Older literature may assign mutants to a linkage group: these were arbitrary and are now allocated to chromosomes on the basis of length. See Green (1966) for details.

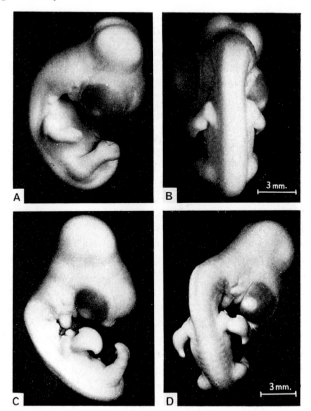

Fig. 2.1. Normal (A,B) and talpid[3] (C,D) chicks aged 6 days. (From Ede and Kelly 1964*b*.)

by Ede and Kelly (1964*a,b*). In the head region abnormalities stem from the fusion of facial rudiments in the midline due to the formation of a single bow-shaped mesenchymal condensation which represents the fused maxillary parts of the first visceral arches; there is also retardation of the development of membrane bone and non-replacement of cartilage by bone. The authors suggested that the basic defect here is the failure of the prechondral mesoderm to split into prechondral plate and lateral strips forming visceral arches and that the failure occurs in the processes by which the mesenchymal condensations of the maxillary rudiments are produced.

  In the trunk and limbs (Ede and Kelly 1964*b*) the vertebral column is shortened and the limbs extremely so, with many digits. The myotomes do not show strict longitudinal arrangement of developing muscle cells, but exhibit a tendency to random orientation. The cells of the outer boundary of the myotomes, which is not regularly curved, tend to mingle with those of the dermal mesenchyme which covers them. Myotome and sclerotome cells are disoriented and mixed and the myocoele enlarged rather than (normally) obliterated. The vertebrae are displaced by an abnormal neural tube and consequently much malformed and fused. In the wing the humerus is small and short, the radius and ulna fused, and metacarpals and carpals represented by a single broad element extending across the whole width of the limb bud. In the hindlimb the metatarsals form a similar band. Beneath the apical ectodermal ridge is a broad band of mesenchyme in which an excessive number of digits form.

In the trunk and limbs, as in the head, the basic defect seems to be one of failure of mesenchyme to form normal condensations; the condensations which are formed are aberrant in number, position, and size and the structures which develop from them are consequently abnormal.

The behaviour of ta cells was studied by Ede and Agerback (1968). They argued that condensation *in vivo* was comparable to reaggregation of dissociated cells *in vitro*. Do talpid cells show abnormal reaggregation behaviour? Ede and Agerback dissociated wing bud mesoderm cells from talpid and normal embryos, suspended them in nutrient solution, and allowed them to reaggregate, whilst being shaken gently, for up to three days. Aggregates were photographed at intervals in long-term culture; in short-term culture (up to 5 hours) samples were removed and the single cells not aggregated counted in a haemocytometer. They found that, although cell numbers in talpid and normal wing buds were equivalent at four days of development, by 5 days the talpid aggregates contained 50 per cent more cells than the normals, although there was no evidence to show more rapid multiplication of talpid cells in culture (Fig. 2.2). However, Ede *et al.* (1975) showed that the mitotic gradient was abnormal in stage-24 talpid embryos. Over their three-day culture period aggregates were formed —a large number of small ones at first, then progressively smaller numbers of larger aggregates. The talpid aggregates were smaller and less numerous throughout. Talpid aggregates were less irregular in shape than normals and more nearly spherical. Ede and Agerback concluded that talpid cells were more adhesive than those of normal chicks and pointed out that adhesiveness and motility were often inversely related.

Bell and Ede (1978) were able to demonstrate the effect of talpid in early limb buds. Small explants of mesoderm from proximal and distal limb buds from chicks at stages 19–21 were grown in single-drop cultures and filmed by a time-lapse technique (Fig. 2.3). The distance between successive positions of the nuclei on each frame of film was recorded, giving a measure of cell mobility. Limb buds were also looked at with a scanning electron microscope (Fig. 2.4) and plastic semithin sections used to measure cell parameters.

They found that even in the earliest stages of limb bud development the talpid cells are abnormal. In the normal limb bud at this stage distal cells

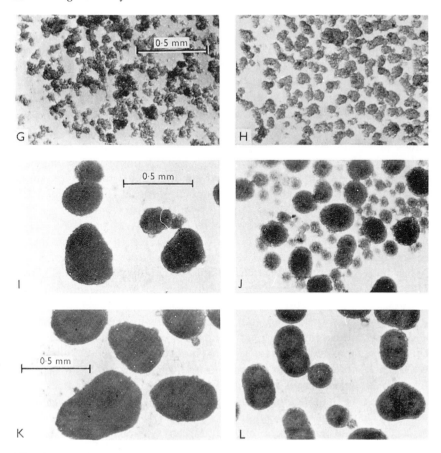

Fig. 2.2. Aggregates of dissociated wing bud mesenchyme cells from 5-day normal and talpid[3] embryos. (G–I) 5-day normal aggregates; (J–L) talpid[3] aggregates at 7, 25, and 50 hours of culture. (From Ede and Agerback 1968.)

are stellate and fairly immobile whilst proximal cells are more polarized and more motile (Cairns 1975, 1977). In talpid, proximal and distal cells did not differ in this way, either in terms of shape or mobility. Talpid cells all resembled distal cells, being stellate and sluggish, but with rather closer contacts (cf. Ede and Flint 1975*a*). Talpid distal mesoderm at this stage had less intercellular space and many short spiky microvillae were present.

The two cell types within the normal limb bud may represent somitic cells destined to form muscular blastemata (Christ *et al.* 1977; Chevallier *et al.* 1977) and the cells which form the other limb tissues, including the skeleton: perhaps talpid blocks the ingress of somitic cells. Alternatively, the distal cells might represent Wolpert's progress zone (Summerbell *et al.*

1973 and see Chapter 8) and a transformation from proximal to distal morphology might be blocked by the mutant. A similar morphological transition is seen in the somites where unsegmented somitic mesoderm cells resemble distal mesoderm but those from segmented mesoderm are more polarized (Bellairs and Portch 1977).

It is interesting to note that Cairns (1977) reported that in talpid[1], a rather milder allele (Abbott 1967), the wing bud is recognizably wider than normal from stage 18; but that cell proliferation was almost normal at stages 21–22, the talpid tending to divide a little more slowly. In talpid[2], distal limb cells adhered at the same rate as normal cells to the glass surface of the culture vessel and proximal cells at the same rate as normal proximal

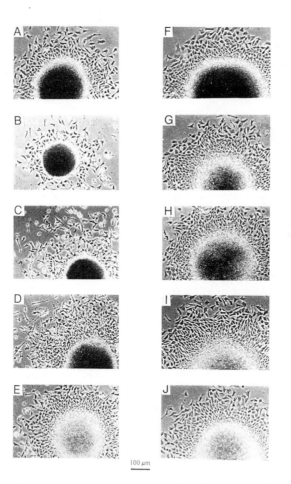

Fig. 2.3. Mesodermal explants from normal (A–E) and talpid[3] (F–J) stage-20 wing buds. (A,F) proximal; all others various distal mesodermal regions. (From Bell and Ede 1978.)

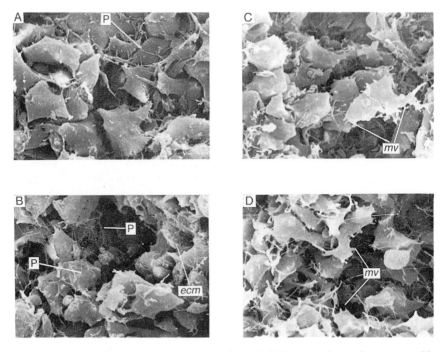

Fig. 2.4. Scanning electron micrographs of limb mesoderm from stage-20 embryos. (A) normal distal mesoderm; (C) talpid³ distal mesoderm; (B) normal proximal mesoderm; (D) talpid³ proximal mesoderm. LP: extracellular process; ecm: extracellular material; mv: microvilli. (From Bell and Ede 1978.)

cells. This is in agreement with the findings of Niederman and Armstrong (1972) who found no difference in adhesiveness between talpid² and normal cells. It may be that the change in adhesion is less in talpid² and was too small to be measured by these investigators.

Since talpid² and talpid³ both produce a fan-shaped limb bud, the suggestion (Ede 1971) that the fan shape is a property of increased cell adhesion is clearly doubtful after the findings of Niederman and Armstrong (1972) and Cairns (1977). But what of the effect on blastema formation?

Ede *et al.* (1974) looked at the light, TEM, and SEM appearance of talpid cells (Fig. 2.5). They described talpid mesoderm cells as having less filopodia than normal giving a characteristic 'empty' appearance to the intercellular spaces: this is especially marked in the case of filopodia less than 0.5 μm in diameter. The talpid filopodia also appeared in bunches at the end of broader cytoplasmic outpushings, reminding the authors of sea anenomes with tentacles. Talpid cells consequently appeared more ragged than normal ones. Ede and Flint (1972) noted that the condensations found in talpid mesoderm in culture were smaller than normal (about half size) but more numerous. Ede and Flint (1975a) used a more rigorous method to confirm the findings of Ede and Agerback (1968). They showed that early talpid cell aggregates were compact, rounded structures and the later larger irregular aggregates were composed of tightly bound clusters of smaller, rounded cells. In the two-cell aggregate, normal cells formed hemispheres meeting in a flat surface, and often less than this was in contact. In talpid, one cell tends to wrap around another making

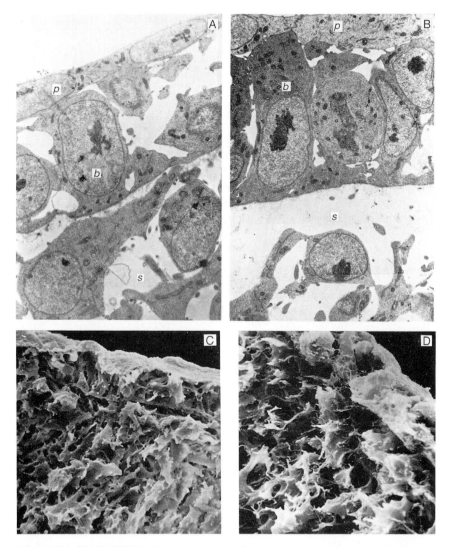

Fig. 2.5. (A,B) TEM views of the ectoderm and underlying mesoderm just proximal to the AER of normal and talipd[3] embryos respectively. (C,D) SEM views of similar areas. b: basal cell; p: periderm; s: intercellular space. (From Ede *et al.* 1974.)

extensive contacts (Fig. 2.6). This tendency persists in larger aggregates. In time-lapse cinematographic studies, the same authors (Ede and Flint 1975*b*) found that the distance moved by cells and the time spent at rest before moving were both significantly greater for normal than for talpid cells cultured in plastic petri dishes. Talpid cells in this situation were more flattened than normal, with extensive ruffled cell membranes and short spiky microvillae all round the cell periphery.

The picture in talpid is clearly very complex indeed, but equally clearly we can equate abnormal blastema formation with changes in cell surface adhesion. Are similar changes to be found in other mutants affecting skeletal blastemata?

Brachypodism (bp, Chr 2)

Brachypodism (Landauer 1952) and an apparent remutation brachypod[H] (Green 1966) affect only the limbs of the mouse: metacarpals and metatarsals are reduced in length, each digit lacks a phalanx, and those phalanges which are present are shortened. Carpals and tarsals are irregular and small extra ones are found. All the long bones of the limb are reduced, the effect being greater in the hind- than the forelimbs. Milaire (1965) noted anomalous precartilaginous blastemata in brachypod and

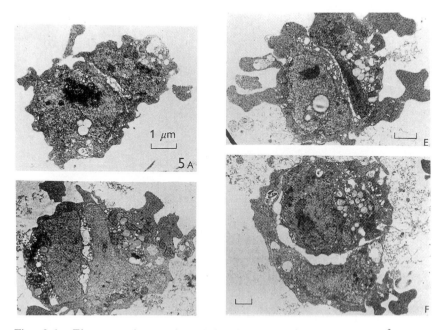

Fig. 2.6. Electron micrographs of 2-cell normal (A,B) and talpid[3] (E,F) aggregates. (From Ede and Flint 1975*a*.)

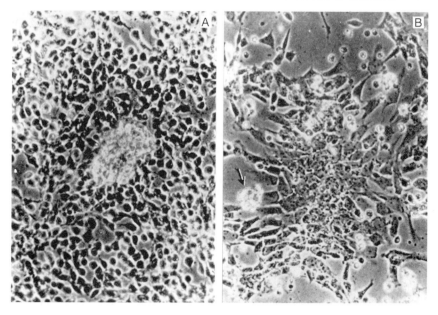

Fig. 2.7. Phase contrast micrographs of (A) normal 12-day mouse hindlimb bud mesenchyme cells and (B) corresponding cells from a brachypod mouse. (From Elmer and Selleck 1975.)

Grüneberg and Lee (1973) described the condition in some detail. The limbs can be seen to be abnormal by 13.5 days and a close inspection reveals abnormalities of the blastemata from 12 days. In the forelimb the brachypod blastemata are slimmer, particularly distally; the bulge which in the normal corresponds to the metacarpo-pharyngeal joint is absent and the blastema of digit II is interrupted basally due to the curving of the blastema. In the hindlimb, brachypod blastemata are also slimmer than normal. The later consequences of these blastemal abnormalities will be dealt with in the section on limb development (Chapter 8).

Elmer and Selleck (1975) dissociated cells from 12-day-old brachypod mice and normal controls (since bp cannot be distinguished from normals at this stage without histological preparations these could not be littermates: in fact they were produced by mating mice from the bp strain known to carry +/+ at the bp locus). Dissociated normal and bp/bp cells behaved very differently. Normal cells, over the first 6 hours, remained rounded, some developing long tapered cytoplasmic processes, many becoming tear-shaped. As incubation proceeded they moved over the surface to aggregate into compact multicellular clumps. Brachypod cells attaching to the culture flask quickly became large flattened cells with conspicuous pseudopodia; many cells did not settle but remained floating in the medium. Settled cells moved towards centres of aggregation where they formed loose aggregates (Fig. 2.7). In many cases monolayers of epithelioid cells were seen, containing cells in mitosis. This abnormal behaviour may be mediated by a growth inhibitor. Ginter and Konyukhov (1966) reported that extracts from 13-day-old bp/bp embryos significantly reduced the growth of 13-day-old normal limb elements in organ culture. Extracts from heterozygotes (+/bp) suppressed growth by approximately half as much (Bugrilova and Konyukhov 1971). A protein of molecular weight 76 000 was isolated (Pleskova *et al*. 1974) to which they ascribed the effects of the gene.

Duke and Elmer (1977) looked at brachypod cells in rotation culture. They found that mesenchymal cells from brachypod hindlimbs, although indistinguishable at the light microscope level, aggregated faster with a decrease in the number of single cells left in solution at any given time. The shape, size, and number of cell aggregates all appeared normal. Brachypod cells were not more tightly packed in these aggregates; histologically no differences in cell-packing density could be seen and the number of cells lysed enzymatically from the aggregates was not significantly different from normal. Brachypod and normal cells were of the same size (Hewitt and Elmer 1976) and no increase in cell lysis was seen.

These results seem to indicate that brachypod cells are more sticky than normal with either more sites for adhesion, or the sites present more favourably arranged. Brachypod and normal cells attach to the surface of culture flasks at the same rate and bind ^{125}I in the presence of lacto-peroxidase in a manner indistinguishable from normal (Hewitt and Elmer 1978*a*), indicating that the number of cell-surface proteins is similar.

In another series of experiments (Duke and Elmer 1978), cell adhesiveness was assessed by studying the process of fusion of fragments of normal and brachypod limb bud mesenchyme. Blocks of tissue from the postaxial regions of normal and bp hind limb buds aged 12 days were placed in contact in hanging drop culture for a period of up to three days. Combinations of tissue tested were +/+:+/+, +/+:bp/bp, and bp/bp:bp/bp. One half of the tissues of each genotype was labelled with methyl-^{14}C-thymidine: any combination, of course, being of a labelled and an unlabelled tissue. Serial autoradiographs were then prepared to see if the junctional surface was flat or curved, implying different adhesiveness. Brachypod rudiments partially enveloped normal fragments, forming a convex bonding (Fig. 2.8), i.e. brachypod cells were more adhesive.

Hewitt and Elmer (1976, 1978*a*) looked at the affinity of normal and brachypod limb bud cells for the plant lectin concanavalin A, since lectin-induced agglutinability had been shown to change with the stage of differentiation of various cell types. They suggested that an aberrant phenotype such as bp/bp might show a changed pattern of carbohydrate components on the cell surface. Both normal and abnormal cells varied in

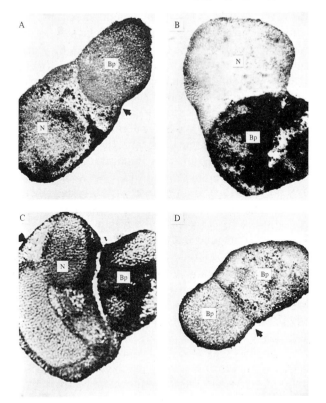

Fig. 2.8. Autoradiographs of normal labelled (N) and brachypod unlabelled (Bp) fragments fused for (A) 24 hours; (B) 30 h; (C) 72 h. (D) shows labelled and unlabelled mutant fragments fused as a control. (From Duke and Elmer 1977.)

agglutination according to age, but in a different temporal sequence. At 11 days agglutination was similar in normal and brachypod cells; at 12 days it was markedly reduced in normals, but persisted at the 11-day level in bp/bp. At 13 days the bp agglutination had dropped in a similar way to that seen in 12-day normals. Trypsinization of the 12-day-old cells abolishes the difference, suggesting that possible adhesive sites are masked rather than missing in normals.

The normal decrease in lectin-induced agglutination with age (i.e. decreased stickiness) may well be due to altered mobility of the binding sites (Martinozzi and Moscona 1975; Paulsen and Finch 1977). Hewitt and Elmer (1978a) demonstrated a strong correlation between both Con A and wheat-germ agglutinin (WGA) and the number of cells exhibiting lectin-induced redistribution of binding sites. Hewitt and Elmer (1978b) investigated the possible role of microtubules and microfilaments in the attachment of Con A and WGA attachment sites. Interference with microtubules by treating cells with 5 μg ml^{-1} colchicine did not affect the distribution of either Con A or WGA binding sites (in the normal, with or without colchicine 60 per cent of cells clustered, in brachypod around 85 per cent). This seems to suggest that microtubules were not involved in the brachypod abnormality. Cytochalasin B showed different effects on Con A and WGA binding sites. Incubation with WGA after Cytochalasin B treatment showed little effect in the redistribution of binding sites or on the number of cells showing such a redistribution. In contrast the redistribution of Con A binding sites was significantly impaired. As a result only 15 per cent of +/+ and bp/bp cells showed clustering, a reduction from 62 per cent (+) and 92 per cent (bp). Eighty per cent of trypsinized cells normally have clustered binding sites: this was reduced to an equal extent (19 per cent +, 16 per cent bp) by cytochalasin B treatment.

Hewitt and Elmer concluded that the differences in clustering between bp and normal cells are not related to the attachment of binding sites *per se* but due to the restraint of binding-site mobility by another mechanism. Microfilaments are seen to be associated with Con A binding sites but not with those associated with WGA. The authors suggest that preliminary work indicates the presence on normal, but not on bp cell surfaces, of a trypsin-sensitive glycopeptide.

Owens and Solursh (1982) used micromass culture of brachypod limb bud mesenchyme from 10–10.5 days to 13 days. Under these conditions where a large number of cells (1 x 10^5) is inoculated into a small amount of medium (5 μl), aggregations occur within 24 hours. These aggregations later form cartilaginous nodules. Owens and Solursh found that 10–10.5-day bp limb bud cultures were most like those produced by normal littermates and that after 72 hours of culture the number of nodules formed was normal and the mean nodule size not significantly different from normal. In older explants the bp cultures become progressively more abnormal, with fewer nodules or aggregates per unit area. This suggests that the initiation of aggregate formation *in vitro* is not affected by the gene and is in fact normal at early stages but affected in older explants. The explants formed also produced a lower proportion of cartilaginous nodules.

The addition of small quantities of other cells (1:10), however, improved aggregation. Stage 10–10.5 day wild-type cells had an effect at a dilution of 1:20.

It seems likely that the effect of bp on cartilage formation is secondary to a defect of condensation: we shall return to this in Chapter 3. A primary defect in chondrogenesis would not account for earlier defects seen in blastemal cells.

Duke and Elmer (1979) looked at the ultrastructure of the cells in brachypod and normal 12–13-day-old limb buds. They found that in the tibial condensation at 12 days, although grossly normal in appearance, cells were separated by 1.5–4.0 μm as opposed to 2.5–6.0 μm in the normal. In the fibula normal-looking cells were again separated by 2.5–5.0 μm and brachypod cells were once more appreciably closer. This close spacing resembles that seen by Thorogood and Hinchliffe (1975) in the early stages of the normal condensation process. Cell processes and filopodia were reduced with cell contacts occurring over a broad area in bp: there were 8–10 cellular contacts between brachypod fibular cells compared to 4–6 in normal littermates. Cell size and shape were also subtly different, brachypod cells being more rounded. Cells not taken from condensations appeared normal as far as intercellular features are concened.

Amputated (am, Chr 8)

The amputated mouse (Flint 1977, 1980; Flint and Ede 1978*a,b*) has effects on somite number and facial development and causes cleft palate. These aspects will be dealt with in more detail in subsequent chapters, but cell culture studies (Flint and Ede 1982) show that amputated resembles brachypodism and talpid in affecting cell adhesion. Flint and Ede cultured somite tissue from 9.5-day-old amputated and normal embryos and studied the results using a variety of methods including time-lapse cinematography, light microscopy, and scanning electron microscopy (Fig. 2.9). Amputated cells in culture dispersed from the cultured explant as a series of small cell

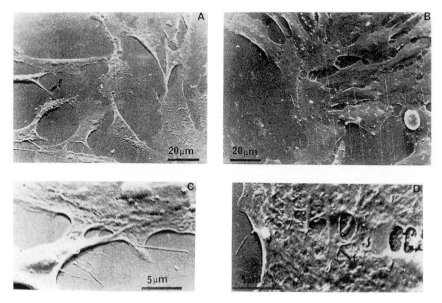

Fig. 2.9. SEM of normal (A,C) and amputated (B,D) cells at the edge of somites explanted from 9.5-day embryos. (From Flint and Ede 1982.)

clumps which remained in contact with each other over a long period. They were bound together by a large number of adhesive contacts, including large numbers of short filopodia amounting to a meshwork and preventing cells from moving apart. In this respect cultured amputated cells resemble those seen *in vivo* in sclerotome, facial mesenchyme, and palate.

Amputated cells in culture move at the same speed as normal cells but do not travel so far in a given time since they spend longer at rest. This is the familiar story that has already been seen in talpid and brachypod and is also found in t^9/t^9 embryos (Spiegelman and Bennett 1974; Yangisawa and Fujimoto 1977; see Chapter 6).

The talpid series of mutations, brachypodism, and amputated lead us to suppose that there is some fundamental link between cell adhesion, or cell surface properties, and blastemal formation. For other mutations little or no experimental work has been done, and we must rely on histological investigations at a difficult stage when the histological differences between cells in condensations and those outside are only just becoming apparent. However, several genes have been investigated by conventional means and traced back to the condensation stage.

Congenital hydrocephalus (ch, Chr 13)

This recessive condition (Grüneberg 1943, 1953) is lethal at or around birth consequent on a striking internal hydrocephalus. This was at first considered to derive from shortening of the chondrocranium but many additional effects of the gene (Green 1970; Grüneberg 1971, 1973) demanded a modified interpretation (Grüneberg and Wickramaratne 1974).

The points which concern us here revolve around the effects on the skeleton, which is affected in its entirety. Grüneberg (1963) gives a summary of the effects on the blastemata which form cartilage and points out that the whole of the cartilaginous skeleton, including the visceral cartilaginous skeleton, is involved. He points to four classes of abnormality: delay in chondrification, absence of cartilage formation, dyssymphsis of cartilages, and anomalous fusions between cartilages and gives extensive examples from chondrocranium to larynx by way of the limb skeleton and vertebral column.

In contrast to this group of abnormalities which essentially depend on the small size or absence of chondrogenic blastemata, those which give rise to membrane bone appear earlier and are larger than normal. This is especially so in the mandible and the zygomatic process of the maxilla. The zygomatic blastema is detectable in 13-day-old ch/ch embryos, a day earlier than in normals and by 14.5 days considerable ossification is seen which is not yet present in the normal. The mandibular ossification is similarly precocious and mandible and maxilla undergo osseous fusion (Fig. 2.10).

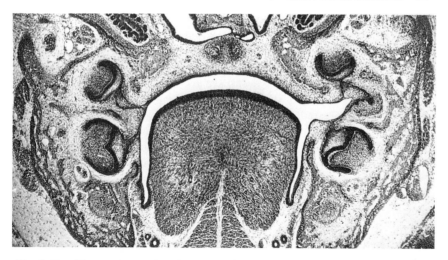

Fig. 2.10. Transverse section through the region of the first molars in an 18-day ch/ch fetus. Note bilateral fusion of maxilla and mandible. (From Grüneberg and Wickramaratne 1974.)

Around the vomero-nasal organ (Jacobson's organ) is a large mesen-chymal condensation which produces both the vomer (a membrane bone) and a cartilaginous cradle which helps support the organ. In ch/ch this blastema is of almost normal size, but gives rise to a massive vomer and almost no cartilaginous elements.

We are reminded here of Hall's work on the induction of cartilage or bone in the skeletal blastemata. It seems that in ch the chondrogenic stimulus is either lacking or unheeded and that the ossification stimulus is present and correct and perhaps taking over mesoderm which is normally destined to chondrify, but has not.

We are helped in the interpretation of this syndrome by the non-skeletal effects of ch. These include an increase in the number and size of nasal glands, the presence of large numbers of mesonephric tubules, and a tendency to accessory ureteric buds. Grüneberg attributed all effects, skeletal and otherwise, to a mesenchymal defect (or more accurately to a mesenchymal and ectomesenchymal defect). We may perhaps guess (in the absence of hard fact) that the defect lies in the ability of the mesenchyme to respond to varied ectodermal or other stimuli.

Short ear (se, Chr 9)

Short ear (Lynch 1921; Green and Green 1942; Green 1951) is so named because of the absence of the cartilaginous scapha of the pinna. Closer examination adds to this a whole series of smaller abnormalities including bifid sternum (a character affected by the genetic background), reduction in the number of ribs, and a whole series of small bones and processes reduced or absent. Cartilaginous bones are affected because their blastemata are reduced in size discouraging or inhibiting chondrification; membrane bones (such as the nasal) are also affected but it is not known if the effect here is direct on the blastema or a secondary effect of the reduction of surrounding cartilaginous bones. Green and Green (1942) pointed out that the effect was not seen at a particular time, but at a particular stage of development in each affected structure. Green (1958) showed that short-ear mice are still inferior to normal mice as far as dealing with fractures is concerned at 30–40 days post partum. Fracture callus is small, and cartilage is formed more slowly and in less quantity. Short ear and a closely linked coat colour mutation dilute (d) form a very useful genetic tool, and over 200 radiation-induced mutations have been generated in this region of chromosome 9. One hundred of these were used by Russell (1971) to construct a complementation map of this region. (Complementation: when a series of alleles se^a–se^z occurs at one locus it often transpires that the combination se^a/se^b has an effect not shared by se^a/se^a. By considering all the genotypes available a complementation map charting the probable order of the alleles on the chromosome can be constructed.) Amongst the radiation-induced mutants was se^1 (Dunn 1972) which complements several pre-implantation se deletions to provide viable young and is thus potentially useful in determining the action of short ear.

Dunn found that se^1/se^1 had a primary defect in the extra-embryonic mesoderm and trophectoderm, which resulted in the overproduction of trophectodermal giant cells (thought to break down decidual tissue), massive haemorrhage, and death. Mesoderm was virtually absent, but this was seen as a secondary defect of the overgrowth of extraembryonic ectoderm. He could not, however, rule out the hypothesis that an extra functional unit, controlling mesoderm proliferation and differentiation, was included in the se^1 deletion. Madison (1952) showed that the rate of mitosis is lower in se than in normal cartilage. In the light of this and the finding of Green (1958) on healing of fractures in se, Green (1968) proposed that the basic defect in se was concerned with the proliferation of cartilage cell forebears. If this is so, then the many soft-tissue effects of se should be secondary. Green studied visceral defects and came to the conclusion that the primary se defect probably did lie in the rate of differentiation of pre-skeletal cells. The alternative, that se affects proliferation of all cells, was felt to be negated by results suggesting that other organs smaller that normal in se regenerated normally. Most visceral defects in se were thought to arise from overcrowding in the abdominal and thoracic cavity consequent upon a reduced rate of skeletal growth.

Droopy ear (de, Chr 3)

Droopy ear (Curry 1959) produces bones of 'poor definition'. Long bones are reduced in length (although Curry gave no quantitative data) and the shoulder girdle and occipital region of the skull are affected. The cartilaginous blastemata of the basicranium and scapula are thinner than normal; the osseous scapula has a perforation in the infraspinous fossa which can be identified at the precartilaginous stage. Curry also considered the cartilage to be histologically abnormal (see Chapter 3).

Phocomelia (pc, Chr ?)

Phocomelia (Gluecksohn-Waelsch, *et al.* 1956) causes disproportionate dwarfing, the effects being most marked in the skull and limbs. In the skull both cartilage replacement and membrane bones are involved: nasals are absent, premaxillae reduced and part of the maxilla usually missing, and the mandible is reduced in size. The forelimbs are less shortened than the hindlimbs and the proximal elements less than the distal—hence radius and ulna, tibia and fibula are most affected.

Blastemal condensations in the limb develop 24 hours later than in normal littermates but are not reduced in size (perhaps because the footplates are a little larger than normal). Perhaps the most interesting feature of the phenotype is the presence of cartilage in unexpected places. This occurs as preaxial polydactyly (again a possible sequel to the over-large limb bud) in the tarsus, near the deltoid crest of the humerus, and on each side of the nasal capsule where a long bar of cartilage (Fitch 1957) near the palatal processes probably interferes with palatal fusion and leads to cleft palate.

Shorthead (sho, Chr ?)

Shorthead (Fitch 1061*a,b*) also has a median cleft palate associated with a broad and short skull. Fore and hindlimbs are shortened, the former more than the latter. The disproportion of the skull is present in 12-day-old embryos, before cartilage matrix can be demonstrated. Fitch (1961*a*) regards the changes as due to the inhibition of the growth of the facial processes; presumably the limb buds are similarly involved but the usual sequels, in terms of interruption of the pentadactyl limb pattern, are not seen.

Shaker with syndactylism (sy, Chr 18)

The abnormalities seen in this syndrome form a good example of the unity of gene action. Hertwig (1942, 1951, 1956) thought that the defects of the inner ear (which are responsible for the shaker part of the name) were consequent upon an insufficient production of endolymph by the stria vascularis. Deol (1963) re-examined the developing ear and was able to demonstrate that the abnormalities described by Hertwig were pre-dated by three days by abnormalities of the periotic labyrinth. The first anomaly seen was in the size of the mesenchymal condensation between saccule and utricle: this is larger and more mature than normal, and consequently the cavitation for the periotic cavity occurs sooner than normal. Further anomalies in the ear (for instance the lack of periosteal lining to the floccular fossa and irregular ossification of the semicircular canals) also suggest that the mesenchyme may be qualitatively as well as quantitatively abnormal. The abnormal periosteum seen in the ear could be systemic, resulting in the less dense than normal skeleton and other osseous abnormalities (Grüneberg 1962)

The syndactylism refers particularly to the hindfeet (although forefeet may also be affected). Syndactylism in this mutant is due to a combination of two events. Union between the phalanges of the feet is primary, the fused phalanges being represented by a single blastema; between carpals and tarsals it is secondary at the cartilaginous stage. The origins of these fusions may well lie in the narrowing of the limb buds which is seen at 12.5 days (Grüneberg 1956).

BLASTEMATA: AN OVERVIEW

It is quite clear from the above that the skeleton can often be seen to depart from normal very early in its formation, at the time when blastemal cells can first be distinguished from non-blastemal ones. The property which we can measure and which distinguishes these cells is mutual adhesion. In at least three cases which have been investigated (am, bp, ta^3) there is a clearly a change in mutant cell adhesiveness. Even though this has not been demonstrated in ta^2 and even though the various systems used to demonstrate adhesiveness give rather complex results, it seems that this parameter may lie behind, or at least give some indication of, the type of anomaly occurring in the first stage of skeletal development.

The classification of such anomalies as defects of condensation is, of course, a difficult one and possibly erroneous. The abnormal condensations of several of the mutants described above are known to produce abnormal cartilage or bone in subsequent stages of development. We must distinguish carefully here between abnormal cartilage and abnormal cartilages. If a blastema is in the wrong place, or of the wrong shape or size, we may expect it to produce a cartilage which is similarly misplaced, misshapen, or which is too small to chondrify; blastemata too close together (although we cannot say exactly what this means) may well fuse. This type of behaviour is the essence of the genes described above, or at least some of them. But in others it seems that the cartilaginous matrix itself may be involved, and be abnormal. We are treading here on rather thin ice. Cartilage matrix is usually recognized by its staining reaction or perhaps by its ultrastructure. In either case there must be a lag between the switching on of the cartilage matrix-specific genes and the appearance of the matrix between the cells. In practice a more reliable method of marking of cartilage formation is the onset of uptake of ^{35}S by blastemata (Searls 1965; Hinchliffe 1977). In fact the increase in sulphate uptake can be mapped slightly before the increased cell packing that we recognize as a condensation can be seen. Is the blastemal skeleton as a skeletal stage an illusion? A mutant producing abnormal cartilage and requiring more or less sulphate than normal to do so might, in fact, be abnormal at a stage before its blastemata are visible; having a defect in chondrogenesis it would be assigned to our category of cartilage-defective mutants; the blastema proper, as seen by the light microscope, would undoubtedly be affected also. Thus the distinction dims; some of our mutants will no doubt prove to be misclassified and the borders between membranous and cartilaginous (and, no doubt membranous and osseous) skeletons must be taken with a pinch of salt as no more than a useful scheme of classification.

Whether or not blastema formation can be separated from matrix production in cartilaginous blastema, amputated, brachypod, and talpid^3 (with supporting data from t-alleles) give us a pertinent clue to what may go wrong. Quite clearly all these mutants differ in their morphological effects, but equally clearly they all suffer from an excess of cell adhesion. The exact nature of this defect is still unknown and will very probably differ amongst them, and amongst any other mutants shown to have a similar phenotype. This is beside the point; the general effect is clearly there and we may pencil in a connection between cell adhesion and blastema formation. The morphology of amputated, brachypod, and talpid is clearly dissimilar. It seems that the effect is localized to a particular group of cells or to particular areas of the body; perhaps the acceptable normal range of cell adhesion for the production of normal structures varies according to region, or more likely according to how much action is taking place in a given region at a given time. But we may also draw distinctions between the morphology and behaviour of, say, talpid and amputated cells. Bellairs *et al.* (1980) showed something of the subtle way in which the behaviour of normal cells changes with time in the course of normal development with each type of mesoderm having a characteristic *in vivo* pattern of behaviour. We can expect that subtle differences between mutants may well lead to very different phenotypes.

The blastema, even if it only gives us *post hoc* notice of an event is nevertheless a useful landmark and the blastemal skeleton a useful concept. Grüneberg (1963) pointed out the consequences which follow blastemal disruption. This usually manifests as a reduction in blastemal size, although this cannot be the only factor as in mutants where abnormalities are widespread (ch, se, de) the reduced blastemata are arranged in characteristic patterns. Each gene reduces blastemal size, but in a specific way so that the skeleton produced is clearly different in every mutant. Brachypod, which affects only the limbs, is an even clearer example of localized blastemal upset.

Certain consequences follow from reduction of blastemal size. It seems in some cases that the reduction in size is due to a delay in blastema formation. If this occurs the formation of the skeleton is delayed: chondrification is retarded, cartilage histogenesis late, and ossification delayed. This may put an affected region out of step with development in other areas and other systems and lead to developmental appointments being missed; many developmental mechanisms depend on the arrival of two tissues in the right place at the right time. If the blastema is too reduced (below what threshold we do not know), chondrification will fail altogether, and so will ossification. If a blastema is too large, then the structure formed from it will also be oversized.

What controls blastemal size is an open question. Implicit in the idea of correct size is that of correct cell number (Hall 1982*b*). In some lower organisms (rotifiers, copepods) cell number is fixed. In mammals this is less true, but it would seem that cell number is nevertheless important, if a little less critical. Unfortunately there seems to be no data on mammalian condensations and cell numbers, although Moss (1972) has shown that growth of skeletal rudiments in the left and right wings of developing chicks is accurate to within one in 20 cells. Even if we knew the number of cells per condensation, we would still need to know the properties of interactions between cells in the blastema with each other. The second of these problems involves relative adhesiveness, mobility, and patterning of cells (which we have already discussed in relation to blastemal formation) and the rate, amount, and quality of the intracellular matrix produced. This has largely to be studied with regard to chondrogenesis, and forms the basis of the next chapter.

REFERENCES

Abbott, U. K. (1967). Avian developmental genetics. In *Developmental biology* (ed. F. H. Wilt and N. K. Wessels), p. 13–52. Thomas Y. Crowell Co., New York.
—— and Holtzer, H. (1966). The loss of phenotypic traits by differentiated cells. III. The reversible behaviour of chondrocytes in primary cultures. *Journal of Cell Biology* **28**, 473–8.
—— Taylor, L. W., and Abplanalp, H. (1959). A second *talpid* like mutation in the fowl. *Poultry Science* **38**, 1185.
Anikin, A. W. (1929). Das morphogene Feld der Knorpelbildung. *Archiv für Entwinklungsmechanik der Organismen* **114**, 549–78.
Bee, J. and Thorogood, P. V. (1980). The role of tissue interactions in the skeletogenic differentiation of avian neural crest cells. *Developmental Biology* **78**, 47–60.

Bell, D. A. and Ede, D. A. (1978). Regional differences in the morphology and motility of mesodermal cells from the early wing bud of normal and talpid[3] mutant chick embryos. *Journal of Embryology and Experimental Morphology* **48**, 185–203.

Bellairs, R. and Portch, P. A. (1977). Somite formation in the chick embryo. In *Vertebrate limb and somite morphogenesis* (ed. D. A. Ede, J. R. Hinchliffe, and M. Balls), p. 449–63. Cambridge University Press.

—— Sanders, E. J., and Portch, P. A. (1980). Behavioural properties of chick somitic mesoderm and lateral plate when explanted *in vitro*. *Journal of Embryology and experimental Morphology* **56**, 41–58.

Berrill, N. J. (1955). *The origin of the vertebrates*. Oxford University Press, London.

Bradamante, Z. and Hall, B. K. (1980). The role of epithelial collagen and proteoglycan in the initiation of osteogenesis by avian neural crest cells. *Anatomical Record* **197**, 305–15.

Bugrilova, R. S. and Konyukhov, B. V. (1971). Dose effect of brachypodism-H gene in mouse. *Genetika* **7**, 76–83.

Cairns, J. M. (1975). The function of the apical ectodermal ridge and distinctive characteristics of adjacent distal mesoderm in the avia wing bud. *Journal of Embryology and experimental Morphology* **34**, 155–69.

—— (1977). Growth of normal and talpid[2] chick wing buds: an experimental analysis. In *Vertebrate limb and somite morphogenesis* (ed. D. A. Ede, J. R. Hinchliffe, and M. Balls), p. 123–138. Cambridge University Press.

Chevallier, A., Kieny, M., and Mauger, A. (1977). Limb-somite relationship: origin of the limb musculature. *Journal of Embryology and experimental Morphology* **41**, 245–58.

Christ, B, Jacob, H. J., and Jacob, M. (1977). Experimental analysis of the origin of the wing musculature in avian embryos. *Anatomy and Embryology* **150**, 171–86.

Cole, R. K. (1942). The 'talpid' lethal in the domestic fowl. *Journal of Heredity* **33**, 82–6.

Coulombre, A. J., Coulombre, J. L. and Mehta, H. (1962). The skeleton of the eye. I. Conjunctival papillae and scleral ossicles. *Developmental Biology* **5**, 382–401.

Curry, G. A. (1959). Genetical and developmental studies on droopy-eared mice. *Journal of Embryology and experimental Morphology* **7**, 39–65.

Curtis, A. S. G. (1961). Timing mechanisms in the specific adhesion of cells. *Experimental Cell Research* (Supp.) **8**, 107–22.

Denison, R. H. (1963). The early history of the vertebrate calcified skeleton. *Clinical Orthopaedics and Related Research* **31**, 141–52.

Deol, M. S. (1963). Development of the inner ear in mice homozygous for shaker with syndactylism. *Journal of Embryology and experimental Morphology* **11**, 493–512.

Duke, J. and Elmer, W. A. (1977). Effect of the brachypod mutation on cell adhesion and chondrogenesis in aggregates of mouse limb mesenchyme. *Journal of Embryology and experimenatal Morphology* **42**, 209–17.

—— —— (1978). Cell adhesion and chondrogenesis in brachypod mouse limb mesenchyme: fragment fusion studies. *Journal of Embryology and experimental Morphology* **48**, 161–8.

—— —— (1979). Effect of the brachypod mutation on early stages of chondrogenesis in mouse embryonic hind limbs: an ultrastructural analysis. *Teratology* **19**, 367–76.

Dunn, G. R. (1972). Embryological effects of a minute deficiency in linkage group

11 of the mouse. *Journal of Embryology and experimental Morphology* **27**, 147–54.

Durston, A. J. (1973). *Dictyostelium discoideum* aggregation fields as excitable media. *Journal of theoretical Biology* **42**, 483–504.

Ede, D. A. (1971). Control of form and pattern in the vertebrate limb. In *Control mechanisms of differentiation and growth*. Society for Experimental Biology symposium 25 (ed. D. Davies and M. Balls), pp. 235–54. Cambridge University Press.

—— (1976). Cell interactions in vertebrate limb development. In *The cell surface in animal embryogenesis and development* (ed. G. Poste and G. L. Nicholson), pp. 495–543. Elsevier, Amsterdam.

—— and Agerback, G. S. (1968). Cell adhesion and movement in relation to the developing limb pattern in normal and *talpid*[3] mutant chick embryos. *Journal of Embryology and experimental Morphology* **20**, 81–100.

—— and Flint, O. P. (1972). Patterns of cell division and chondrogenesis in cultured aggregates of normal and talpid[3] mutant chick limb mesenchyme cells. *Journal of Embryology and experimental Morphology* **27**, 245–60.

—— —— (1975a). Intercellular adhesion and formation of aggregates in normal and talpid[3] mutant chick limb mesenchyme. *Journal of Cell Science* **18**, 97–111.

—— —— (1975b). Cell movement and adhesion in the developing chick wing bud: studies on cultured mesenchyme cells from normal and talpid[3] mutant embryos. *Journal of Embryology and experimental Morphology* **18**, 301–13.

—— Bellairs, R. and Bancroft, M. (1974). A scanning electron microscope study of the early limb bud in normal and talpid[3] mutant chick embryos. *Journal of Embryology and experimental Morphology* **31**, 761–85.

—— —— and Teague, P. (1975). Cell proliferation in the developing wing bud of normal and talpid[3] mutant chick embryos. *Journal of Embryology and experimental Morphology* **34**, 589–608.

—— and Kelly, W. A. (1964a). Developmental abnormalities in the head region of the talpid[3] mutant of the fowl. *Journal of Embryology and experimental Morphology* **12**, 161–82.

—— —— (1964b). Developmental abnormalities in the trunk and limbs of the talpid[3] mutant of the fowl. *Journal of Embryology and experimental Morphology* **12**, 339–56.

—— Flint, O. P, Wilby, O. K., and Colquhoun, P. (1977). The development of precartilage condensations in limb bud mesenchyme *in vitro* and *in vivo*. In *Vertebrate limb and somite morphogenesis* (ed. D. A. Ede, J. R. Hinchliffe, and M. Balls), pp. 161–80. Cambridge University Press.

Elmer, W. A. and Selleck, D. K. (1975). In vitro chondrogenesis of limb mesoderm from normal and brachypod mouse embryos. *Journal of Embryology and experimental Morphology* **33**, 371–86.

Fitch, N. (1957). An embryological analysis of two mutants in the house mouse, both producing cleft palate. *Journal of experimental Zoology* **136**, 329–61.

—— (1961a). A mutation producing dwarfism, brachycephaly, cleft palate and micromelia. *Journal of Morphology* **109**, 141–9.

—— (1961b). The development of cleft palate in mice homozygous for the shorthead mutation. *Journal of Morphology* **109**, 151–7.

Flickinger, R. A. (1974). Muscle and cartilage differentiation in small and large explants from the chick embryo limb bud. *Developmental Biology* **41**, 202–8.

Flint, O. P. (1977). Cell interactions in the developing axial skeleton in normal and mutant mouse embryos. In *Vertebrate limb and somite morphogenesis* (ed. D. A. Ede, J. R. Hinchliffe, and M. Balls, pp. 465–84. Cambridge University Press.

—— (1980). Cell behaviour and cleft palate in the mutant mouse amputated. *Journal of Embryology and experimental Morphology* **58**, 131–42.

—— and Ede, D. A. (1978*a*). Cell interactions in the developing somite: in vivo comparisons between amputated (am/am) and normal mouse embryos. *Journal of Cell Science* **31**, 275–92.

—— —— (1978*b*). Facial development in the mouse: a comparison between normal and mutant (amputated) mouse embryos. *Journal of Embryology and experimental Morphology* **48**, 249–67.

—— —— (1982). Cell interactions in the developing somite: in vitro comparisons between amputated (am/am) and normal mouse embryos. *Journal of Embryology and experimental Morphology* **67**, 113–25.

Fyfe, D. M. and Hall, B. K. (1981). A scanning electron microscope study of the developing epithelial skeletal papillae in the eye of the embryo chick. *Journal of Morphology* **167**, 201–9.

Ginter, E. K. and Konyukhov, B. V. (1966). Pleiotropic effect of gene brachy-podism-H in the mouse. *Genetika* **3**, 83–92.

Gluecksohn-Waelsch, S., Hagedorn, D., and Sisken, B. F.(1956). Genetics and morphology of a recessive mutation in the house mouse affecting head and limb skeleton. *Journal of Morphology* **99**, 465–79.

Godman, G. C. and Porter, K. R. (1960). Chondrogenesis studied with the electron micropscope. *Journal of Biophysical and Biochemical Cytology* **5**, 221–55.

Goedbloed, J. F. (1964). The early development of the middle ear and the mouth cavity. A study of the processes in the epithelium and the mesenchyme. *Archives of Biology* **75**, 207–44.

Gould, R. P., Day, A., and Wolpert, L. (1972). Mesenchymal condensation and cell contact in early morphogenesis of the chick limb bud. *Experimental Cell Research* **72**, 325–36.

—— Selwood, L., Day, A., and Wolpert, L. (1972*b*). The mechanism of cellular orientation during early cartilage formation in chick limb and regenerating amphibian limb. *Experimental Cell Research* **83**, 287–96.

Green, E. L. (Ed.) (1966). *Biology of the laboratory mouse*, (2nd ed). McGraw-Hill, New York.

—— and Green, M. C. (1942). The development of the manifestations of the short ear gene in the mouse. *Journal of Morphology* **70**, 1–19.

Green, M. C. (1951). Further morphological effects of the short ear gene in the house mouse. *Journal of Morphology* **88**, 1–21.

—— (1958). Effects of the short ear gene in the mouse on cartilage formation in healing bone fractures. *Journal of experimental Zoology* **137**, 75–88.

—— (1968). Mechanism of the pleiotropic effects of the short ear mutant gene in the mouse. *Journal of Experimental Zoology* **167**, 129–50.

—— (1970). The developmental effects of congenital hydrocephalus (ch) in the mouse. *Developmental Biology* **23**, 585–608.

Grüneberg, H. (1943). Congenital hydrocephalus in the mouse, a case of spurious pleiotropism. *Journal of Genetics* **45**, 1–21.

—— (1953). Genetical studies on the skeleton of the mouse. VII. Congenital hydrocephalus. *Journal of Genetics* **51**, 327–58.

—— (1956). Genetical studies on the skeleton of the mouse. XVIII. Three genes for syndactylism. *Journal of Genetics* **54**, 113–45

—— (1962). Genetical studies on the skeleton of the mouse. XXXII. The development of shaker with syndactylism. *Genetical Research* **3**, 157–66.

—— (1963). *The pathology of development*. Blackwell, Oxford.

—— (1971). Exocrine glands and the Chievitz organ of some mouse mutants. *Journal of Embryology and experimental Morphology* **25**, 247–61.

—— (1973). A ganglion probably belonging to the n. terminalis systems in the nasal mucosa of the mouse. *Zeischrift für Anatomie Entwicklungsgeeschichte Gesellschaft* **140**, 39–52.

—— and Lee, A. J. (1973). The anatomy and development of brachypodism in the mouse. *Journal of Embryology and experimental Morphology* **30**, 119–41.

—— and Wickramaratne, G. A. de S. (1974). A re-examination of two skeletal mutants of the mouse, vestigal tail (vt) and congenital hydrocephalus (ch). *Journal of Embryology and experimental Morphology* **31**, 207–22.

Hale, L. J. (1956). Mitotic activity during early skeletal diffentiation of the scleral bones of the chick. *Quarterly Journal of the Microscopical Society* **97**, 333–53.

Hall, B. K. (1971). Histogenesis and morphogenesis of bone. *Clinical Orthopaedics and Related Research* **74**, 249–68.

—— (1978a). Initiation of osteogenesis by mandibular mesenchyme of the embryonic chick in response to mandibular and non-mandibular epithelia. *Archives of oral Biology* **23**, 1157–61.

—— (1978b). *Developmental and cellular skeletal biology.* Academic Press, New York.

—— (1981). The induction of neural crest-derived cartilage and bone by embryonic epithelia: an analysis of the mode of action of an epithelial–mesenchymal interaction. *Journal of Embryology and experimental Morphology* **64**, 305–20.

—— (1982a), Distribution of osteo— and chondrogenic neural crest cells of osteogenically inductive epithelia in mandibular arches of embryonic chicks. *Journal of Embryology and experimental Morphology* **68**, 127–36.

—— (1982b). The role of tissue interactions in the growth of bone. In *Factors and mechanisms influencing bone growth* (ed. A. D. Dixson and B. G. Sarnat), pp. 205–16. Alan R. Liss, New York.

Ham, A. W. (1969). *Histology*, J. B. Lipincott, Philadelphia.

Hamburger, V. and Hamilton, H. L. (1951). A series of normal stages in the development of the chick embryo. *Journal of Morphology* **88**, 49–92.

Hertwig, P. (1942). Neue Mutationen und Koppelungsgsgruppen bei der Hausmaus. *Zeitschrift für induktive Abstammungs-u. Vererbungslehre* **80**, 220–46.

—— (1951). Entwicklungsgeschichtliche Untersuchungen über Bewegungsströrungen bei Mausen. *Verhandlunger der Anatomischen Gesellschaft* **49**, 97–107.

—— (1956). Erblicke Missbildungen des Gehörorgans bei der Maus. *Verhandlunger der Anatomischen Gesellschaft* **53**, 256–69.

Hewitt, A. T. and Elmer, W. A. (1976). Reactivity of normal and brachypod mouse limb mesenchymal cells with ConA. *Nature, London* **264**, 177–8.

—— —— (1978a). Developmental modulation in lectin-binding sites on the surface membranes of normal and brachypod mouse limb mesenchymal cells. *Differentiation* **10**, 31–8.

—— —— (1978b). The involvement of microfilaments and microtubules in the lateral mobility of lectin binding sites of normal and brachypod mouse limb mesenchyme. *Cell Differentiation* **7**, 295–303.

Hinchliffe, J. R. (1977). The chondrogenic pattern in chick limb morphogenesis: a problem of development *and* evolution. In *Vertebrate limb and somite morphogenesis* (ed. D. A. Ede, J. R. Hinchliffe, and M. Balls), pp. 293–310. Cambridge University Press.

—— and Johnson, D. R. (1980). *The development of the vertebrate limb.* Clarendon Press, Oxford.

Holtfreter, J. (1968). Mesenchyme and epithelia in inductive and morphogenetic processes. In *Epithelial-mesenchymal interactions* (ed. R. Fleischmajer), pp. 1–30. Williams & Wilkins, Baltimore.

Hunton, P. (1960). A study of some factors affecting the hatchability of chicken eggs, with special reference to genetic control. MSc thesis, Wye College, University of London.

Jacobson, W. and Fell, H. B. (1941). The developmental mechanics and potencies of the undifferentiated mesenchyme of the mandible. *Quarterly Journal of Microscopical Science* **82**, 563–86.

Janners, M. T. and Searls, R. L. (1970). Changes in the rate of cellular proliferation during the differentiation of cartilage and muscle in the mesenchyme of the embryonic chick wing. *Developmental Biology* **23**, 136–65.

Johnson, L. G. (1973). Development of chick embryo conjunctival papillae and scleral ossicles after hydrocortisone treatment. *Developmental Biology* **30**, 223–7.

Johnston, M. C. and Listgaren, M. A. (1972). Observations on the migration interaction and early differentiation of the orofacial tissues. In *Developmental aspects of oral biology* (ed. H. C. Slavkin and L. A. Bavetta), pp. 56–80. Academic Press, New York.

Kollar, E. J. and Baird, G. R. (1969). The influence of the dental papilla on the development of tooth shape in embryonic mouse tooth germs. *Journal of Embryology and experimental Morphology* **21**, 137–48.

Landauer, W. (1952). Brachypodism, a recessive mutation of house mice. *Journal of Heredity* **43**, 293–8.

Lynch, C. J. (1921). Short ears, an autosomal mutation in the house mouse. *American Naturalist* **55**, 421–6.

Madison, C. R. (1952). The growth and differentiation in vitro of tissue from the normal and short ear mouse. PhD dissertation, Ohio State University.

Martinozzzi, M. and Moscona, A. A. (1975). Binding of ^{125}I-concanavalin A and agglutination of embryonic neural retina cells. *Experimental Cell Research* **94**, 253–66.

McLean, F. C. and Urist, M. R. (1968). *Bone—fundamentals of the physiology of skeletal tissues* 3rd edn. University of Chicago Press.

Milaire, J. (1965). Étude morphogénétique de trois malformations congénitales de l'autopode chez la souris (syndactylisme–brachypodisme–hémimélie dominante) par des methods cyctochimiques. *Mémoires de l'Académie r.de Belgique Classe des Sciences*. Collection in 4° Ser. **16**.

Minor, R. (1973). Somite chrondrogenesis. A structural analysis. *Journal of Cell Biology* **56** 27–50.

Montagna, W. (1945). A re-investigation of the development of the wing in the fowl. *Journal of Morphology* **76**, 87–113.

Moscona, A. A. (1960). Patterns and mechanisms of tissue reconstruction from dissociated cells. In *Developing cell systems and their control* (ed. D. Rudnick), pp. 45–70. Ronald Press, New York.

Moss, M. L. (1963). The biology of acellular teleost bone. *Annals of the New York Academy of Science* **109**, 337–50.

—— (1964). The phylogeny of mineralised tissues. *International Review of General and Experimental Zoology* **1**. 297–331.

—— (1972). The regulation of skeletal growth. In *Regulation of organ and tissue growth* (ed. R. J. Goss). Academic Press, New York.

Newman, S. (1977). Lineage and pattern in the developing wing bud. In *Vertebrate limb and somite morphogenesis* (ed. D.A. Ede, J. R. Hinchliffe, and M. Balls), pp. 181–197. Cambridge University Press.

Newsome, D. A. (1976). In vitro stimulation of cartilage in embryonic chick neural crest cells by products of retinal pigmented epithelium. *Developmental Biology* **49**, 496–507.

Niederman, R. N. and Armstrong, P. B. (1972). Is abnormal limb bud morphology in the mutant talpid^2 chick embryo a result of altered intercellular adhesion? Studies employing cell sorting and fragment fusing. *Journal of experimental Zoology* **181**, 17–32.

Orvig, T. (1967). Phylogeny of tooth tissues: evolution of some calcified tissues in early vertebrates. In *Structural and chemical organisation of teeth* (ed. A. E. W. Miles, pp. 45–110. Academic Press, New York.

Owens, E. M. and Solursh, M. (1982). Cell–cell interactions by mouse limb cells during in vitro chondrogenesis: analysis of the brachypod mutation. *Developmental Biology* **91**, 376–88.

Palmoski, M. J. and Goetinck, P. F. (1970). An analysis of the development of conjunctival papillae and scleral ossicles in the eye of the scaleless mutant. *Journal of experimental Zoology* **174**, 157–64.

Paulsen, D. F. and Finch, R. A. (1977). Age and region dependent concanavalin A reactivity of chick wing-bud mesoderm cells. *Nature* **268**, 639–41.

Pleskova, M. V., Rodinov, V. M., Bugriva, R. S., and Konyukhov, B. V. (1974). The partial purification of growth inhibiting factor of the brachypodism-H mouse embryos. *Developmental Biology* **37**, 417–21.

Reinbold, R. (1968). Rôle du tapetum dans la differéntation de la sclérotique chez l'embryon de poulet. *Journal of Embryology and experimental Morphology* **19**, 43–7.

Romer, A. S. (1942). Cartilage an embryonic adaptation. *American Naturalist* **76**, 394–404.

—— (1963). The ancient history of bone. *Annals of the New York Academy of Science* **109**, 168–76.

—— (1964). Bone in early vertebrates. In *Bone biodynamics* (ed. H. M. Frost), pp. 13–37. Little, Brown & Co., Boston.

Russell, L. B. (1971). Definition of functional units in a small chromosomal segment of the mouse and its use in interpreting the nature of radiation induced mutations. *Mutation Research* **11**, 107–23.

Saxen, L. (1977). Directive versus permissive induction: a working hypothesis. In *Cell and tissue interactions* (ed. J. W. Lash and M. M. Burger), pp.1–9. Raven Press, New York.

Schaeffer, B. (1961). Differential ossification in the fishes. *Transactions of the New York Academy of Science* **23**, 501–5.

Schmidt, A. J. (1968). *Cellular biology of vertebrate regeneration and repair.* Chicago University Press.

Searls, R. L. (1965). An autoradiographic study of the uptake of S35 sulphate during the differentiation of limb bud cartilage. *Developmental Biology* **11**, 155–68.

—— (1967). The role of cell recognition in the development of the embryonic chick limb buds. *Journal of experimental Zoology* **116**, 39–50.

—— (1973). Newer knowledge of chondrogenesis. *Clinical Orthopaedics* **96**, 327–44.

—— Hilfer, S. R., and Mirow, S. M. (1972). An ultrastructural study of early chondrogenesis in the chick wing bud. *Developmental Biology* **28**, 123–37.

Smith, H. W. (1961). *From fish to philosopher.* Doubleday, New York.

Spiegleman, M. and Bennett, D. (1974). Fine structural study of cell migration in the early mesoderm of normal and mutant mouse embryos (T-locus t^9/t^9). *Journal of Embryology and experimental Morphology* **32**, 723–38.

Stewart, P. A. and McCallion, D. J. (1975). Establishment of the scleral cartilage in the chick. *Developmental Biology* **46**, 383–9.

Summerbell, D., Lewis, J. H., and Wolpert, L. (1973). Positional information in chick limb morphogenesis. *Nature, London,* **224**, 492–6.

Tarlo, L. B. H. (1964). The origin of bone. In *Bone and tooth* (ed. H. J. J. Blackwood). Pergamon Press, Oxford.

Thorogood, P. V. (1972). Patterns of chondrogenesis and myogenesis in the limb buds of normal and talpid[3] chick embryos. Ph.D. thesis, University College of Wales, Aberystwyth.

—— and Hinchliffe, J. R. (1975). An analysis of the condensation process during chondrogenesis in the embryonic chick limb bud. *Journal of Embryology and experimental Morphology* **33**, 581–606.

Toole, B. P. (1972). Hyaluronate turnover during chondrogenesis in the developing chick limb and axial skeleton. *Developmental Biology* **29**, 321–9.

Tyler, M. S. and Hall, B. K. (1977). Epithelial influences on skeletogenesis in the mandible of the embryonic chick. *Anatomical Record* **188**, 229–40.

Urist, M. R. (1962). The bone–body fluid continuum: calcium and phosphorus in the skeleton and blood of extinct and living vertebrates. *Perspectives in Biology and Medicine* **6**, 75–115.

—— (1963). The regulation of calcium and other ions in the serums of hagfish and lampreys. *Annals of New York Academy of Science* **109**, 294–311.

—— (1964). Further observations bearing on the bone–body fluid continuum: composition of the skeletons and serums of cyclostomes, elasmobranchs and bony vertebrates. In *Bone biodynamics* (ed. H. M. Frost), pp. 151–79. Little, Brown & Co., Boston.

Van de Kamp, M. (1968). Fine structural analysis of the conjunctival papillae in the chick embryo: a reassessment of their morphogenesis and developmental significance. *Journal of experimental Zoology* **169**, 447–62.

Wessels, N. K. (1977). *Tissue interactions and development.* W. A. Benjamin, Menlo Park, California.

Yangisawa, K. O. and Fujimoto, H. (1977). Differences in rotation mediated aggregation between wild type and homozygous brachyury (T) cells. *Journal of Embryology and experimental Morphology* **40**, 277–83.

Zwilling, E. (1961). Limb morphogenesis. *Advances in Morphology* **1**, 301–30.

3. The cartilaginous skeleton

CARTILAGE

The onset of cartilage formation can be accurately determined by the first appearance of a specific extracellular matrix. Mutants affecting cartilage formation are obviously likely to affect this matrix and so we must pay attention to the signals which determine the onset of matrix formation, i.e. what governs the initiation of matrix formation, and to the nature of the matrix itself. Mutants may delay matrix formation, reduce it, or qualitatively or quantitatively change matrix components. We might also suspect, with foreknowledge of the blastemal condensations, that cellular morphology and behaviour, measured in adhesiveness, motility, and perhaps mitotic rate may be involved, since cartilage is unique amongst skeletal tissues in growing interstitially.

MATRIX

The formation of cartilage matrix involves both the activation of genes associated with specific matrix components and the repression of genes associated with chronologically earlier extracellular products produced by embryonic mesenchyme. The differentiated chondrocyte synthesizes type II collagen, cartilage-specific proteoglycan, proteoglycan link protein, and chondronectin which constitute a unique array of matrix-forming macromolecules.

Collagen

Chondrocyte collagen is different from that synthesized by other tissues (Miller 1977). Type II collagen from cartilage is made up of three identical polypeptide chains ([alpha 1(II)], Fig. 3.1). Type I collagen from skin and bone consists of two [alpha 1(I)] and one alpha 2 chain (Miller 1976; Miller and Mutukas 1969; Trelstad *et al.* 1970). In the developing limb bud Von der Mark (1980) has shown that the precartilaginous blastema synthesizes type I collagen; when the cells become chondrocytes type II collagen only is synthesized. Later, of course, when the cartilage is replaced by bone type I, collagen will again be found.

Biosynthesis of collagen is complex (reviewed by Minor 1980; Olson 1981; Mayne and Von der Mark 1982) and a number of inherited diseases of collagen synthesis are known: since none specifically affects cartilage (although cartilage may be secondarily involved) a detailed description of this process is unnecessary here.

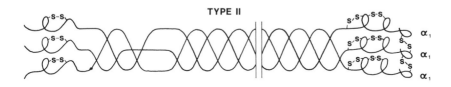

Fig. 3.1. Type I and II collagen.

Proteogylcans

Our current concept of cartilage proteoglycans comes from work on the cow, pig, rat, and developing chick (Hascall 1977; Lash and Vasan 1982). Proteoglycan is assembled from a monomer consisting of core protein (m.w. 2.0×10^5) with side chains formed of two types of sulphated glycosaminoglycans (GAGs), chondroitin sulphate and keratan sulphate, and two oligosaccharides (Fig. 3.2). Chondroitin sulphate is made up of alternate residues of glucoronic acid and N-acetylgalactosamine linked to the core protein via a short region glucoronic acid–galactose–galactose– xylose. The core protein attaches via an O-glycoside bond between xylose and the hydroxy group of serine in the protein backbone. Sulphation is in either the 4 (chondroitin 4-sulphate) or 6 (chondroitin 6-sulphate) position of the N-acetylgalactosamine. Keratan sulphate is a disaccharide of N-acetylglycosamine and galactose units also linked to the core protein by an O-glycosidic bond to serine or threonine. A third sulphated GAG, dermatan sulphate, is a minor constituent of some cartilages (Kimata *et al.* 1978).

The oligosaccharide components are a group of three related oligo-saccharides of different sizes forming neuraminie-rich oligosaccharides linked to the core protein by the same linkage as keratan sulphate, and a mannose-rich oligosaccharide apparently linked to asparagine in the core protein.

The proteoglycan is visualized as consisting of a core protein which binds hyaluronic acid at one end, and which is relatively free from carbohydrates.

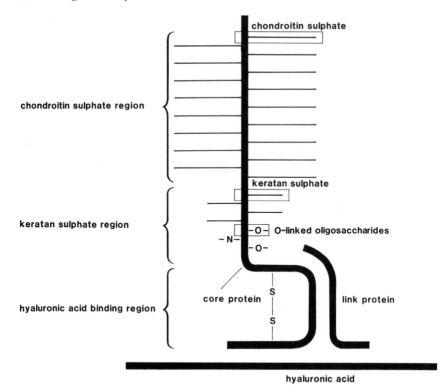

Fig. 3.2. Proteogylcan structure.

In the middle of the molecule is a zone with approximately 50 keratan sulphate side chains and terminally a chondroitin-sulphate-rich zone of approximately 100 side chains (Hunter and Caplan 1982; Hascall and Heinegard 1979). The monomer is assembled into a polymer by the association of many units with hyaluronic acid; this process is aided by a specific link protein which has affinities for both hyaluronic acid and the hyaluronic acid linking zone of the core protein.

Proteoglycans can be extracted from cartilage after pulse labelling with $^{35}SO_4$ and/or a labelled amino acid (for instance ^3H serine) while guanidinium chloride and density gradient centrifugation will give details of physical and chemical properties (DeLuca *et al.* 1977, 1978, 1980; Hascall and Sajdera 1970). The first proteoglycan monomer produced by cultured chondrocytes has very long chondroitin sulphate chains, very short keratan sulphate chains, and an ineffective binding region. After further culture, and coinciding with the maximal uptake of ^{35}S, cells produce shorter chondroitin sulphate chains, longer keratan sulphate chains, and a higher proportion of chondroitin-4-sulphate. After further culture 'senescent' cells produce even shorter chondroitin sulphate chains, longer keratan sulphate chains, and even more chondroitin-4-sulphate. The properties of the type II collagen produced by these cells do not change during culture (Hascall *et al.* 1976; De Luca *et al.* 1977; Caplan and Hascall 1980). The maturity of the proteoglycan produced depends, amongst other things, on position; proximal limb regions, for example, have a more mature proteoglycan profile than distal ones (Ovidia *et al.* 1980).

The synthesis of these polymers is obviously complex. The core protein is transcribed at a m.w. of 340 000 (Uphold *et al.* 1979); we have seen that the core protein has a m.w. of 200 000, so some unknown modification must occur between synthesis and assembly of monomers. Chondroitin sulphate chains are formed by the addition of the basal xylose–galactose–galactose–glucaronic acid residues followed by repeating disaccharide units. Specific glycosyltransferases govern the addition of each type of monosaccharide. Keratan sulphate and the similar group of O-linked oligosaccharides are also built under the control of glycosyltransferases covering N-acetylgalactosamine, galactose, and neuraminic acid. The N-linked oligosaccharides also have specific enzymes.

Sulphation of the glycosaminoglycans involves synthesis of PAPS (3'-phosphoadenosine 5'-phosphosulphate) and the transfer of sulphate from this to an appropriate acceptor.

Link protein

One or more link proteins (Hardinham 1979; Tang *et al.* 1979) stabilize the union of the protein cored monomer to hyaluronic acid (Caterson and Baker 1978). The number and quality of the link proteins seem to depend on their origin; three have been described in bovine nasal septum (Caterson and Baker 1978) and in xiphoid process (McKeown-Longo *et al.* 1982).

Chondronectin

This is a cartilage-specific glycoprotein (Hewitt *et al.* 1980) of m.w. 180 000 which splits into subunits of m.w. approximately 80 000 (Kleinman *et al.* 1981) and plays a role in attaching chondrocytes to collagen, a process also involving cartilage proteoglycans. Details of this attachment have yet to be described.

Maturation of the matrix

Cartilage-specific chondroitin sulphate proteoglycan (Levitt and Dorfmann 1973; Palmoski and Goetinck 1972) is easily located by uptake of ^{35}S sulphate in autoradiographic studies (Janners and Searls 1970) where incorporation is first seen towards the centres of chondrogenic blastemata.

Palmoski and Goetinck (1972) and Levitt and Dorfman (1973) have shown that embryonic chick cartilage synthesizes two species of chondroitin sulphate, which are also synthesized by limb mesenchyme and that chondrogenesis is marked by a preferential increase in the synthesis of one of these species (Goetinck *et al.* 1974) accompanied by a change from synthesis of type I collagen to type II. Later work (Okayama *et al.* 1976; Kitamura and Yamagota 1976; De Luca *et al.* 1977) suggests that neither of these is identical to the chondroitin sulphate synthesized by early limb mesenchyme.

The elaboration of cartilage macromolecules in the extracellular matrix involves the interaction of proteoglycans and hyaluronic acid to form huge complexes (Hascall and Heinegard 1974*a,b*; Heinegard and Hascall 1974) and the assembly of collagen molecules into fibrils. There is also interaction between acid mucopolysaccharides and collagen and this may influence fibril formation (Mathews 1965; Wood 1960; Mathews and Decker 1967; Toole and Lowther 1968). Specifically, it has been suggested that the ·presence of large amounts of mucopolysaccharide–protein complexes prevents the assembly of large-diameter collagen fibres. The production of extracellular matrix alters the environment of the cells which produce it, and this could lead to a feedback effect on the cells. During cytodifferentiation of cartilage cells, chondroitin sulphate synthesis increases, hyaluronate production decreases, and hyaluronidase increases, suggesting that the production and subsequent removal of hyaluronidate is an essential event in cartilage matrix synthesis (Toole 1972; Toole *et al.* 1972). Hyaluronic acid is known to be important in amphibian regeneration (Toole and Gross 1971). Extraneous hyaluronidase will inhibit proteoglycan synthesis in chondrocyte cultures (Wiebtain and Muir 1973). On the other hand extraneous chondroitin sulphate stimulates chondroitin sulphate production (Nero and Dorfman 1972).

Does the composition of the surrounding matrix feed back to the cartilage matrix producing cells? Caplan (in Hunter and Caplan 1982) tested this by exposing high-density cultures of chondrogenic tissue to trypsin, chondroitinase ABC, and hyaluronidase which remove 80–95 per cent of the extracellular proteoglycan and protein without removing cells from the culture vessel. After this treatment cells were pulsed with ^{35}S sulphate and ^{3}H-serine and were shown to produce normal proteoglycans. Sequential collagenase digestion allowed another approach, the removal of chondrogenic and non-chondrogenic cells from their culture plates (Lennon 1982). After pulsing and replating, the matrix formed was again normal. A third approach was to use methylxyloside. This produces complete chondroitin chains (although reduced in number and size) which are not associated with the proteoglycan core protein and hence not incorporated into the matrix. Keratan side chains are almost unaffected

(Lohmander *et al*. 1979*a*,*b*) The effect of this treatment is that cartilage cells are surrounded by considerably less matrix than in normal controls, and are hence closer together. Again treated cells produce normal collagen and normal proteoglycan monomer (Von der Mark *et al*. 1982), indicating that qualitative synthesis of collagen and proteoglycan is unchanged. Washing out the methylxyloside followed by pulse labelling shows that cells sitting in an abnormal matrix still produce normal proteoglycan. These results seem to indicate that the variation in matrix production associated with maturity is programmed into the cell, and not a property of its surroundings.

CELL–CELL INTERACTION

Holtfreter (1968) noted that, once induced by pharyngeal ectoderm to produce cartilage, newt ectomesenchyme increased in size as additional cells were recruited. The effect could be due to an assimilative stimulus provided by the cartilage cells themselves. Ahrens *et al*. (1977, 1979) also noted that in high-density cultures of dissociated avian limb mesenchyme the cartilaginous aggregates rapidly increased in size mainly due to assimilation of adjacent mesenchyme cells, although the increase in size might also have a component due to mitosis (Epperlein and Lehmann 1975). Moscona (1956) noticed that there was a lower limit of 30–40 to the number of limb bud cells which would form cartilage and that smaller aggregates would not (equivalent to Grüneberg's (1963) threshold for chondrification?). Umansky (1966) found that ability to form cartilage in dissociated limb bud mesenchyme was dependent upon cell density. Cartilage will form in conditions favouring cell association, i.e. in densities greater than confluity (Caplan 1970), in pellets (Karasawa *et al*. 1979), or in suspension, particularly when shaken (Matsutani and Kuroda 1980). Aggregates of mesenchymal cells from chick wing can be found in culture from stage 17 onwards (Ahrens *et al*. 1977) but will only form cartilage if implanted between stages 20 and 26 (Solursh *et al*. 1981). In mouse cell cultures progressively smaller aggregates are formed with increasing age, suggesting that a decreasing proportion of cartilage-forming cells is present as development proceeds.

Differentiated cartilage cells require cell contacts in order to maintain their phenotype (Holtzer and Abbott 1968); dispersed chondrocytes become fibrogenic (Abbot and Holtzer 1966). Non-cartilage cells are also able to block chondrogenesis (interference). The presence of a small number of non-chondrogenic cells in mass cultures can result in loss of phenotype. Once non-chondrocytes have been eliminated, a second type of interaction can be seen, this time amongst the chondrocytes and stimulating the production of cartilage matrix (Solursh and Meier 1974). We have already seen that this is not due to matrix components. Solursh and Meier (1973) noted that this effect is due to a diffusible factor present in the conditioned medium.

These results suggest that cells which are beginning to develop into cartilage cells interact with nearby mesenchyme cells in some way (Chiakulas 1957). We do not know the mechanism of this interaction. Cell contact may be important.

Conditioned medium from sternal cell cultures stimulates proteoglycan synthesis in limb mesenchyme (Solursh and Rietter 1975a) but does not appear to enhance the conversion of mesenchyme into cartilage. But concentrated conditioned medium from limb mesenchyme culture does seem to accelerate cartilage differentiation in quail limb buds (Matsutani and Kuroda 1978). Close cellular association results in the production of more precellular matrix than usual, which also increases matrix component synthesis.

Exposure to dibutyrl CAMP or related drugs will promote cartilage clone formation (Solursh et al. 1981) and stimulate cell aggregates to form cartilage (Ahrens et al. 1977; Solursh et al. 1983). Even mesenchyme cells lying between aggregates which would normally form loose connective tissue form cartilage instead (Owens and Solursh 1981). The mode of action of this process is unknown. The chemical might inhibit cell division (Solursh and Rieter 1975b) alter cell shape, or increase intracellular levels of CAMP (Solursh et al. 1979).

MORPHOGENESIS OF CARTILAGE

Thorogood (1982) pointed out the important distinction between cartilage cytodifferentiation and cartilage morphogenesis. Morphogenesis of cartilage is fundamental because it underlies, and determines, much of the form of the vertebrate endoskeleton. Bones arising by endochondral and perichondral ossification depend on a cartilaginous template to produce a scaffold for subsequent ossification. Cartilage even affects, albeit indirectly, the form of membrane bones; Diewert (1980) showed the effects of Meckel's cartilage on the morphogenesis of the mandible. Articulations between the bones of the skeleton are also dependent upon cartilage and require very precise morphogenesis of joint form.

Morphogenesis is brought about by controlled growth. We may therefore expect mutants which affect the shape and size of cartilaginous skeletal elements to do so through the medium of growth either directly, or indirectly by affecting the control of growth. In either case we must be clear about the ways in which cartilage can grow, and the factors controlling this growth.

Cartilage is unique amongst skeletal tissues in being able to grow interstitially: a skeleton formed in cartilage is thus able to grow *en masse* by division of its chondrocytes without disturbing the tissues which surround it or attach to it. This also allows cartilage to grow very rapidly, much more so than bone which grows slowly by apposition and is remodelled from the surface. But, besides interstitial growth, cartilage may grow in other ways—by apposition of cells on its surface, by deposition of matrix, and by enlargement of cartilage cells.

Interstitial division of chondrocytes in the matrix-secreting blastema seems to be random (although zones of division have been reported in relation to the epiphyses). The products of division move apart, probably because they are secreting matrix, but the dynamics of the process is such that it is common to find two, three, four, or more chondrocytes, presumably the products of recent division, occupying a single lacuna in the matrix. The contribution made by cell proliferation seems to be higher in the early stages of cartilage growth, to be superseded by matrix secretion as the tissue becomes more mature. These factors are difficult to separate since newly divided cartilage cells are active secretors of matrix. Mitoses can be seen in immature and mature cartilage and the classic textbook picture is one of decreasing mitotic rate with increasing age. Stockwell (1979) considered this to be inaccurate. In areas where cartilage growth has been extensively studied (the growth plate, the epiphysis, and mammalian articular cartilage) studies with ^{3}H thymidine show that mitoses are commoner within the mass of tissue than subperiostially, contrary to what might be expected if growth were largely subperiostial (Messier and Leblond 1960; Fitzgerald and Shtieh 1977). In contrast Searls (1973) maintained that 75 per cent of chondrocytes in chick long-bone cartilage have dropped out of the cell cycle, whereas at the limb bud stage all cells are involved in division. It is possible that the mitotic region of the epiphysial growth plate might become exhausted, and this could become a growth-limiting mechanism. The contribution of mitosis to growth may be quite small: suppression of mitosis by X-irradiation in the embryonic chick tibiotarsus (Biggars and Gwatkin 1964) causes only a 10 per cent reduction in subsequent increase in length.

Appositional growth dates from the appearance around the cartilaginous blastema of a bilaminar perichondrium with an outer fibroblastic and inner chondroblastic layer. The latter seeds chondrocytes into the growing cartilage. Apposition can continue as long as the perichondrium is active, that is throughout the life of the organism. Different amounts of appositional growth at various points would, of course, lead to shape change but the morphogenic importance of this mechanism seems not to have been investigated.

Matrix secretion is also important in cartilage growth. Stocum *et al.* (1979) wondered if the different phases of cartilage-cell natural history were accompanied by different rates of matrix secretion. They were able to measure the lengths of various histological zones in the tibiotarsus of the developing chick embryo and obtain mitotic indices. As the cells progress from flattened to hypertrophic, well-marked peaks occur for both chondroitin sulphate and collagen synthesis; these peaks largely correspond, although the chondroitin sulphate peak leads that for collagen over the period 10–14 days. At 8 days the peaks are located in the hypertrophic zone and by 12 days have moved to the least differentiated proximal cells, showing that peak synthesis cannot be equated with a particular histological appearance.

Hicks (1982) found that matrix synthesis (measured by the uptake of ^{35}S sulphur into chondroitin sulphate) in the large tibial and small fibular chick limb primordia was very similar when corrected for unit volume. Instead of differential secretion, polarized secretion might, however, occur, i.e. all the cells comprising a population could secrete more matrix on one aspect than another. This would require that the cells were oriented with respect to each other—commonplace in differentiated cartilage (the growth plate for example). Gould *et al.* (1974) turned this argument on its head and proposed the reverse, that cellular orientation in developing long bones is brought about by the forces of matrix secretion from an initially unpolarized and unoriented cell population within the primordium. The radial whorls of chondrocytes seen within the blastema arise, according to Gould, by progressive accumulation of matrix produced by central cells which pushes them apart and compresses more peripheral cells including those of the perichondrium. This hypothesis is not in line with the findings of Ede *et al.* (1977), Trelstad (1977), and Holmes and Trelstad (1980) who have shown that orientation and polarity exist within the developing blastema before matrix secretion commences.

Cell enlargement begins to play a part in cartilage growth as cartilage matures. Even in cartilages not destined to ossify considerable enlargement of cells near the centre has been noted (Stockwell 1979). Along the length of a cartilaginous bone primordium a variety of cell shapes can be seen. The epiphyses consist largely of small rounded cells which merge towards the diaphysis into a region of flattened cells oriented at right angles to the long axis of the rudiment. These cells will later become organized into the epiphysial growth plates, but at this stage simply merge into a mid-diaphysial region of large rounded chondrocytes.

These large central cells will later hypertrophy if the cartilage rudiment is to ossify; a rather poorly understood process involving a great increase in cell volume and commonly death of the enlarged chondrocytes. In the mandibular condyle, however, hypertrophy is not accompanied by cell death and the chondrocytes survive (Silbermann and Frommer 1972a); cell survival may also occur at other sites. Biggars (1957) blocked hypertrophy in the chick tibiotarsal primordium by placing the tissue for 1.5 hours in glycerol saline at −79°C, a treatment not severe enough to inhibit mitosis on subsequent thawing and culturing. Diaphysial growth was much reduced by this process, but ephiphysial morphogenesis was almost indistinguishable from that seen in normal controls. Wolpert (1981) suggested that the degree of hypertrophy might be a contributing factor to the non-equivalence of limb bone primordia. Hicks (1982) experimented on chick tibia and fibula cartilages aged 6 days. At this stage the tibia was four times the volume of the fibula, but after a further 6 days in culture it was 12 times as large. The relative contributions of cell division, matrix deposition, and cell-volume increase to this differential growth were all measured, and the greatest effect found to be due to the greater expansion of tibial cells during hypertrophy.

The ordered arrangement of blastemal cells has fundamental consequences for cartilage morphology. In longitudinal sections of long-bone blastemata polarity is already evident (Fell 1925; Thorogood and Hinchliffe 1975) and becomes more obvious with further development. Trelstad (1977) suggested that blastemal cells are polarized just as epithelial cells are, although in the latter the polarization is much easier to see. The axis running through the nucleus and the Golgi apparatus can be used to assess cell polarity, and ordered patterns have been located in this way in long-bone blastemata (Trelstad 1977 in chick; Holmes and Trelstad 1980 in mouse), vertebral blastemata (Trelstad 1977), and metacarpal blastemata (Ede *et al.* 1977) as well as in non-cartilaginous tissue (Holmes and Trelstad 1980). These patterns correspond with the piezoelectric polarity seen in adult skeletal elements (Athenstaedt 1968, 1970, 1974): this is a property of bone generated by the fibrillar components of the matrix, principally by collagen, and suggests a fundamental role for collagen in cell polarization.

THE LONG-BONE MODEL

Perhaps the relative importance of these components of growth and morphogenesis can best be understood by looking at the growth of a familiar system, the mammalian long bone.

The mammalian long bone develops from a cartilaginous blastema, the potential methods of growth of which have already been discussed. Most of the cell division responsible for the elongation of the blastema occurs towards the ends of the model. The chondrocytes in the midsection mature and hypertrophy and eventually the surrounding matrix calcifies. The chondrocytes of this region are widely believed to die (but cf. the mandibular condyle).

The primary centre of ossification

A bud of osteogenic cells, capillaries, and pericytes derived from the periostium enters the degenerating cartilage at a site near the middle of the cartilaginous model. Bone begins to be deposited upon calcified cartilage matrix. Once the primary centre of ossification has developed, the bone primordium is divided into three distinct regions, the central diaphysis and two cartilaginous epiphyses.

Secondary centres of ossification

Within the cartilaginous epiphyses secondary centres of ossification may appear, by a mechanism similar to that which forms the primary centre. Cartilage is retained as an articular covering surrounding the secondary centre of ossification over the joint surface and as a growth plate, a transverse disc of cartilage extending the full width of the bone and separating epiphysis from diaphysis.

THE GROWTH PLATE

Although the active growth plate varies in detail amongst mammals, it conforms to a recognizable basic plan. The chief feature of this plan is that chondrocytes at various stages of maturation may be found; these have been divided arbitrarily, although really a merging continuum, into six zones. From the epiphysis to the metaphysis these are (Fig. 3.3):

1. Inert zone;
2. Zone of reserve cells;
3. Zone of proliferation;
4. Zone of maturation;
5. Zone of hypertrophy;
6. Zone of vascular invasion.

Inert zone

This is a wide expanse of cartilage in which chondrocytes are apparently randomly arranged, separated by large amounts of matrix. The epiphysial blood vessels terminate on the epiphysial aspect of this zone. Mitosis are rare, and the inert zone probably contributes little to growth through cell division.

Zone of reserve cells

The cells of this zone are nearly spherical in cross-section and again irregularly scattered in the matrix. The reserve cell zone is variable and may be absent or indistinguishable from the inert zone. In the rabbit (Rigal 1962) this zone labels well with ^{3}H thymidine suggesting that its function is as a stem-cell zone, but in the rat the labelling index is much lower.

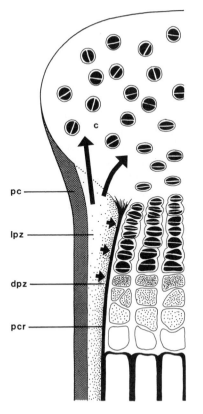

Fig. 3.3. Cell columns and the ossification groove. The bone shaft is covered by a perichondrium (pc). The bone end is capped by cartilage (c). Cells of the loosely packed zone (lpz) are thought to be cartilage precursors, those of the densely packed zone (dpz) osteocyte precursors contributing to the perichondrial ring of bone (pcr).

Zone of proliferation

The chondrocytes of the proliferative zone are typically discoidal and (in mammals) arranged into regular columns. Each chondrocyte is separated from its neighbour above and below only by a thin tranverse septum of matrix, while rather more matrix separates individual columns. Column formation is partly due to the longitudinally running collagen fibres and partly due to characteristic division of the chondrocytes (Dodds 1930), the axis of whose mitotic figures lies perpendicular to the long axis of the bone. Following division, two hemispherical daughter cells lie close together, then become flattened as each grows past its neighbour. Finally the two daughter cells become separated by a thin septum of cartilage matrix; this last stage may be delayed so that two, four, or eight cells may be found in a single matrix lacuna. Cell proliferation is maximal in this zone (Kember 1960; Messier and Leblond 1960).

Zone of maturation

The chondrocytes in this zone enlarge and lose their characteristic discoidal shape; this increase in size continues throughout the zones of maturation and hypertrophy until, when the matrix surrounding a cell calcifies, its vertical height has increased some five times (Brighton 1978). In this zone cytoplasmic glycogen increases and can often be seen by suitable staining under light or electron microscopy. Calcium also accumulates on mito-chondrial and cell membranes (Brighton and Hunt 1974, 1976) and many enzymes, particularly alkaline phosphatase, are at their highest concentrations.

Zone of hypertrophy

The sequence of degenerative changes which occur in this zone is classically thought to lead to the death of most chondrocytes. The cytoplasm becomes vacuolated, the nuclear envelope fragments, and cell organelles are lost. Calcium is released from mitochondria and cell membranes.

The matrix and chondrocytes in maturation and hypertrophic zones undergo a series of changes. It appears that calcium accumulation and ATP production are mutually exclusive in cartilage cells (Lehniger 1970). Calcium accumulation is energy dependent and in the absence of ATP production and in low oxygen tension, glycogen reserves are utilized. Once glycogen is exhausted the system breaks down and calcium is liberated. The proteoglycan in the hypertrophic zone becomes disaggregated (Howell 1971; Howell *et al.* 1969); aggregated proteoglycan is believed to inhibit calcification. Hypertrophic zone chondrocytes are also rich in lysosomes (Pita *et al.* 1975).

Calcification

This process may be thought of as fixing the growth which has occurred due to proliferation and expansion of the chondrocytes. (See Stockwell (1979) for a recent review.) Once glycogen reserves have been depleted from the hypertrophic chondrocytes, calcium is liberated. Matrix vesicles are at their highest concentration in the zone of hypertrophy (Pita *et al.* 1975) and are thought to act as nucleating agents on which crystals can be seeded, as the first identifiable apatite crystals are seen in apposition to these vesicles (Bonucci 1969). This process will be dealt with more fully in Chapter 4 when we consider bone.

Zone of vascular invasion

The growth plate does not increase in thickness, despite the continual proliferation and growth of chondrocytes, because of the resorption of mineralized and unmineralized cartilage matrix by the metaphysial vessels. The transverse septa of the lower hypertrophic zone are rarely mineralized (Schenk *et al.* 1967) and only two-thirds of the longitudinal septa are mineralized. The transverse septa are eroded by the endothelium of the capillary loops, followed by the remainder of the uncalcified cartilage. Calcified cartilage is removed by chondroclasts. The longitudinal bars which remain project into the diaphysis of the bone and later have bone matrix deposited on their surface.

Closure of the epiphysis

Epiphysial closure is a gradual process resulting in a progressive diminution in the growth plate until it is reduced to a slow-growing remnant or is completely resorbed. Each epiphysis will close at a specific time. The number of chondrocytes diminishes, leaving very short columns with few hypertrophic cells separated by large deposits of matrix. At some stage the growth plate will be pierced by capillaries uniting metaphysial and epiphysial blood vessels. More perforations appear and the whole plate then usually disappears. Both peripheral and central invasions by blood vessels have been described (Hains 1975; Hains *et al.* 1967).

CARTILAGE GROWTH AND BONE GROWTH

In the mammalian long bone three processes of growth are important and depend upon cartilage: elongation of the diaphysis; increase in width of the growth plate (though not increase in width of the diaphysis which is due to intramembranous ossification); and growth of the articular surfaces. These can be considered separately.

Elongation of the diaphysis

This is often measured by the longitudinal growth of the bone which replaces the cartilage, a relatively precise technique. Sissons (1956) pointed out that cartilage growth in this dimension was a product of cell growth rate and the length of the column of hypertrophied cells; as the latter is sensibly constant, the rate of bone growth is an index of the rate of production of chondrocytes in the growth plates. These rates vary widely according to species, age, and which end of a particular bone is being measured (Hinchliffe and Johnson 1982). Serafini-Francassini and Smith (1974), for instance, noted that the rate of elongation of the upper and lower ends of the human femur were the same for the first year of life, but subsequently 70 per cent of the growth was at the lower end. Closure of one or other epiphysis may also lead to growth solely at one end of a long bone; if an epiphysis is accidentally or experimentally damaged, compensatory growth will occur at the other end of the bone (Hall-Craggs 1969).

The cell kinetics of cartilage growth were reviewed by Kember (1982). We have seen that the elongation of the diaphysis is by the division of chondrocytes which form into columns of flattened cells which have been likened to piles of coins. In mammals (and in various lizards) the mitotic spindles are transverse, i.e. at right angles to the length of the longitudinal columns. Sissons (1956) estimated 7.5 new cells per column per day in the tibial growth plate of four-week-old rats; Kember (1960) found five new cells per day in his 'standard' 25-cell column. During maturation this rate falls in parallel with the decrease in elongation of the shaft (a 50 per cent decrease in mouse or rat, Tanner 1962; Walker and Kember 1972a,b) accompanied by a decrease in the number of cells in the proliferating zone and a decrease in the diameter of hypertrophic cells (35–21 μm, Walker and Kember 1972a,b; Thorngren and Hansson 1973).

Increase in width of the growth plate

The growth plate also increases in width to maintain the proportions of a growing bone. This involves an increase in the number of cell columns (Dodds 1930) which is believed to occur at the margins of the plates. Lacroix (1951) suggested that central growth is less likely because the columns are held parallel by calcified cartilage matrix at their lower ends. Growth in this dimension is now thought to be centred in the ossification groove (Tonna 1961) where fibrocellular tissue is associated with growth and mitotic figures. Shimomura *et al.* (1973) noted that new cells here orientate in the same direction as pre-existing cartilage cells, so contributing to growth in both length and diameter. Rigal (1962) suggested that growth here is interstitial, and Shapiro *et al.* (1973) suggested that the fibrocellular tissue contains separate groups of progenitor cells. Densely packed cells near the ossification groove are progenitors of osteoblasts, more loosely packed ones chondrogenic. The transverse growth of the epiphysis has been estimated (Stockwell 1979) as around 1/10 of the increase in length per day.

Articular surfaces

Articular surfaces must also grow, and we know that both interstitial and appositional growth are important, but few data on relative importance exist. The epiphysial region is perhaps best looked at in three chronological stages, the early chondroepiphysis, the pre-ossification, and post-ossification stages. The early chondroepiphysis comprises the period from the first deposition of matrix to the formation of the joint cavity. The two cartilaginous ends of adjacent long bones are at first separated by a fibrous interzone which will give rise to the joint cavity. At first appositional growth takes place over both interzonal and perichondral surfaces of the cartilaginous bone *anlage*, then the interzone differentiates into three; the central part develops a joint cavity and the two outer parts continue, for the moment, as chondrogeneous tissue contributing to the ends of the bone. A little later these layers cease to produce new cartilage and become embedded in the smooth matrix of the joint surfaces. The chondroepiphysis now depends on interstitial growth plus apposition from the fibrous perichondrium. Interstitial mitoses are confined to the mitotic annulus, a band of tissue lying below the articular surface. After ossification at a secondary centre most of the chondroepiphysis is replaced by bone. Slow growth continues by proliferation of the mitotic annulus.

SYNCHONDROSES AND SECONDARY CARTILAGE

In the mammalian skull rather special cartilaginous features exist. The cranial base is made up of a series of midline cartilaginous plates meeting in a series of synchondroses resembling epiphyses but with ossification zones on either side.

Cleall *et al.* (1968) and Hoyte (1971) have shown that growth in man, guinea pig, and rat is by equal increments on both sides of the joint but De Brul and Laskin (1961) and Vilmann and Moss (1980) suggest that local differences in growth may occur within a single synchondrosis, either between sides or within a single side.

Secondary cartilage (see Beresford 1981 for review) has features in common with embryonic cartilage (large haphazard chondrocytes, sparse matrix, and certain functional and pathological properties—Durkin 1972). The major secondary cartilage, that of the mandible, is an important centre for mandibular growth. A plate of hyaline cartilage is here sandwiched between a fibrous articular zone and the bone of the mandible. Blackwood (1966) described an articular zone and an intermediate zone producing cells which pass as chondrocytes through the underlying cartilaginous zone in 5–6 days to emerge in the marrow cavity beneath. Silbermann and Frommer (1972*a,b*; 1974) confirmed these findings in the mouse and suggested that the surviving cells might form multinucleate chondroclasts or become osteoprogenitor cells.

Many factors therefore influence cartilage growth. We may expect to find mutations which disrupt any or all of these. The matrix may be formed of unusual components, both the mucopolysaccharides and the protein forming the mucopolysaccharide backbone being suspect as well as collagen. Or there may be a specific increase or decrease in one component, thus upsetting the balance. Or the matrix may be balanced but simply sparse. We can expect such matrix defects to originate in morphologically abnormal cells; we might also expect to see morphologically normal cells producing a morphologically normal matrix at a reduced rate, perhaps due to a decrease in the rate of cell division.

Perhaps the commonest reported type of cartilage mutant is a chondrodysplasia producing a characteristic disproportionate dwarf, probably because this phenotype is so easy to recognize. Chondrodystrophies act on the type of growth seen in the elongation of longbones, but not on increase in width. This produces a disproportion seen in every cartilage replacement bone in the body, although the effect is most marked in the limbs. We must ask ourselves if the mutations producing this phenotype are all similar, or if they merely attack a vulnerable tissue, or perhaps a vulnerable process.

THE MUTANTS

Mov 13 (Chr ?)

Perhaps the most interesting of the mutations affecting the deposition of extracellular matrix is also one of the most recently described. Schnieke *et al.* (1983) applied to the mouse a potentially revolutionary technique, previously worked out in *Drosophila* but unknown in mammals, of inducing a mutation by introducing a virus into the genome rather than waiting for a spontaneous mutation that might be of interest or using a mutagen.

Schnieke was able to introduce Moloney leukaemia virus into the germ line of mouse strains by exposing early embryos to the virus or by microinjection of virus particles into zygotes. She produced 13 strains of mice each with virus integrated at a different Mendelian locus. Twelve of these showed no change in phenotype but the thirteenth—Mov 13—proved to be a recessive lethal causing early embryonic death, which occurred between 13 and 14 days of gestation. Further investigation showed that the Mov 13 gene was expressed only by mesoderm-derived fibrogenic or myoblast cell lines, and transcribed during the second half of embryogenesis. Hybridization studies showed that the Mov 13 gene represented the structural gene for the alpha 1(I) collagen chain. The mutation thus effectively blocks the formation of type 1 collagen and presumably interferes with the formation of extracellular matrix. To date no studies on the morphology of Mov 13 embryos have appeared.

Brachypodism (bp, bp^H, Chr 2)

We have already seen that the brachypod mutation affects membranous condensations. It also affects the next stage in the process of skeletogenesis

(Milaire 1965; Grüneberg and Lee 1973) delaying the production of metachromatic cartilage matrix. Although perhaps atypical in that the effect is seen only in some of the limb bones, there is no doubt that the cartilage produced is abnormal.

Krotoski and Elmer (1973) looked at the alkaline phosphatase content of brachypod and normal 17-day hindlimbs. Alkaline phosphatase has been implicated in the cellular hypertrophy of chondrocytes. Krotoski and Elmer found that the amount of alkaline phosphatase was reduced, but that in all respects tested the enzyme appeared to be normal, and suggested that they were looking at a secondary effect due to the immaturity of the brachypod limbs at this stage; the number of cells containing the enzyme seemed to be the crucial difference between normal and mutant. This in turn seemed to stem from the inability (or delay) of the small number of cartilage cells present to hypertrophy. Rhodes and Elmer (1972) suggested that decrease in the deposition of ground substance by the chondrocytes was due to the inability of the cells to hydroxylate proline, causing a reduction in the intra- and intercellular levels of collagen.

Rhodes and Elmer (1975) found newborn bp fibulae histologically abnormal, shorter in length than normal, and, as expected, with a homogeneous chondrocyte population (Fig. 3.4). Using the whole bp fibula as an experimental tool was clearly not comparing like with like (since the normal fibula at birth is much larger and has zoned cartilage). To overcome this, measurements were related to DNA content. Collagen, as estimated by peptide-bound hydroxyprolene, was about twice as plentiful in bp as in normal and the amount of acid mucopolysaccharides increased 2.8 times. The ratio hexose:hydroxyproline was 3.85 (cf. 2.62 in normals). Since all these figures are based on unit DNA one explanation would be that the brachypod tissue was not dividing at the same rate as normal, or that excessive cell death was occurring. Measurement of DNA synthesis by methyl ^{3}H thymidine incorporation showed that in fact the brachypod had a higher value (at 48 hours + = 72.3, bp = 147.2 cpm/μg DNA/10^{-3}), suggesting a higher rate of cell division. This was also noted by Elmer and Selleck (1975)—see below. The authors suggested that the increase was either due to the whole of the chondrocyte population being in a proliferative state, or to the cell cycle being reduced in bp, but do not distinguish between the possibilities. As normal cartilage matures, so cell division rate decreases. These results could therefore reflect the immaturity of the bp fibula.

Rhodes and Elmer argued that normal differentiation into proliferating, maturing, and hypertrophying chondrocytes (not seen in the bp fibula, Krotoski and Elmer 1973) might depend on matrix secretion. Diegelmann and Petrofsky (1972) suggested that critical levels of collagen are necessary for the initiation of osteogenesis.

The high collagen content of mutant fibulae could be due to a different

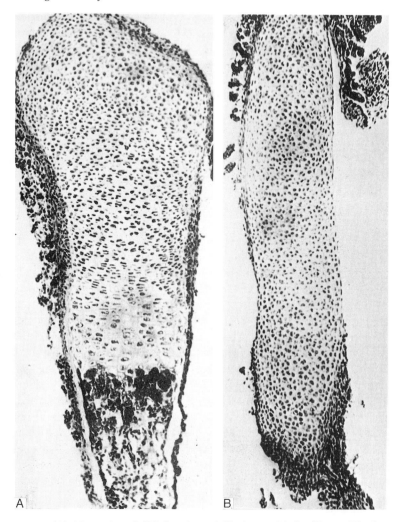

Fig. 3.4. (A) Normal and (B) brachypod fibulae at birth. (From Rhodes and Elmer 1975.)

rate of synthesis or degradation, or the production of a different collagen molecule. To test this, proline incorporation was measured, and found to occur at 54 per cent of the normal rate. Incorporation of ^{3}H-tryptophan (an amino acid not found in cartilage) was found to be at 56 per cent of the normal level, suggesting a generalized decrease in protein synthesis. The stability of the collagen formed in brachypod was assayed by following the loss of radioactivity (due to degradation or turnover of collagen) following a pulse of ^{3}Hproline. In the normal there was 40 per cent loss after three days, in the mutant a 25 per cent drop. The normal degradation of collagen at this time is associated with the initiation of ossification, a process not occurring in the brachypod fibula. The type of collagen synthetized by bp cells was shown to be normal, of the type expected from cartilage.

Similar experiments on synthesis and turnover of acid mucopolysaccharides were performed. Incorporation of ^{14}C-galactosamine was normal in bp, but a pulse-labelling experiment showed that there was virtually no turnover of acid mucopolysaccharides (AMPS) in mutants, against a 60 per cent loss in three days in normal littermates. The change from mesenchyme to cartilage is normally accompanied by a loss of hyaluronic acid and an increase in chondroitin sulphate synthesis (Toole 1972). To characterize the type of AMPS present a double label of ^{3}H galactose and ^{35}S sodium sulphate was used. In the normal, electrophoresis showed two peaks of activity. One band contained ^{3}H label only, and corresponded to hyaluronic acid standards; the second contained ^{3}H and $^{35}SO_4$ and corresponded to chondroitin sulphate standards. In brachypod only one peak was seen, containing both ^{3}H and $^{35}SO_4$ and coinciding with the chondroitin sulphate marker. In fact the normal peak corresponding to hyaluronic acid most probably represented unsulphated chondroitin, as it was removed by chondroitinase but not by hyaluronidase.

Rhodes and Elmer suggested that the delayed transition from cartilage to bone (and indeed from early undifferentiated to differentiated cartilage) is not due to abnormal collagen or chondroitin synthesis, but based on uncoordinated rates of synthesis and degradation.

Elmer and Selleck (1975), besides noting differences in the morphology of cultured brachypod cells (see Chapter 2), also carried out biochemical analyses. The DNA content of 12-day-old normal explants increased by 50 per cent during the first 24 hours of incubation; then no further increase was noted. In bp cultures the DNA content increased linearly over the first three days of culture. The significance of this finding is not further discussed, although it agrees with the distribution of mitoses seen in the cultures. More interestingly collagen synthesis was also measured as hydroxyproline content. At 12 days, on explanting, both normal and bp cells contained small and equal amounts of collagen. Both cell types synthesized the same amount of collagen during the first 12 hours of culture, and over three days brachypod produced an (insignificantly) reduced amount.

Elmer and Selleck also looked at the qualitative composition of the collagen formed. They performed two analyses under slightly different conditions. In the first half the culture medium was replaced with fresh medium supplemented with ^{14}C proline and in the second all the medium was replaced by ^{14}C glycine tagged medium in the case of normal explants and 5–^{3}H proline labelled medium in the case of brachypod tissue. In this series of experiments the medium was serum free. From the elution profiles of cells grown in serum based media it was clear that both normal and brachypod tissue synthesized type 1 collagen predominantly, with a little type 2 suggesting some fibroblast involvement. Overall the ratio of type 1 to type 2 was 9:1, characteristic of cartilage collagen. In the serum-free conditions both normal and abnormal cells produced an alpha-1:alpha-2 ratio of 4:1, i.e. both responded to the changed culture conditions by transcriptional and translational changes in the collagen synthesis pathway.

Duke and Elmer (1977) studied chondrogenesis in brachypod by means of rotation culture. Normal and mutant aggregates formed from mesenchyme of 12-day normal and brachypod hindlimbs were of similar size, shape, and cell number. Would such apparently similar aggregates show the abnormal chondrification typical of brachypod *in vitro*? Sections through normal aggregates after 24 h in culture showed several metachromatically staining areas (averaging 6.6 per aggregate, with a mean diameter of 183 μm), with cells close together, oriented in a 'circular manner', and surrounded by a perichondrium-like layer of spindle-shaped cells. Cartilage matrix accumulation was greatest in the centre of each condensation. Chondrogenic regions accumulated more ^{35}S than non-chondrogenic areas. After 24 h new condensations were not formed but each existing one increased in size by matrix deposition and annexation of the cells between condensations, so that the latter tended to fuse. After four days in culture the aggregates of cells usually contained only one condensation.

In brachypod chondrogenesis was strikingly different. The average number of condensations was 2.3 per aggregate and their average diameter 80 μm. Even after seven days incubation, 50 per cent of the brachypod aggregates remained non-metachromatic and showed no ^{35}S incorporation. Cartilage was first seen after 48 h (24 h later than in normals), although ^{35}sulphur accumulation was seen after 24 h. The cells in bp condensations did not assume the typical circular outline with attendant perichondrium. When distinct condensations were recognizable, their small size and numbers and random distribution in the medium mitigated against fusions.

In a later paper (Duke and Elmer 1978) the same authors studied the fusion of fragments of normal and brachypod limb-bud mesenchyme. Once again condensations in brachypod were small and failed to fuse. The combination of explants from normal and brachypod mice failed to influence cartilage differentiation in either: this is of interest in the light of Konyukhov's claim (Konyukhov and Ginter 1966; Konyukhov and Bugrilova 1968; Pleskova *et al.* 1974) that a diffusible proteinaceous inhibitor influences brachypod growth.

Duke and Elmer's (1979) study of the ultrastructure of brachypod tibial and fibular blastemata showed some differences in matrix-secreting cells. The tibial condensation at 12 days appeared 'quasi-normal' with a slight decrease in the distance between cells. Structures typical of precartilaginous cells were noted, with some endoplasmic reticulum expanded by granular matrix components; a small Golgi apparatus was present. The matrix was composed of small fibrils of collagen 15–20 nm in diameter and granules of proteoglycan, free or attached to the fibrillar collagen. Matrix material was scattered rather homogeneously between cells. In the fibula, however, there were extensive cell–cell contacts and many cell junctions (8–10 per cell, cf. a control value of 2–4). Extracellular material was in the form of regional deposits on the cell surface, and much less than in normal limbs; the fibrillar component of the matrix appeared to be coalescing into thicker fibres, although thinner fibres were still in the majority. Brachypod showed a deficiency of matrix granules (mucopolysaccharides) and had both normal 10–15 nm diameter fibres of collagen and much larger 45–95 nm diameter fibres. These thicker, banded collagen fibres have also been reported elsewhere in abnormal cartilage (Seegmiller *et al.* 1971—see below; Levitt *et al.* 1975; Kochhar *et al.* 1976); their appearance is thought to be linked to decreased mucopolysaccharide content of the matrix.

Shambaugh and Elmer (1980) looked at the development of glycosaminoglycans (mucopolysaccharides) in developing brachypod mesenchyme, measuring radioactive incorporation rates and steady-state levels of hyaluronic acid and chondroitin sulphate between 11 and 14 days of gestation. They measured total glycosaminoglycans (essentially hyaluronic acid plus chondroitin sulphate) colorimetrically, and used ^{3}H-glucosamine and ^{35}S-sulphate to assay hyaluronic acid, chondroitin sulphate, and glycoprotein synthesis.

They found that the rate of incorporation of ^{3}H-glycosamine into chondroitin sulphate in organ culture was constant from 11.5 to 12.5 days and then increased rather rapidly; this correlates with the visible appearance of cartilage matrix deposition. Normal and abnormal were almost indistinguishable at this point; the abnormals had less glucoronic acid per limb bud as would be expected from the smaller size of their condensations. In cell culture, however, normal cells incorporated significantly more ^{3}H-glucosamine than mutant cells.

Use of ^{35}S in organ culture was equivocal, since the variance in uptake was very great; in cell culture again uptake was signficantly lower in mutants than in controls. Taken together these results suggested reduced synthesis of chondroitin sulphate rather than reduced sulphation.

Hyaluronic acid incorporation increased up to day 12.5, then decreased. In the brachypod limb buds hyaluronic acid levels were greater than normal over the whole period of the experiment (11.5–14.5 days). The decrease at 12.5 days has been reported as due to an increase in hyaluronidase (Toole 1972); this was found in both normal and brachypod limb buds, with a slight delay in bp, perhaps due to the higher initial concentration of hyaluronidate. Loss of hyaluronidate in cell culture is also slower in bp than in normal.

Owens and Solursh (1982) suggested that bp cells may provide a defective chondrogenic stimulus. Stage-16–17 bp cells (10–10.5 days) had more normal levels of chondrogenesis than cultures of any other age, and were thus used in experiments designed to test the relative abilities of brachypod and normal cells to promote aggregate and nodule formation by stage-21 brachypod cells. Owens and Solursh found that there was a clear difference between the capacities of normal and bp stage-16 cells to induce cell aggregates and cartilage nodules in stage-21 normal cells, both in the number of aggregates per unit area and in the number of cartilage-containing nodules.

Owens and Solursh outlined evidence to suggest that cell–cell interaction is necessary for the formation of cellular condensations and suggested that the specific defect here is in the ability of the bp cells to induce chondrogenesis. This is consistent with the idea that there is a specific subpopulation of cells present in the limb which are responsible for production of a chondrogenic stimulus. bp affects both the initiation of cartilage nodule formation and cartilage nodule growth. Owens and Solursh suggested at least three stages may be present in the onset of chondrogenesis: (1) an inductive signal; (2) recognition of the signal; and (3) active response to the signal. They suggested that bp limb mesenchyme is unable to provide the inductive signal.

Obviously much work has gone into the elucidation of the brachypod problem and we now have many data on the mutant. There seem to be no chemical differences in the matrix components (if we discount the small amount of broad-banded collagen seen by Duke and Elmer) nor should we expect any. It would be a very unlikely mutant which governed the production of an aberrant collagen or an undersulphated or unusual chondroitin sulphate in such a localized area of the body, for brachypod, although considered here for its effects on the skeleton as a whole because of the thoroughness with which it has been investigated, is very much a gene with a local action, affecting only some parts of the limb buds; the tibia for instance is virtually unaffected, while effects on the fibula are large. Instead we see a much more plausible arrangement, a local delay in chondrogenesis brought about by a change in cellular adhesiveness which seems to become critical only in certain blastemal areas. The delay in chondrogenesis, dependent upon a delay in blastemal formation, leaves a legacy of tardiness, a temporal upset in the process of chondrogenesis which is never overcome.

The talpid chick

The talpid chick shows a similar but more widespread change in cell adhesion to that seen in brachypod. Is the cartilaginous matrix affected here?

Hinchliffe and Ede (1968) noted that in talpid[3] cartilage replacement bones failed to form, yet membrane bone ossified normally from skeletal blastemata. They followed the differentiation of the shoulder girdle as chorio-allantoic grafts. This region contains the membranous clavicle, which developed with striking normality in talpid and the cartilaginous scapula, coracoid, and proximal humerus, which did not. After explanting, at 4.5 days *in vivo*, normal and talpid chondroblasts were similar with cell diameter just beginning to increase, but not yet orientated with their long axis at right angles to the developing rudiment except in small scattered areas. These unorientated cells had a round nucleus and a circular outline in section. Later the centre of the talpid rudiment began to hypertrophy with the chondrocytes staining positive for glycogen but negative for phosphatase activity, these hypertrophic areas being surrounded by still disorientated 'young' chondrocytes. Later still, almost all of the scapular cartilage hypertrophied, to a stage where it stained for both glycogen and alkaline phosphatase, i.e. the last state of hypertrophy. These cells were swollen, possessed vacuolated cytoplasm, had an alkaline phosphatase positive cytoplasm, and sat in an alkaline phosphatase positive matrix. Other rudiments remained in 'arrested hypertrophy' with small cells with no alkaline phosphatase activity. The talpid perichondrium did not differentiate normally, remaining as a layer of elongated fibroblasts to the outside and a region of undifferentiated mesenchymal cells to the inside.

We are obviously dealing here with a problem of bone formation, and will return to this topic in Chapter 4. But Hinchliffe and Ede asked the interesting question whether the lack of bone formation is due to a failure of its induction by talpid cartilage, implying a defect in the cartilage itself rather than a hangover from the abnormal blastemal formation. Talpid cartilage seems to fare better than the brachypod fibula since it hypertrophies, but this may be purely a size effect as the hypertrophy is patchy.

Ede and Flint (1972) looked at reaggregations of talpid and normal limb mesenchyme in culture. They found 'precartilage areas' which stained metachromatically with alcian blue. Within these areas were one or more condensations with cells orientated in concentric rings about a central group of cells. Mucopolysaccharide staining was deepest at the centre of these condensations. In talpid cultures condensations were less well defined (Fig. 3.5) though of similar structure, but towards the centre cells of adjacent rings were less clearly separated by layers of matrix. Talpid condensations, although occupying more area than normals initially, tended to fall behind (77 per cent increase in normals, 56 per cent in ta at four days of culture). Talpid aggregates also tended to have more condensations than normal; in talpid the greater area of condensation was composed of large numbers of small aggregates. ^{35}S uptake was no different from normal in talpid cultures.

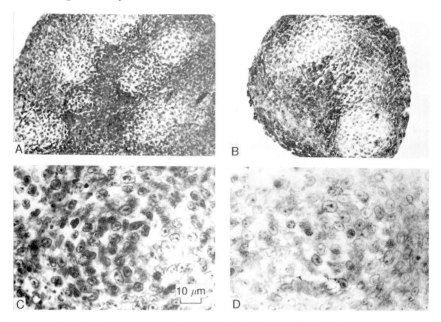

Fig. 3.5.   Two-day aggregates of normal (A,C) and talpid³ (B,D) cells. (From Ede and Flint 1972.)

Congenital hydrocephalus (ch, Chr 13)

One other of the genes described in Chapter 2 as affecting blastemal formation is also known to affect cartilage matrix. Breen *et al.* (1973) found that sternal cartilage from ch/ch mice aged 15–18 days, 19–21 days post partum had only around 40 per cent of the acid glycosaminoglycans of normal littermates. The ratios of chondroitin 4 sulphate to total glycosaminoglycans and sulphate to galactosamine were constant in all genotypes. Cartilage from ch/ch was also easier to cut than that from normal littermates: calcification is delayed by about seven days in ch heterozygotes. Again later stages of skeletogenesis are deranged by earlier events, and again the defect seems to be quantitative rather than qualitative.

CHONDRODYSTROPHIES

We now come to a group of abnormalities typifying what we normally think of as achondroplasia, i.e. disproportionate dwarfing. However these conditions are better referred to as chondrodystrophies, of which achondroplasia is but one. In man this imprecision in terminology has led to confusion; in laboratory animals conditions labelled achondroplasia (the achondroplastic mouse, rabbit, etc.) may or may not correspond to achondroplasia in man. It is wisest to assume that they do not; often the first described chondrodystrophy in a particular species is labelled achondroplasia (as it was in man) and later discoverers have to hunt for synonyms.

In this group of chondrodystrophies all the cartilage replacement bones in the body are affected, membrane bone is only secondarily involved, and the effects seem to make themselves obvious rather late in fetal development, or after birth, affecting the elongation of long bones through action in the epiphysial growth plate. Many conditions like this have been described, but the greatest information comes from beyond the level of the light microscope which merely reveals a deranged epiphysis; we shall deal mainly with a group of chondrodystrophies, one from the chick, one from the rabbit, and the remainder from the mouse where ultrastructural and/or biochemical observations have been made.

The nanomelic (nm) chick

The nanomelic chick (Landauer 1965) is a recessive, lethal at hatching, or just before at 18–21 days *in ovo*, with a large brachycephalic head and greatly reduced limbs. All long bones, including carpals and tarsals, are reduced in length but quite broad. There are no consistent abnormalities other than those arising from disproportion in the wings and feet, although a few embryos have been described with sporadic defects, e.g. absence of wings and sirenoidea.

Mathews (1967) looked at the acid mucopolysaccharides in sternal and limb cartilage from 17-day-old nanomelic chicks. He found that normal cartilage contained 60 per cent of its acid mucopolysaccharide as chondroitin sulphate A, 30 per cent as chondroitin sulphate C, and 10 per cent as unsulphated chondroitin. In nanomelic cartilage this distribution was unaltered, but the total was reduced to 10 per cent of its normal value. Nanomelic bone had a rather smaller defect of acid mucopolysaccharides. In skin acid mucopolysaccharides were normal and the absence of chondroitin sulphate breakdown products in the amniotic fluid led Mathews to suppose that the defect in nanomelia lay in the synthesis of chondroitin. Collagen levels were normal in cartilage but decreased in bone, perhaps as a result of defective mineralization.

Fraser and Goetinck (1971) looked at sternal cartilage and were able to demonstrate first that mutant and normal sternae contained equivalent amounts of protein and DNA, and secondly that in culture nanomelic chick cartilage accumulated ^{14}C and ^{35}S at about one-third of the normal rate over a six-hour period. This accumulation, of course, measures the balance between synthesis and degradation; pulse-labelling experiments showed an equivalent loss or degeneration of labelled products to the culture medium in normal and mutant cultures, confirming Mathews' suggestion that the defect was in synthesis. In another experiment they plotted the concentration of tricholoracetic acid precipitable (that is protein linked) material produced from ^{14}C-labelled glucose. This fraction was accumulated at 43 per cent of the normal rate by nanomelic sterna and chondroitinase ABC removed 43.3 per cent of the label from nanomelic cultures, as against 69.8 per cent from normals. The amount of chondroitinase ABC insensitive label was similar in both cultures, indicating a specific defect in protein-linked chondroitin sulphate metabolism. This was confirmed by measuring the hexosamine content of hydrolysed polysaccharide which proved to be 23 per cent in nanomelics and 39 per cent in normals. What is being measured here is hyaluronidase-sensitive galactosamine-containing poly-saccharide, i.e. chondroitin or chondroitin sulphate. Sephadex gel filtra-tion of radioactive polysaccharides suggested, since there was no difference in elution profiles between normal and nanomelic cartilage, that the latter was producing chondroitin sulphate at one-third of the normal rate, rather than making a normal number of chains of one-third of the normal length. The similarity of uptake rate of ^{14}C-labelled glucosamine and ^{35}SO$_4$ in normals and nanomelics suggests that the chains that are produced are normally sulphated.

Palmoski and Goetinck (1972) studied the synthesis of proteochondroitin sulphate complex in cell culture. Nanomelic cells incorporated 9 per cent of the control value over 48 hours of culture, close to the original 10 per cent value obtained by Mathews (1967). Chromatography of the culture medium (where 90–5 per cent of the proteochondroitin sulphate was found) showed that normal proteochondroitin sulphate had two peaks, whereas nanomelic had only one, corresponding to peak II of the control which normally represents 14 per cent of the whole. This suggested that normal and nanomelic cartilage produce equal amounts of the peak II proteochondroitin, whilst nanomelic cartilage produces no peak I type. A similar result was obtained by treating normal cells with bromodeoxy-uridine, a teratogen thought to inhibit the synthesis of cell-specific (luxury) molecules. Bourett *et al.* (1979) noted that cyclic AMP levels are reduced in nanomelia.

Pennypacker and Goetinck (1976) thought that the absence of cartilage-specific proteochondroitin sulphate might affect collagen biosynthesis, Mathews' (1967) results notwithstanding. A continued synthesis of the earlier Type I collagen type, rather than a switch to cartilage-specific Type II collagen would not necessarily affect the total amount of hydroxyprolene present. In a proteoglycan separation based on a 1 per cent agarose column they found that normal 14-day chick sterna produced three peaks, 90 per cent of incorporated radioactivity residing in peaks Ia and Ib, and 10 per cent in peak II. In nanomelia the chromatographic technique disclosed a small synthesis of peak I material (10 per cent of the total) and normal quantities in peak II. This confirms the reduced synthesis of proteoglycans found by other workers.

Pulse labelling with ^{3}H-glycine was followed by collagen extraction and carboxymethylcellulose chromatography. Chromatographs from normal and mutant sterna were virtually identical, suggesting that collagen type was not affected by the mutant. The percentage of lysine residues in collagen synthesized by the mutant was also normal.

McKeown and Goetinck (1979) posed the interesting question of whether the cartilage derived from neural crest was affected by the nanomelia gene in the same way as that derived from mesenchyme. They carried out studies of synthesis rate and analysed proteoglycans from sternal and Meckel's cartilage. These results confirmed earlier studies on the sternum and showed that the ectomesenchymal cartilage of the mutant was identical to that derived from mesenchyme.

Pennypacker and Goetinck (1976) also looked at the ultrastructure of cartilage to assess the effect of the biochemical changes seen in the matrix (Fig. 3.6). The nanomelic chondrocytes appeared normal, although there was a significant reduction in the amount of extracellular space, with an increase of 50–60 per cent in the number of chondrocytes per unit area. The intracellular matrix was notable for a dense mesh of collagen fibres and a severe reduction in the number of matrix granules (corresponding to the reduction in AMPS). The collagen fibres in the matrix are of normal diameter, and fail to display 64-nm periodicity.

If nanomelic chondrocytes are exposed to para-nitrophenol-beta-D-xyloside (a stimulant of mucopolysaccharide synthesis) chondroitin synthesis levels are raised to normal values. The chondroitin sulphate synthesized in the presence of the xyloside is the same in size and composition as that produced by normal cells in similar conditions (Stearns and Goetinck 1979).

The normality of xylosyltransferase and of proteoglycan link protein in nanomelia (McKeown-Longo 1981) taken with the xyloside studies suggests that the defect may be at the level of proteoglycan core protein. Direct evidence in support of this hypothesis was provided by Argraves *et al.* (1981) who radioactively labelled proteins from normal and mutant chondrocytes then separated them by polyacrylamide gel electrophoresis. Normal extracts carried a labelled band of m.w. 246 000 which was absent in the mutants. This protein could be precipitated in normals by an antiserum directed against normal cartilage proteoglycan monomer.

McKeown and Goetinck (1979) showed that nanomelia produced only 10 per cent of proteoglycan which sediments in the same position as cartilage-specific proteoglycans after sucrose density treatment. This minor amount may represent a small population of proteoglycans present in both normal and mutant but masked in the former, or there may be a little synthesis in nanomelia of normal or abnormal core protein. But the data strongly suggest that in nanomelia core protein is essentially absent.

It seems from both biochemical and ultrastructural evidence that the

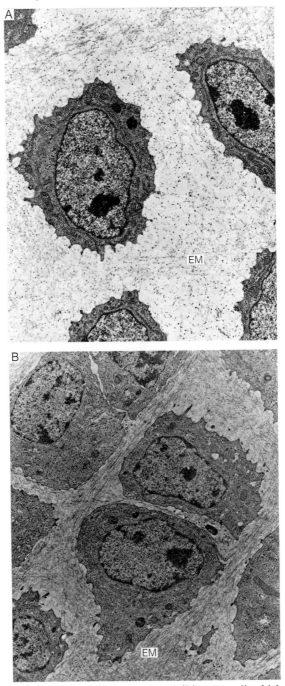

Fig. 3.6.   Sternal cartilage of (A) normal and (B) nanomelic chick embyros aged 14 days. EM: extracellular matrix. (From Pennypacker and Goetinck 1976.)

nanomelic chick gene disrupts protein-mucopolysaccharide synthesis while leaving collagen unscathed, suggesting that the two main components of the matrix are not co-regulated.

Cartilage matrix deficiency (cmd, Chr ?)

Cmd homozygotes are usually liveborn, but succumb almost immediately (Rittenhouse *et al.* 1978). They have a short snout, short extremities, and a cleft palate. At birth long bones are rather less than half normal length. Calcification appears to proceed normally; membrane bones are not affected. cmd tracheal rings show tightly packed chondrocytes with little matrix and many pyknotic cells often containing conspicuous vacuoles (Fig. 3.7). cmd cartilage stains with toluidine blue and also with picofuschin, a collagen stain which does not normally stain cartilage matrix. The authors infer from this that cmd cartilage matrix must contain unusual amounts or types of collagen. This is borne out by electron microscopy where cmd matrix is seen to be crisscrossed by broad collagen fibres in a sparse background of mucopolysaccharides (Fig. 3.8); chondro-

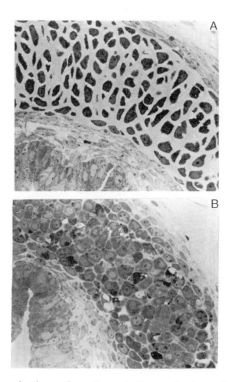

Fig. 3.7. One μm plastic sections through the tracheal cartilage of (A) normal, (B) cmd/cmd 17–18-day-old fetuses. (From Rittenhouse *et al.* 1978.)

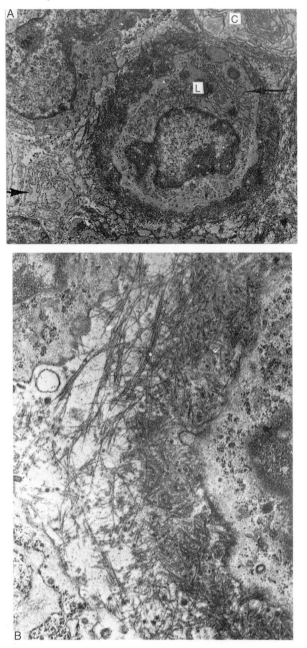

Fig. 3.8. (A) The cmd chondrocyte has lipid inclusions (L), abundant endoplasmic reticulum (arrow), and is invested by a dense collagenous mesh. C is a probable pycnotic chondrocyte. (B) At higher magnification this mesh is seen to be made up of an orderly arrangement of thick collagen fibres. (From Rittenhouse *et al.* 1978.)

cytes are characteristically surrounded by a dense ring of collagen fibres. These do not show 64-nm banding. Mucopolysaccharide granules in the matrix are large and often clustered.

Kimata *et al.*(1981) found that the dry weight of cmd 18-day fetal knee joints was less than half the normal value, but that there was no decrease in amino acid content per mg dry weight. Hydroxyproline content was normal in cmd, but hexosamine and hexuronic acid content was markedly reduced (glucosamine 60 per cent, galactosamine and hexuronic acid 20 per cent normal values).

Incubation with ^{3}H- and ^{35}SO$_4$-labelled precursors showed no proportionate difference (i.e. the H:SO$_4$ ratio was normal) but the incorporation of ^{3}H was 44 per cent and ^{35}SO$_4$ was 18 per cent of the control value. In the controls only small amounts of radioactivity were released into the medium from incubated knees, but in cmd one-third of the label leaked out into the medium and a further third was found in the first extraction fraction, suggesting increased solubility.

^{35}S radioactivity was followed in glycerol density gradient centrifugation. In the normal the first and second extracts gave a typical pattern of cartilage, with proteoglycan segregating into two distinct fractions. The faster sedimenting proteoglycans accounted for 90 per cent of the total ^{35}S sulphate incorporation. No corresponding peak was seen in the mutant: most of the ^{35}S radioactivity (90 per cent) was seen in the slower sedimenting fractions common to both normal and mutant. Sephadex 2B column chromatography and caesium chloride density gradient centrifugation confirmed that the high molecular weight cartilage proteoglycan was absent from cmd cartilage, although low molecular weight proteoglycans were present.

The glycosaminoglycan samples prepared from the glycerol density gradient centrifugation were further analysed by Bio-Gel column chromatography: in the normal all the S-labelled material was retained by the column; in the abnormal almost all was not. The slower sedimenting fractions for the normal were resolved into three components with respect to ^{3}H label, the middle peak also carrying substantial amounts of ^{35}S label. In the mutant there were two peaks labelling with ^{3}H, one coinciding with the ^{35}S label. These peaks were sensitive to chondroitinase ABC and ACII. Both mutant and normal therefore contained mostly 4-sulphated chondroitin sulphate. Half the ^{3}H radioactivity in the first normal peak was sensitive to hyaluronidase, and therefore represents hyaluronic acid. The first peak of the mutant, however, contained a mixture of chondroitin sulphate and hyaluronic acid, suggesting that the chondroitin sulphate chains of the mutant are larger than normal, and are excluded from the column with hyaluronic acid.

The synthetic potential of mutant cartilage was tested with beta-D-xyloside: normal and mutant reacted identically, suggesting that this exogenous stimulant for the initiation of chondroitin sulphate synthesis was able to affect both mutant and normal cartilage equally. Transfer of xylose on to core protein was also normal. Collagen synthesis was almost normal, except that again labelled protein was present in greater quantity in mutant than in normal culture medium.

These experiments give similar results to the picture seen in the nanomelic chick and suggest that the defect in cmd might also be the absence of core protein from cartilage proteoglycan. To test this, rat chondrocarcinoma was used to raise specific core protein antibody in rabbit. Control cartilage stained strongly with this antibody, but no reaction was seen in the mutant indicating the absence of core protein. Both types of cartilage stained strongly with type II collagen antibody; significantly this fluorescence—usually seen only after hyaluronic treatment (since the proteoglycan core protein blocks the reaction—von der Mark *et al.* 1976) was present in the mutant without hyaluronidase pretreatment. In the mutant there was a dense type II collagen staining surrounding certain cells—equivalent to the dense collagen seen in light and electron microscopy.

It seems clear that the defect in cmd relates to the absence of core protein from cartilage proteoglycan, and that cmd and nanomelia thus represent similar defects in two different experimental animals.

Spondylo-metaphysial chondrodysplasia (smc, Chr ?)

Johnson (1984) recently described a condition which in some ways resembles cmd. The recessive gene for spondylo-metaphysial chondro-dysplasia produces fertile males and non-fertile females which are dwarfed, with thick tails (sometimes kinked) and short limbs. Long bones are bowed and epiphyses enlarged. Ribs are shortened and costo-chondral junctions splayed.

In light microscope sections smc cartilage shows clear vacuoles in the chondrocytes, and the matrix stains unevenly (Fig. 3.9). In semithin

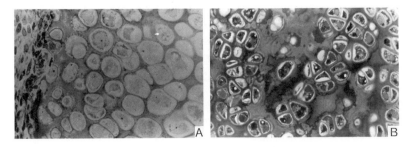

Fig. 3.9.   Condylar cartilage from normal (A) and spondylo-metaphysial chondro-dysplasia (B) mice aged 18 days. Plastic 1μm sections stained with methylene blue.

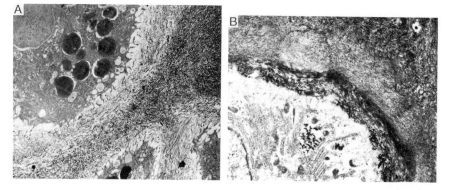

Fig. 3.10.   smc cartilage consists of close set chondrocytes with lipid droplets in the cytoplasm set in a dense matrix (A). Around each chondrocyte (B) there is often a dense collagenous mesh (cf. Fig. 3.8). Pale circular patches (arrow) are often visible in the matrix.

sections stained for collagen with toluidine blue the matrix stains much more deeply around the chondrocytes. The vacuoles within the cells stain as if they contain lipid.

Electron microscopy shows cartilage with cells relatively closer together than normal, with a matrix rich in dense collagen fibres, rings of which, as in cmd, surround the chondrocytes (Fig. 3.10). In the matrix circular pale patches are visible criss-crossed by collagen fibres. In cells which are surrounded by collagen meshes the space between cytoplasmic processes and matrix is occupied by aggregations of large, broad collagen fibres. The relationship, if any, between these deposits and the circular area within the matrix, but distant from the chondrocytes, is not clear.

Thurston *et al.* (1985*b*) noted that the long bones of smc mice were very abnormal. At 10 days post-partum the long bone still resembles the much earlier cartilaginous model. The epiphyses are small and the shaft bowed by a large mass of cartilage which projects into the diaphysis. The cartilage cells undergo hypertrophy but not degeneration; mitotic figures are found deep in the hypertrophic zone. A large acellular area is usually present in the centre of the cartilage. Secondary centres of ossification are absent in mice aged up to 40 days, although the latter may show limited vascular invasion of the epiphyses. Limited secondary centres of ossification are seen in older mice. Columns of cartilage cells are absent: at 18–21 days chondrocytes appear to be arranged in rows at 45° to the normal columnar orientation.

Kinetic studies are difficult in the absence of cell columns. However at 10 days the zone showing mitoses is wide. At 14 days the labelling is more haphazard with proliferating cells scattered throughout the middle third of the cartilaginous epiphysis. Most labelled cells are peripheral at this age.

By 16–21 days the number of labelled cells has decreased. Hypertrophic cell height at 16 days is normal. Glycogen content of hypertrophic smc cells is increased; calcification is not seen in the epiphysis, and is limited in the diaphysis.

Clearly the phenotypes of cmd and smc have some similarities in that they both produce an apparent surfeit of collagen: they are also clearly different since smc survives, and there are differences in the matrix.

The cartilage anomaly rat

Grüneberg's cartilage anomaly in the rat (Fell and Grüneberg 1939) is mentioned here because of similarities to both cmd and smc. Grüneberg and Fell describe the formation of heavily staining capsules around the chondroblasts. These show linear concentric layers of deposit and central degenerative areas (Fig. 3.11). Although the mutant is long extinct, and cannot therefore be tested it seems to resemble the conditions described above quite closely.

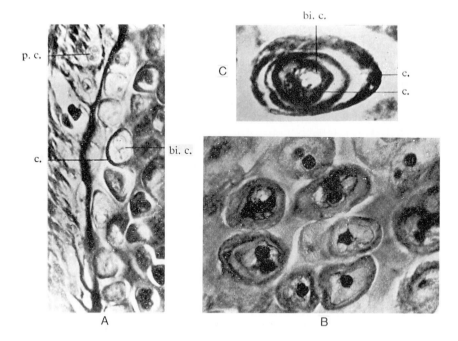

Fig. 3.11.   The cartilage anomaly rat. (A) Transverse section through the costal cartilage. (B) Tracheal cartilage. (C) A binucleate chondrocyte (bi.c) with laminal capsule (c). (From Fell and Grüneberg 1939.)

Brachymorphic (bm, Chr 19)

Brachymorphic, first described by Lane and Dickie (1968), is a typical chondrodystrophy classifiable 4–5 days after birth by its short domed skull and short thick tail. A little later the limbs can also be seen to be affected. Lane and Dickie measured bones from the skull and axial skeleton and found typical shortening of the long bones (femur 61 per cent, humerus 72 per cent, radius 69 per cent, ulna 71 per cent) with no reduction in width, and a similar picture in the skull. The epiphysial plates were thinner than normal, and proliferating cell columns shorter, as was the zone of hypertrophy. This is typical in chondrodystrophy and was shared by the cn and stb mutants which they also investigated. Johnson (1978) found, in a study of papain-cleaned bones aged 6–100 days, that the subnormal rate of bone growth was present from the earliest time that it could be measured by this method, that the faster growing end of the bone was most disturbed by the mutation, and that growth was interrupted by the trauma of weaning. Miller and Flynn-Miller (1976) noted that brachymorphic tail vertebrae and femora were especially robust and the other long bones wider than normal. Lane and Dickie (1968) in fact obtained some A:N ratios (for radius width 1.03, 1.02, ulna width 1.05, 1.01, and femur width 1.01) which were greater than unity in bm, but also in stubby (stb) where ulna rose as high as 1.07–1.10: Miller and Flynn-Miller also examined stubby, but did not comment on this point.

Miller and Flynn-Miller found that brachymorphic cartilage showed disrupted growth. Chondrocyte columns were irregular and individual cell clusters abnormal in arrangement and number. The normal regularity of cellular maturation was lost, so that the cartilage plates were abnormal in outline and thickness. This was noted both in the synchondroses in the base of the skull (Fig. 3.12) and in the tail vertebrae. In the basispheno-occipital synchondrosis the cartilaginous plate was so wide that it protruded through the bone and perichondrium to produce a very abnormal joint. Miller and Flyn-Miller suggested that this joint thickening is due to failure in resorption of abnormal cartilage; perhaps the bm cartilage does not allow mineralization.

Orkin *et al.* (1977) examined the brachymorphic growth plate further and noted that the epiphyses were reduced in size, that the extracellular matrix reacted poorly with stains for sulphated glycosaminoglycans, and that under the electron microscope the collagen fibres appeared normal, but matrix granules of proteoglycan aggregates were not visible in the columnar and hypertrophic zones of mutant growth plates, whilst the number of granules present in reserve zones was smaller than normal (Fig. 3.13).

Orkin *et al.* (1976) examined both glycosaminoglycans and collagen in brachymorphic cartilage matrix by incubation with radioactive precursors.

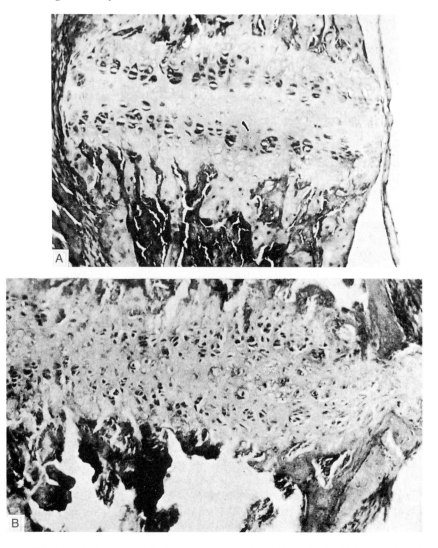

Fig. 3.12. Basispheno-occipital synchondroses from 5-week-old (A) normal and (B) brachymorphic mice. (From Miller and Flynn-Miller 1976.)

They found that both hydroxyproline and hydroxylysine levels were identical in normal and brachymorphic collagen, and corresponded well to the expected values for type II collagen. Incubation with a cross-linkage inhibitor followed by cellulose column chromatography also gave normal results, as did incubation with cyanogen bromide and electrophoresis to reveal the pattern of constituent peptides.

bm matrix contains the same amount of glycosaminoglycans as normal (estimated by uronic acid content of pronase and NaOH extractions) but the amount of $^{35}SO_4$ incorporated *in vitro* and *in vivo* was sharply reduced ($^{35}SO_4$:^{3}H-glucosamine uptake = 1.3 bm; 2.8 normal), and bm cartilage contained 26 per cent less sulphate than normal. DEAE cellulose chromatography of glycosaminogylcans revealed one major component in the normals, which eluted with the chondroitin sulphate standard and labelled with ^{3}H and $^{35}SO_4$. Mutant cartilage showed two peaks both labelling with ^{3}H but only of which contained any $^{35}SO_4$. These peaks also eluted in abnormal places. Chromatographic properties of skin glycosaminoglycans were normal. The two peaks of mutant cartilage were unaffected by hyaluronidase, suggesting that neither was hyaluronic acid. Sephadex gel chromatography showed that the normal and mutant glycosaminoglycans were of similar size. Digestion of the matrix by chondroitinase ABC produced, in the normal, disaccharides corresponding to chondroitin 4 sulphate (the majority) with a little corresponding to chondroitin 6 sulphate. In the mutant a little of the former was obtained, but most of the disaccharide corresponded to unsulphated disaccharide derived from unsulphated chondroitin.

Greene *et al.* (1978) looked at the incorporation of sulphate autoradio-

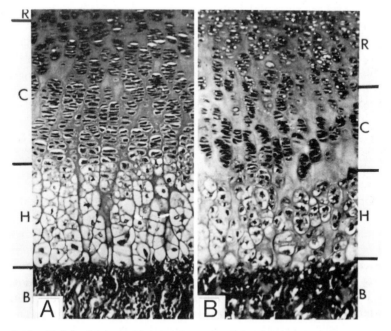

Fig. 3.13. Epiphysial growth plates from the tibia of (A) 5-day-old normal and (B) 5-day-old bm/bm mouse. R, resting zone; C proliferative/columnar zone; H, hypertrophic zone; B, bone. (From Orkin *et al.* 1977.)

graphically using $Na_2{}^{35}SO_4$ and ^3H glucosamine. They found that the incorporation of these precursors was markedly reduced, although incorporation of ^3H-leucine and ^3H-proline was not. The reduction in incorporation was general, although more marked in the proliferative than resting zones.

Sugahara and Schwartz (1979) suggested that the reduction in sulphation could be due to a deficiency in the conversion of adenosine 5'-phosphate (APS) to 3'-phosphoadenosine 5'-phosphate (PAPS), a sulphate donor, and demonstrated a 93 per cent decrease in APS kinase in newborn mice. This clearly implicated APS-kinase. More recent work however (Schwartz *et al.* 1982) reported that both APS-kinase and ATP-kinase are affected. Schwartz suggested that the mutation may be in a subunit common to both these enzymes. The related compound cyclic AMP (cAMP) was altered in palatal shelves of bm mice, susceptible to cortisone-induced cleft palate (Pratt *et al.* 1980). Since cyclic AMP levels in mesenchyme are known to trigger chondrogenesis (Solursh *et al.* 1978; Kosher *et al.* 1979), the cAMP levels of the bm were obviously of interest. Hindin and Erickson (1981) found between 13 and 15 days cAMP levels in hindlimbs increased, then decreased by day 17. bm showed a nearly identical pattern, but one day ahead of normal littermates. Hindin and Erickson suggested that these findings may be related to the undersulphated matrix in bm.

Pennypacker *et al.* (1981) cultured limb mesenchyme at high density and observed the formation of cartilage nodules after three days in culture. These cultures, as expected, incorporated ^3H-glucosamine at normal rates, but brachymorphic cultures, incorporated less $^{35}SO_4$. The glycosamino-glycans produced were fractionated by DEAE cellulose chromatography: initially hyaluronic acid was synthesized, but by day 6 chondroitin sulphate was the main component. The difference between incorporation of ^3H and $^{35}SO_4$ was also at its greatest in mutant cultures at this time. In mutants hyaluronic acid and chondroitin sulphate peaks were broader.

The achondroplastic (ac) rabbit

The brachymorphic mouse obviously has a generalized biochemical defect which is primarily expressed in cartilage. A similar situation, albeit with a different defect, occurs in the achondroplastic rabbit. This (Brown and Pearce 1945; Pearce and Brown 1945*a,b*) is a typical chondrodystophy with shortened limbs, a domed skull, and irregular metaphysial line. Brown and Pearce reported that ac/ac individuals were almost invariably born dead (the longest survival being 12 hours). Affected individuals had markedly shortened extremities, a broad short head with prominent calvarium, a shortened spine, a short thorax, and a short broad tail. In some cases the palate was cleft.

Shepard *et al.* (1969) looked at the cartilage from these affected individuals. They found that in the resting zone semithin plastic sections of osmium-fixed material showed an excess of dead cells, the highest concentration of these being in the centre of the cartilaginous masses (Fig. 3.14), with none found in the perichondrium. Dead cells were not seen elsewhere (muscle, liver, thymus) in such profusion or distribution. The number of cells within a single matrix capsule was reduced in the centre of achondroplastic cartilage, but not near the edges. Less matrix was present in abnormals than in controls; cartilage columns were irregular (Fig. 3.15). Histochemical studies for a number of enzymes were indistinguishable in normal and achondroplastic cartilage. Electron microscopy showed only the increased number of dead and dying cells; at the periphery of the tissue the dwarf cartilage cells were normal. Shepard *et al.* supposed that the dead cells in the centre of cartilaginous masses were those furthest away from a blood supply; perhaps they were unable to metabolize at low oxygen concentrations?

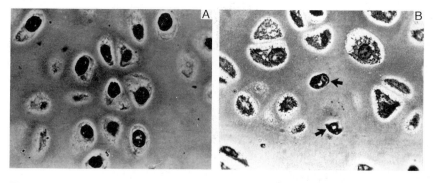

Fig. 3.14. Electron micrograph of (A) normal resting cartilage; (B) achondro-plastic rabbit cartilage. Arrows indicate dead cells. (From Shepard *et al.* 1969.)

Shepard and Bass (1971) used organ culture methods to study the uptake of ^{14}C from glucose and galactose, ^{35}S from sulphate, and ^{3}H from thymidine. The uptake of the last two labels was normal but the amount of ^{14}C incorporated was significantly increased (in fact almost doubled). Heart and kidney controls did not show this increased incorporation. Glucose labelled in position 1 was incorporated faster than that labelled in position 6. Autoradiography of ac cartilage showed an increase in grains from uniformly labelled glucose and galactose, with most of the increase being peripheral. Cells identified as dead naturally did not incorporate label. Sulphate incorporation was uniform over all the growth plate in normal and abnormal animals and grain counts were similar.

Shepard and Bass pointed out that glucose metabolism in cartilage is

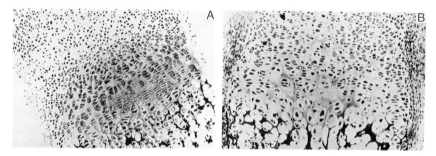

Fig. 3.15. Metaphysial growth plate of (A) normal; (B) achondroplastic newborn rabbit. Note the reduced number of cells per cartilage capsule and irregularity of the plate in (B), especially towards the centre. (From Shepard *et al.* 1969.)

unique. Glucose acts as a source for both acid mucopolysaccharides and collagen via conversion through prolene: but also relatively anaerobic cartilage relies heavily on glycolysis for respiratory needs. They suggested that the problem in rabbit achondroplastic cartilage may be in aerobic glycolysis.

Bargman *et al.* (1972) surmised that the defect might be in oxidative energy metabolism. They used liver mitochondria as a substrate. When preparations from achondroplastic rabbits were placed in a Warburg apparatus, P:O ratios were uniformly low, averaging 1 (normal = 2) with succinate as substrate and 2 (normal = 3) for pyruvate-malate. This demonstrated the absence or inactivity of one phosphorylation site in ac mitochondria. Although phosphorylation was decreased oxygen utilization was normal, suggesting that terminal oxygen transport steps in metabolism could be at fault. This was supported by the normality of the relevant enzymes (DPNH oxidase, succinic oxidase, cytochrome oxidase) in related studies of heart muscle. Disrupted mitochondria had normal levels of ATPase and were inhibited to an equal degree by oligomycin and specific beef heart inhibitor. Further experiments showed that phosphorylation at site III (the cytochrome oxidase region) was essentially absent in ac rabbits.

Webber *et al.* (1981) grew slices of ac/ac rabbit cartilage *in vitro* for 24 hours and looked at the labelling index of cells after exposure to tritiated thymidine. They found that ac labelling index was 4.3 per cent (control 21.7 per cent, $P < 0.01$) but that the paucity of proliferating cells in the ac growth plate made the recording of a meaningful generation time impossible. The histology of the growth plates agreed with that seen by Shepard and Bass (1971). Secondary centres of ossification were absent in ac, perhaps due to a generalized retardation of the mutant animals. Alternatively, the authors raised the interesting point that cartilage canals and accompanying blood vessels may not be able to penetrate into ac cartilage. The general resistance of cartilage to vascular invasion is well known and has been ascribed to anti-invasion factor (AIF) produced by chondrocytes (Kuettner and Pauli 1978). Overproduction of AIF, if it indeed exists, may upset the formation of secondary centres of ossification.

Here we see another condition, clinically described as a chondro-dystrophy and obviously affecting cartilage and little else at the light and electron microscopic level, defined as a specific biochemical effect present in all cells in the body but showing only in cartilage. We should perhaps consider cartilage as a tissue running near the ragged edge, with high synthetic rates, a high growth rate, and a poor supply of nutrients obtained by diffusion.

Cartilage anomaly (can, Chr ?)

This recessive gene (Johnson and Wise 1971) produces a typical chondro-dystrophic phenotype, homozygotes being smaller than normal littermates at birth and gaining weight slowly. Skull length, body length, tail length, femur length, tibial and middle metatarsal length are all reduced. Interestingly skull width, femur, and tibial width are also somewhat reduced (cf. bm). Death usually occurs at around 10 days, but, excep-tionally, individuals have survived longer, the oldest recorded being 38 days old. The cause of death is unknown, but Johnson and Wise ruled out cyanosis, a common problem in chondrodystrophies; cn blood is normally oxygenated. The domed skull and shortened limbs of the homozygote can be recognized at birth and these defects can be traced back with certainty to 17-day and, in some cases, 15-day fetuses. Malocclusion of the incisors is present due to a shortening of the upper jaw. The junction of osseous and cartilaginous ribs is typically spatulate and ossification is reduced in epiphysial plates. Chondrocytes in the head of the femur are closely packed with less matrix than normal (about 50 per cent increase in cells per unit area). The epiphysial plate is thin and cartilage columns poorly aligned, with a narrow zone of hypertrophy and an irregular line of calcification (Fig. 3.16). These abnormalities were seen in 17-day-old fetuses; a litter of 14-day-old fetuses from known heterozygous parents contained no histologically discernible abnormality. The cartilaginous matrix in all cases stained poorly. In electron micrographs can chondrocytes show regular bipolar deposits of glycogen. The matrix appears more fibrillar than usual; in fact the number and diameter of collagen fibres appears normal, but the mucopolysaccharide fraction of the matrix appears depleted (Fig. 3.17).

Johnson and Hunt (1974) looked at the biochemistry of the matrix. They found that protein-bound radioactivity was incorporated at normal rates by newborn can cartilage, whether the precursor was ^{14}C-glucosamine, ^{14}C-galactosamine, or ^{35}S-sulphate. At three days glucosamine incorporation

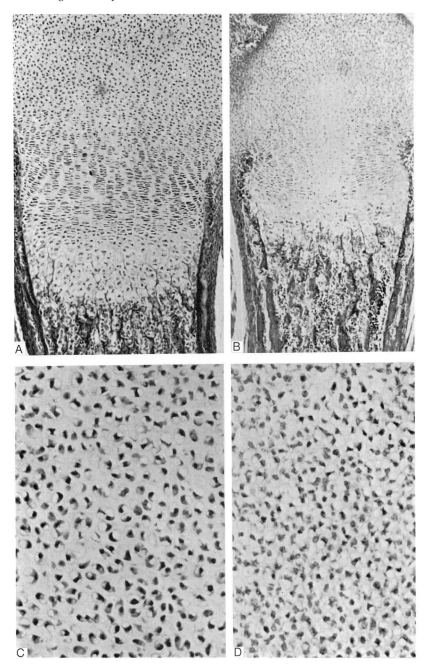

Fig. 3.16. Longitudinal sections through the head of the femur of (A) normal; (B) can/can mice. (C,D) Higher power views of normal and can cartilage. (From Johnson and Wise 1971.)

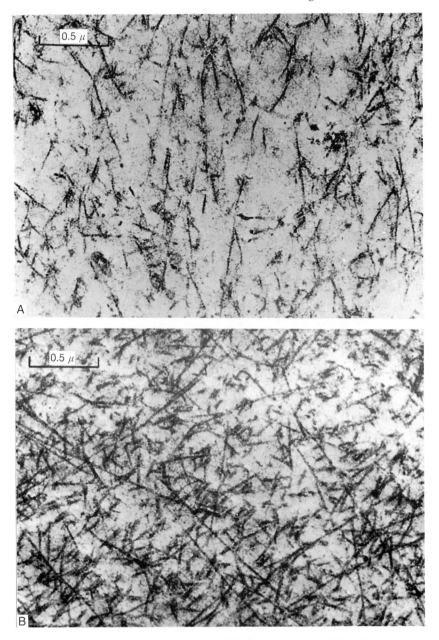

Fig. 3.17.   Cartilage matrix from normal (A) and can/can (B) newborn femur. (From Johnson and Wise 1971.)

was significantly higher in can than in normals, but by five days it had returned to normal levels. The increased incorporation was shown to be due to increased synthesis by chase experiments. The extra glucosamine was incorporated into mucopolysaccharides susceptible to the action of chondroitinase ABC. Non-protein-bound radioactivity showed a similar pattern. At three days and at five days both UDP-glucose dehydrogenase and UDP-glucose-4-epimerase (both concerned with mucopolysaccharide synthesis) were elevated. Non-specific glucose-6-phosphate dehydrogenase levels, however, were normal. Protein synthesis from ^{14}C-glycine was significantly reduced in can cartilage at birth and three days, but increased by five days; in the liver no changes were seen. Little of the protein synthesized by can cartilage at three days was removed by collagenase; at five days both collagen and non-collagenous protein were increased. Oxidative phosphorylation (cf. the ac rabbit) was normal.

In litters of embryos from heterozygous parents sacrificed at 17 days, incorporation of ^{14}C-glucose was normally distributed (66 individuals) but incorporation of ^{14}C-glycine was bimodal (56 individuals) with 13/56 embryos producing a lower peak, and tentatively identified as can homozygotes.

These results are at first sight rather contradictory. The earliest abnormality seen was a decrease in protein synthesis in 17-day fetal cartilage; this continues until 3 days post partum. The period between three and five days marks a change in cartilage metabolism. Protein synthesis (including collagen synthesis) increases, as does the incorporation of galactosamine into mucopolysaccharides and the levels of the synthetic enzymes measured. Johnson and Hunt suggested that these changes perhaps form a belated compensation for earlier deficiencies.

The compensatory effect was further studied by Johnson (1974) who transplanted tail vertebrae from can homozygotes at a time when the compensation was likely to be maximal beneath the kidney capsules of normal sibs from an earlier litter. It was clear from the morphology of the vertebrae recovered 14 days later that no discernible compensation had taken place (Fig. 3.18); the morphology of the cartilage was typical of that seen in non-transplanted material, with little matrix, little bone formation, and a short hypertrophic zone.

Achondroplasia (cn, Chr 4)

Achondroplasia in the mouse was described by Lane and Dickie (1968). cn homozygotes can be distinguished at birth by the usual domed skull and shortened limbs and tail, and reduced birthweight. During the following weeks they become progressively more retarded and never attain adult weight. Malocclusion is common and Miller *et al.* (1974) described slight dental abnormalities. Neither sex breeds well. The usual cause of death is

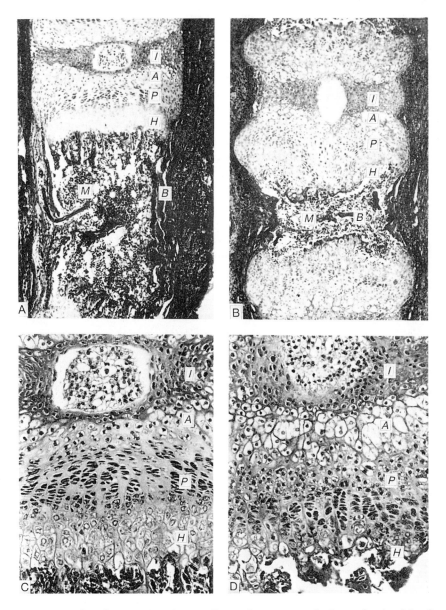

Fig. 3.18. (A,B) Normal and can tail vertebrae, respectively, maintained for 3 weeks in kidney capsule of a 3-month-old normal sib. (C,D) details of cartilage from (A,B). B, bone; M, metaphysis; P, proliferative zone; H, hypertrophic zone; I, intervertebral disc; A, articular cartilage. (From Johnson 1974.)

reported as cyanosis consequent upon a crowded thorax. The standard measures of length (skull, body, tail, long bones) are all reduced to circa 70 per cent of their normal values and bone widths are a little less than normal. Konyukhov and Paschin (1970) looked at growth rates of various organs and reported on the histology of the epiphysial plates. Reserve and columnar zones were normal in size, but the zone of hypertrophy was reduced; individual chondrocytes were less hypertrophied than normal (Fig. 3.19). The number of chondrocytes per lacuna was reduced. Chondrocyte columns, however, were regular in contrast to the more usual deranged picture. The same authors (1967) had already shown that in subcutaneous bone homotransplants aged between 7 and 14 days the cn bones showed some signs of returning to normality after 14 days implantation (cf. findings of Johnson (1974) with can transplants).

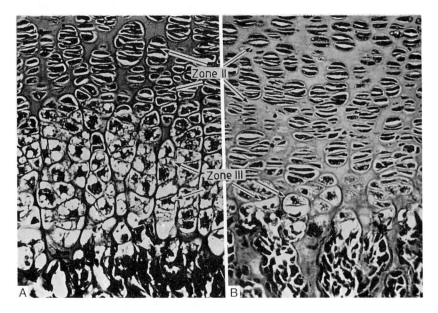

Fig. 3.19.   The proximal epiphysial plates of the humerus of (A) normal; (B) cn/cn 7-day-old mice. (From Konyukhov and Paschin 1970.)

Interestingly normal humeri transplanted into cn sibs showed retarded growth: + humeri transplanted into + recipients grew 11 per cent, + into cn 4 per cent, and cn into cn 8 per cent. Konyukhov and Paschin interpreted these results as being due to the effect of the cn gene on chondrocytes and the presence of a circulating growth retarding substance in the cn blood. This hypothetical factor was not located by serum protein gel electrophoresis and serum blood glucose and blood mucopolysaccharide levels were normal.

Silberberg and Lesker (1975) looked at the histology of the cartilage in more detail and performed some biochemical assays. They found that between two and four weeks hexosamine and hexauronic acid levels decreased in both cn and control animals, and that N- acetyloamino sugar levels dropped in normals but were maintained in cn. Sialic acid levels were high in cn at two weeks, the ratio was reversed at three weeks, and the original differences partially restored at four weeks. Hydroxyproline levels decreased with advancing age, more so in dwarves than non-dwarves. Many of these data were, however, inconsistent.

At two weeks old cartilage columns in the limbs were short and the chondrocytes somewhat deranged (cf. Konyukhov and Paschin 1970). Hypertrophy was reduced and the spicules of bone which had formed were linked transversely. After four weeks two types of abnormal could be distinguished. In less affected individuals growth zones were comparatively regular with small chondrocytes and reduced hypertrophy but with increasing numbers of necrotic cells. In the second group growth zones were strikingly abnormal with short columns of small chondrocytes separated by large amounts of matrix. There were no hypertrophic cells and no primary spongiosa; instead of the latter a bony lamella was present beneath the atrophic cartilage. Widespread resorption of bone had led to perforation of the growth plates. In older animals these perforations widened to form extensive communications between epiphysial and diaphysial cavities. Similar changes were seen in the vertebrae; ribs were on the whole less affected. The authors were unable to account for the presence of two classes of homozygote. The most affected histologically were not, in fact, the smallest, having a mean weight of 14 g against 12 g in less affected individuals; it is thus unlikely that, for instance, starvation due to malocclusion was to blame.

Bonucci *et al.* (1976) looked at the histology and histochemistry of cn ribs, caudal vertebrae, and tibial epiphyses at ages ranging from 1 to 30 days. They confirmed the shortening of the cartilage columns and compression of the hypertrophic zone with a recognizably normal series of proliferation, maturation, and hypertrophy and degenerating chondrocytes. Intracellular matrix stained less well with colloidial iron and alcian blue than in normal littermates, and there was a reduction in PAS staining (Fig. 3.20). These authors considered that the changes they saw in cn cartilage resemble those seen in normal ageing cartilage, and suggested that the dwarfing is due to premature ageing of the chondrocytes. Miller and Flynn-Miller (1976) looked histologically and radiologically at cn mice. They found fairly even chondrocyte columns in the synchondroses of the skull base with irregular organization of the collagen matrix. The synchondroses closed irregularly and early. Jolly and Moore (1975) noted that there were

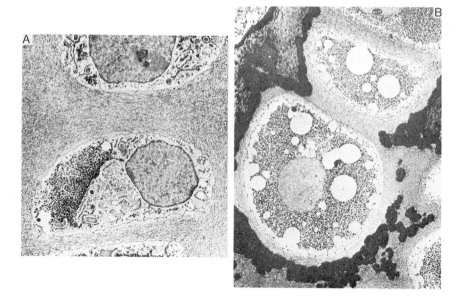

Fig. 3.20. (A) Maturing chondrocytes from 5-day-old cn/cn mouse. Note glycogen deposits. (B) By 20 days there is more accumulation of glycogen, many vacuoles, calcification at the level of the maturing chondrocyte, and a lack of cell degeneration. (From Bonucci *et al.* 1976.)

marked changes in skull proportions in cn. Brewer *et al.* (1977) tried to use the fluorescent dye Procion to measure skull growth, but found that fluorescence was widespread and very complex in disposition; radiography was also unsatisfactory and they suspected that the necessary general anaesthetic was retarding the growth of both normal and abnormal mice. Papain-digested material showed abnormalities from 6 days in the mandible, basioccipital, and basisphenoid bones. The thickness of the mandibular condylar cartilage was reduced, but its histology normal. Growth at the synchondroses in the base of the skull reduced the basisphenoid by 40 per cent (equivalent to the reduction in the humerus of 34 per cent—Lane and Dickie 1968; 38 per cent—Jolly and Moore 1975; 36 per cent—Silberberg and Lesker 1975). The basioccipital (with contributions from two synchondroses) was reduced by 24 per cent. The mandible was reduced in length by only 11 per cent; this relatively minor retardation might be supposed to be secondary to that seen in the basicranium. However the mild derangement of the manibular condyle suggests that secondary cartilage may be less affected by the cn gene.

Silberberg *et al.* (1976) looked at the ultrastructure of cn cartilage from mice aged 3–7 weeks. In 3-week-old animals the chondrocytes appeared normal, save for the presence of a large number of dense bodies resembling primary lysosomes. Glycogen deposits were more abundant than in controls (cf. can). The matrix had closely packed collagen fibres and degenerating and dead cells were more common than usual. One individual had single rather than paired chondrocytes with scanty, short, cytoplasmic footlets, scanty endoplasmic reticulum, and collapsed endoplasmic reticulum cisterns, fragments of which were seen in autophagic vacuoles. Golgi apparatus was delicate, and lipid droplets commonly present. One is tempted to equate this phenotype with the more acute form of cn described by Silberberg and Lesker (1975) in light microscopical sections.

A similar heterogeneity was seen in four- and five-week-old sections. Unfortunately similar defects were seen in normal mice aged 3–5 weeks; the authors suggested implausibly that this was due to the action of the cn gene in heterozygous form.

Bonucci *et al.* (1977) also investigated the ultrastructure of cn cartilage. They found no abnormalities in the matrix, with 10–200 nm diameter collagen fibres and small intrafibrillar granules. Proliferating and maturing chondrocytes resembled those of controls, except for the presence of ample deposits of glycogen (Fig. 3.20). Hypertrophy appeared normal, but growth columns had fewer cells than in controls over the 5–15-day period, a feature which became marked in 20–30-day-old mice. Matrix calcification at these ages was seen around maturing chondrocytes, and hypertrophic cn cells did not seem to be degenerating. No mention of heterogeneity in cn/cn was made by these authors.

Kleinman *et al.* (1977) found that cn and normal mice had similar amounts of hexosamine/mg dry weight and that control and mutant knee joints incorporated equivalent amounts of $^{35}SO_4$ and ^{3}H-glucosamine into a fraction extracted with 4M Guanidine hydrochloride. Five to 10 per cent of the radioactivity introduced was unextracted in both mutants and controls. These findings suggested that synthesis and turnover of proteoglycans was normal in cn, and the similarity of $^{35}SO_4$:^{3}H ratios suggested that the GAGs were not undersulphated. Caesium chloride centrifugation showed that the proteoglycan-rich part of the gradient was essentially similar in normal and cn. The relative amounts of GAGs (by Sephadex G 200 or DEAE cellulose separation) were also similar.

Amino acid synthesis of whole knee joints and purified collagen alpha chains showed no differences in the amount of collagen judged by hydroxyproline content or type of collagen judged by amino acid analysis. Similar proportions of salt extractable (80 per cent) and non-extractable (20 per cent) collagen were found in normals and mutants. The collagen present, as expected, was identified as type II.

Silberberg and Lesker (1975) had reported an increase of 60 per cent in hydroxyproline content of 21- and 28-day-old mice. Kleinman *et al.* were unable to repeat this finding in tissue of a similar age. Moreover material from Kleinman's strain assayed by Silberberg showed normal levels of hydroxyproline.

Thurston *et al.* (1985*a*) also found two classes of cn homozygote in their work on the kinetics of cn cartilage. These could be distinguished on both their histology (Fig. 3.21) and growth potential. Their type I was mild, and type II more extreme. Both types I and II were found in their sample of mice aged 22 days. In type II at this age there was a complete absense of growth plate labelling despite heavy labelling with ^{3}H-thymidine in the epiphysis and metaphysis. Type I dwarves at 22, 26, and 34 days showed a much reduced proliferation zone size, reduced mean labelling index, and reduced cell height. Type II dwarves had normal proliferation zone sizes and labelling indices but much reduced hypertrophic cell heights. Type II homozygotes had a reduced growth rate at all ages: type II had a zero growth rate at 22 days, but some growth was evident at 26 and 34 days.

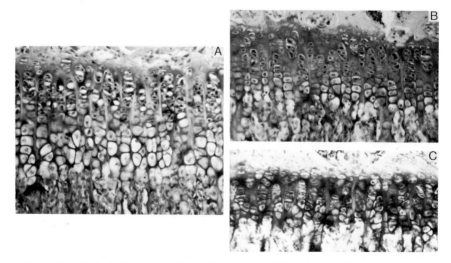

Fig. 3.21. Epiphysial plates from 22-day-old (A) normal; (B) cn/cn type I homozygote; and (C) cn/cn type II homozygote.

Histologically type II homozygotes had a very thin growth plate with a reduced number of cells in every zone: this was especially marked in the hypertrophic zone when the bottom cell of a column was often not hypertrophied. Column formation was quite regular but the cells were packed tightly and often hyperchromatic. Trabecular formation is linked to the presence of short longitudinal septa. Type I homozygotes had a noticeably wide growth plate and more normal trabecular formations.

The pathology of cn is obscured by the apparent presence of two types of homozygote. These two types of homozygote have been reported by Silberberg and Lesker (1975), Silberberg *et al.* (1976), and Thurston *et al.* (1985*a*). Other authors have not reported the heterogeneity, if indeed it was present in their samples. Kleinman *et al.* (1977) disagree with Silberberg and Lesker's biochemical findings.

It seems easier to suppose that the stocks of Silberberg and Thurston *et al.* contain some factor not present in that of Kleinman. In fact the former authors seem to have obtained their stocks of the mutant from the Jackson Laboratory after it was outcrossed to C57BL, C3H and Lg/J (the original mutation occurred in the A/J strain), during a stable period of 20 generations. Kleinman's stock was exported after further outcrosses to the F_1 of C57BL/6J and C3HeB/FeJ made to restore fertility. It seems likely that in the process the factor responsible for the more extreme phenotype was bred out.

It also seems possible that the 'inconsistent' biochemical findings of Silberberg and Lesker (1975) were due to sampling a mixed population.

What is the factor causing this heterogeneity? Could it be that the type II phenotype is more extreme due perhaps to a threshold effect seen in small individuals, i.e. a quasi-continuous variation? Silberberg and Lesker reported that the most extreme phenotype was not seen in the smallest mice; in fact their mean weight was greater than in less affected dwarves. Silberberg *et al.* (1976) however suggested that the most disrupted chondrocytes came from the smallest mice. These authors also reported changes similar to those seen in the more extreme phenotype in normal littermates.

Is it that another factor, probably genetic and not at the cn locus, is present in certain stocks carrying the cn gene? In the absence of two doses of the recessive cn, this is seen only as a mild derangement of chondrocyte morphology; if cn is present then the phenotype of the latter is made more extreme. It would seem sensible that any further work on cn is performed on descendants of Kleinman's stock, which seems to lack this factor.

Even allowing for the possible heterogeneity, cn is still an enigma. Much work has been performed, but there seems to be little biochemical or morphological difference between cn and normal; the only positive evidence seems to be in the abnormal maturation of cn chondrocytes.

Chondrodystrophy (cho, Chr ?)

This recessive mutation (Seegmiller *et al.* 1971) has the short snout and mandible and disproportionately short limbs of many chondrodystrophies. There is a median cleft palate and protruding tongue. The animals die immediately after birth because, according to the authors, the tracheal cartilage does not have sufficient strength to maintain an open airway.

Mutant animals can be distinguished on day 15 of gestation by shorter limbs, short jaw, and cleft palate. Long bones are typically short and wide with wide flared metaphyses; membrane bones appear normal. Meckel's cartilage is reduced in length.

Mutant cartilaginous blastemata appeared normal up to day 13: at day 17, Seegmiller noted that the cartilaginous ends of the femur were enlarged relative to the resting formed bone and that large areas of cartilage (mainly in the proliferation zone) were devoid of chondrocytes. The axial length of mutant and normal cartilages are similar, but the latter greatly increased in width. Mutant chondrocytes from the reserve cell zone appear normal in size, shape, and number but later become variable in size and shape, often resembling proliferation zone cells. These in turn are flattened, irregular in shape and size, and misaligned laterally rather than lying in columns. Cartilaginous matrix is heterogeneous, vesicular, and stains less well than normal with toluidine blue, alcian blue, and PAS. In the mutant calcified bone surrounds the lower part of the epiphysis and extends up the sides of this rather bulbous mass of cartilage. In the diaphysis there is much calcified bone and little formation of marrow cavity; most of the bone seems to be periosteal in origin. In electron microscope sections of costochondral rib junctions, reserve zone cells appear normal. In the proliferative zone, columns are not found; chondrocytes appear in oblique arrays and are fewer than normal. Instead of being spindle-shaped these cells are fibroblastic or amoeboid in appearance. Several cells are often in contact via cytoplasmic processes. In the zone corresponding to the normal hypertrophic zone there is a paucity of chondrocytes and cellular debris is scattered throughout the matrix. Clusters of darkly stained pycnotic cells were observed in all zones of the mutant cartilage. The fibrils of the matrix showed uniform transverse banding: these fibrils were up to 200 nm in diameter (normal 25 nm). The authors suggested that column formation fails to occur due to reduced AMPS in the matrix which renders it physically weak and unable to align daughter chondrocytes after mitosis. The reduced matrix is synthesized by chondrocytes which lose their phenotype and become amoeboid. We have already noted the relationship between paucity of AMPS in the matrix and wide, banded collagen fibres.

A later paper (Seegmiller *et al.* 1972) examined the tracheal cartilage. Here mutant chondrocytes often contained cytoplasmic droplets which stained with toluidine blue (Fig. 3.22). In electron micrographs the cell surface was smoother than normal and large golgi vacuoles were present (Fig. 3.23). Again large banded collagen fibres up to 60 nm wide were seen in the matrix (Fig. 3.24). No accumulations of AMPS were seen in mutant chondrodystrophic skin, muscle, or liver.

Stephens and Seegmiller (1976) investigated the glycosaminoglycans of cho mice in the light of these findings. They found the macromolecular structure of 18-day-old fetal rib and limb cartilage normal with respect to

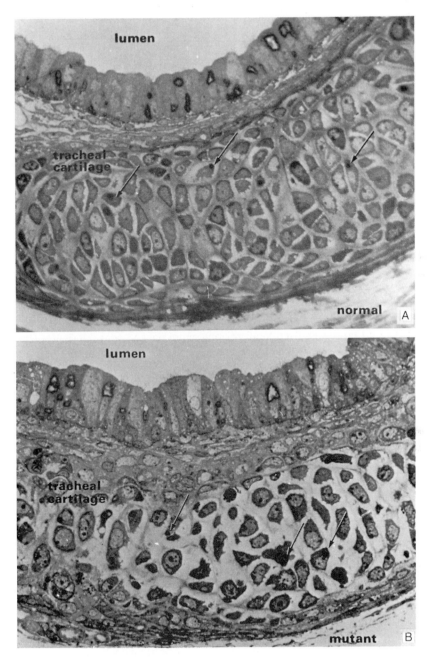

Fig. 3.22. Light micrograph of (A) normal and (B) cho/cho tracheal cartilage. (From Seegmiller *et al.* 1972.)

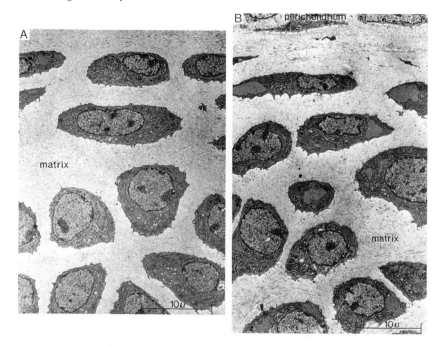

Fig. 3.23.   Low-power electron micrograph of normal (A), and cho (B) chondro-cytes. (From Seegmiller *et al.* 1971.)

protein, uronic acid, and amino sugar levels. Incorporation of ^{35}S, ^{14}C-glucosamine, and ^{3}H-glucose was normal but glycosaminoglycans leached more readily from mutant than from normal tissue. After Sephadex gel electrophoresis no differences were seen in the structure of normal and cho glycosaminoglycans. These findings clearly invalidate the hypothesis of Seegmiller *et al.* (1971) that GAGs in cho are quantitively reduced or qualitatively altered. Stephens and Seegmiller reported preliminary findings that content and synthetic rate of collagen is abnormal in cho. However Seegmiller *et al.* (1981) in a detailed study of cho rib cartilage matrix found that it did not differ from normal in the following parameters: hydroxy-proline content, hydroxyproline content after pulse labelling, percentage hydroxyproline incorporated into collagen, and lysine incorporation into collagen. CMC chromatography showed that the alpha collagen chain was the major form made by mutant and normal cartilage, and that both made predominantly type II collagen. Normal and mutant accumulated the same amount of ^{3}H-glucosamine into glycosaminoglycans which were of the same molecular weight. Gel analysis yielded identical distribution of subunits. Amounts of unsulphated and 4-sulphated chondroitin were similar, with the 4 form predominating. Molecular weight of mutant proteoglycan was normal.

Monson and Seegmiller (1981) confirmed that the ^{35}S-labelling pattern of mutant fetuses was normal at 18 days by autoradiography. They suggested that the lack of rigidity of cho matrix could be due to a defective proteoglycan–collagen interaction. The stain ruthenium red attaches to matrix granules containing proteoglycan; thus the cho matrix might retain less rutherium red in stained vessels associated with collagen fibres. Unfortunately, the number of these stained granules precipitated on to collagen was not significantly different from normal.

All in all, chondrodystrophy is a rather frustrating mutant; we have the light and electron microscopic appearance of the abnormal cartilage, but all biochemical parameters measured are normal. Perhaps the defect lies in whatever mechanism normally fails to align collagen fibres side by side in normal cartilage. It seems that reduction in GAGs is not the only factor which controls this mechanism.

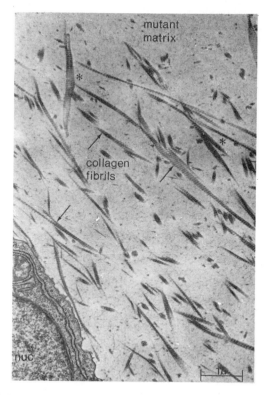

Fig. 3.24. High-power electron micrograph of a small area of cho cartilage showing the broad, banded collagen fibres present. nuc, nucleus. (From Seegmiller *et al.* 1972.)

Stumpy (stm, Chr ?)

Stumpy arose in a mutagenicity experiment (Ferguson *et al.* 1978). All cartilage replacement bones are affected, sometimes grossly, sometimes more subtly. The proximal limb bones are most reduced, being shortened but not narrowed (humerus 58 per cent normal, femur 66 per cent), and distal limb bones less affected. The skull is also shortened, but not narrowed. The growth of stm long bones is rather unusual (Johnson 1978) as growth is almost normal between six and 16 days, then interrupted in the period 16–21 days. Later, there is some catch-up growth over days 21–31. Other chondrodystrophies tend to have reduced growth from the moment that the condition can be identified.

In light histological sections (Johnson 1977) the cartilage matrix shows reduced staining, with an increase in the number of cells per lacuna. In later studies (Thurston *et al.* 1983) 10-day growth plates were seen to be slightly abnormal with smaller hypertrophic cell height, slight irregularity of columns, and chondrocytes not strictly wedge-shaped. This disorganization worsens and by day 21 the plate is very irregular with many pycnotic cells and much nuclear debris (Fig. 3.25). In the manibular condyle (Johnson 1983) dead cells were seen in 16-day-old mice (Fig. 3.26), often in pairs with a single lacuna. Electron microscopy (Johnson 1977, 1983) shows normal-looking chondrocytes set in a normal-looking matrix, but many members of pairs of cells lying in close approximation to their neighbours (Fig. 3.27), and frequent nests of four or eight cells. Adjacent cells have large amounts of cell membane in contact with a narrow intracellular gap or tight junctions (Fig. 3.28). These regions are often complicated by interdigitations of the cell membrane. In the condyle (Johnson 1983) hypertrophic cells were often binucleate or had a dumbell nucleus.

The cell kinetics of stumpy cartilage was studied by Thurston *et al.* (1983). Labelling with radioactive thymidine enables a profile of cell division to be built up along the chondrocyte column. Tritiated thymidine produces autoradiographic grains only over the nucleus, and thymidine is incorporated only into DNA. Thurston found that normal mice aged 10–14 days had a low proportion of labelled cells in the first few positions in the chondrocyte column (the resting zone), a peak in the middle (proliferating zone), and a decrease in the zone of hypertrophy. Stumpy profiles were rather variable; all profiles up to 21 days were obviously abnormal and at 10 days most of the label was in the first 14–15 cell positions. At 16 days the hypertrophic zone abutted directly on to the proliferative zone with no zone of differentiation and maturation of chondrocytes. By 21 days the stumpy plate had made something of a recovery and the results were more nearly normal.

The effective proliferative zone size (Kember 1972) was essentially

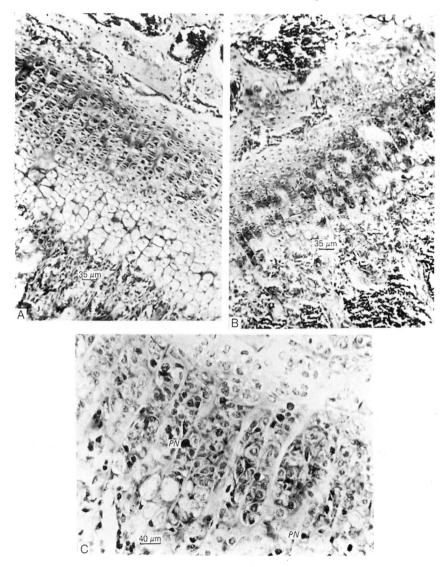

Fig. 3.25. The proximal growth plate of the tibia in (A) normal; (B) stm/stm. (C) At higher magnification stumpy shows a lack of hypertrophy and the presence of pycnotic nuclei (PN). (From Thurston *et al.* 1983.)

normal: but stumpy hypertrophic cell height was reduced throughout. Growth rate calculated on cell proliferation data showed a decrease in stumpy over the period 10–21 days.

It seems clear that the root of the stumpy syndrome lies in the rather odd cell division seen in these mice. Dumbell nuclei could indicate incomplete

Fig. 3.26. The head of the mandibular condyle from a stumpy (stm) mouse.

nuclear division, binucleate chondrocytes nuclear division not followed by cell division and adjacent daughter cells with interdigitations, further problems in cytoplasmic splitting, or the moving apart of daughter cells which usually follows this event. The dead cells seen in the condyle and epiphysial plates might be the results of abortive cell division. However these dead cells, present in the relatively large condyle and limb cartilages, were not reported from rib; like the dead cells seen at the centre of cartilaginous masses in the rabbit achondroplasia by Shepard *et al.* (1969), they may be due to the presence of substandard cartilage cells at the end of a long diffusion gradient. The timing of gene action is also interesting. Although stm can be recognized by a competent observer at around 5 days, the effect of the gene is first seen in the cartilaginous growth plates at 10 days and is maximal, in terms of effect on growth rate, rather later and seems to diminish after 21 days. This type of 'catch-up' effect was also seen in can (Johnson and Hunt 1974). It is interesting to note that the appearance of a pile of cells occupying a lacuna suggestive of cells not moving apart after division has also been reported by Bringel *et al.* (1983) in the Alaskan Malamute. Stumpy also clearly shares some features with cho: the cells closely in contact, the presence of dead cells, and the lipid droplets are all features common to both, yet the most characteristic feature of cho, the wide collagen fibres are not present in stm.

Disproportionate micromelia (Dmm, Chr ?)

Brown *et al.* (1981) described a mouse chondrodystrophic mutant which is unusual in being dominant. The homozygous abnormal, dwarf with a cleft

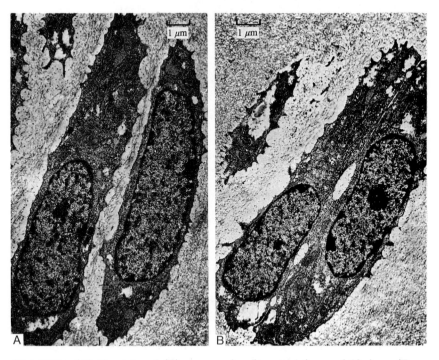

Fig. 3.27. (A) Normal and (B) stumpy chondrocyte pairs aged 10 days. (From Johnson 1977.)

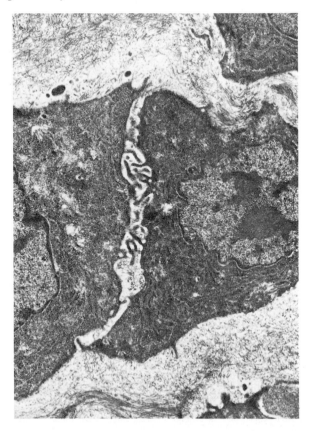

Fig. 3.28. Interdigitations between stumpy chondrocytes. (From Johnson 1983.)

palate dies around birth. The heterozygous dwarf is typical in having a short head and reduced long bones (50–60% normal length, width unaffected). Membrane bones are normal. Skull width and length are both reduced. Some heterozygous Dmm are of normal size at birth, but have cleft palates. Survivors can be classified as dwarves by seven days, and are even more affected by 28 days. Females are slightly more affected than males.

Dmm homozygous epiphyses from 18-day-old fetuses have a 'cystic' appearance to the matrix, especially centrally where large, rounded extracellular spaces are described (Fig. 3.29). Chondrocytes are slightly compressed so that their long axis is parallel to the long axis of the bone. The short calcified metaphysis is abnormally wide. Dmm heterozygotes have a reduction in the height of proliferative and hypertrophic zones but normal organization in the growth plate: matrix stains subnormally with alcian blue and safranin-O.

Biochemical determinations on Dmm homozygotes showed the following. Glucosamine content was normal but galactosamine content reduced to about one-third of its normal value. Decreased galactosamine content correlated with a decrease in proteoglycan synthesis, ^{3}H-glucosamine, and ^{35}S uptake being reduced by 52 per cent and 74 per cent, respectively. Dmm homozygote cartilage appeared to lose its proteoglycans (i.e. leak label) very readily. Structural organization of proteoglycan showed no differences from normal. Type II collagen (as visualized by labelling with anti-type II collagen antibody) was reduced throughout the abnormal

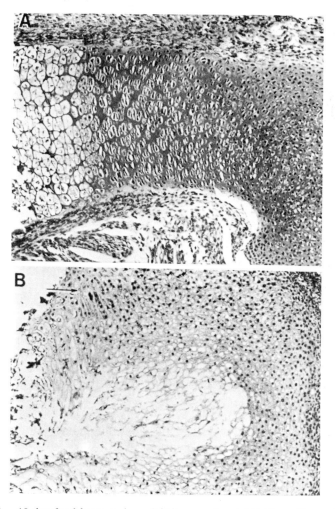

Fig. 3.29. 18-day fetal knee region of (A) normal and (B) Dmm/Dmm showing reduced matrix staining and cystic disorganization. (From Brown *et al.* 1981.)

matrix, and had a pericellular distribution; large areas between the cells lacked type II collagen. The labelling pattern for type I collagen was normal. Incorporation of ^{3}H-proline into collagen showed a 30 per cent reduction, but the distribution in salt extract, medium, and pepsin extract was normal. Column chromatography suggested that both normal and Dmm cartilage contained type I collagen, but that Dmm contains less type II.

The authors suggested that the mutation affects matrix synthesis (both major components being involved) and that the pericellular distribution of type II collagen is a consequence of this.

Lethal chondrodystrophy (ch) in the turkey

Goetinck (1983) reported biochemical data on the ch turkey (Gaffney 1975) which suggested that the mutant may affect glycosylation of cartilage proteoglycans. In this mutant cartilage galactosamine is 32 per cent normal (Leach and Bass 1977). ^{35}S incorporation is similarly reduced, showing that the reduction of incorporation into galactosamine is probably due to reduced synthesis. Para-nitrophenol-beta-D-xyloside stimulates chondroitin sulphate synthesis in the mutant to normal levels; Goetinck suggested that this shows that the mutation does not affect chondroitin sulphate synthesis.

Mutant glycosaminoglycans are similar in size to those found in normals, but sediment slowly on sucrose density gradients suggesting, since most of the molecular weight resides in the carbohydrate component, that there are fewer sulphated GAG chains per proteoglycan molecule.

This could be due to a shortened core protein, or could occur on a core protein of normal length. Goetinck found that the m.w. of the core protein in mutants was 246 000, equivalent to that found in normal chickens and turkeys. The core protein thus seems to bear fewer side chains per unit length, and the synthesis of UDP-xylose or the xylosythyltransferase reaction may be at fault.

Stubby (stb, Chr 2)

The recessive stubby resembles achondroplasia (cn) but is less severe. It was first described by Lane and Dickie (1968) as having a short, domed head and typical short limbs and tail. Miller and Flynn-Miller (1976) found cartilage histology within normal limits and Johnson (1978) found that the growth pattern resembled that of cn.

Bare patches (Bpa, Chr X)

Phillips *et al.* (1973) described a sex-linked mutation, bare patches, which is a male lethal. In the heterozygotes there is a group of abnormalities which includes stippled foci of premature calcification within cartilaginous areas (Happle *et al.* 1983). This lesion was seen in both the cartilaginous areas of the limb bones and in the vertebrae on day 5. Because of associated anomalies (cateracts, ichthyosis, patchy hairlessness, and a linear upset in hair pigmentation) all of which are common to Bpa and chondrodysplasia punctata (Cpx) in man (Happle 1977, 1981) which is also sex-linked, Happle *et al.* (1983) proposed homology for Bpa and Cpx. The symptoms certainly seem to be homologous, but pending an investigation of the basic pathology of the disease (unknown in man or mouse) declaration of homology may be premature.

Paddle (pad, Chr ?)

Paddle is a lethal recessive condition with complete penetrance (Nash 1969). It results in a median cleft palate (see Chapter 7) and chondro-dystrophy. Histology of the growth plate (in Johnson and Nash 1982) shows disorganization in all five zones of growth. Transmission electron microscopy shows the presence of abnormal chondrocytes and matrix in both proliferative and degenerative zones. Freeze-fraction of chondrocytes shows abnormal particle distribution and size on rough endoplasmic membranes and cell membranes in the proliferative zone.

Other conditions

This consideration of chondrodystrophies cannot be exhaustive; Landauer's bibliography (1968) of micromelia lists hundreds of references, many of which fall into the category. Grüneberg (1963) gave a useful résumé of the more interesting mutants described up to that date. Many earlier descriptions do not go beyond genetical data, bone measurement, and perhaps light histology and so do not contribute as much to the understanding of chondrodystrophies as the conditions discussed above.

THE DEFECTS SEEN IN ANIMAL MODELS: AN OVERVIEW

We now have the necessary methods, by the use of specific antigens and radiolabelled precursors, backed up by enzyme assays, to attack the problems of chondrodystrophy at the biochemical level. What conclusions can we draw from the mutants described here?

Let us first dispose of brachypod, talpid[3], and congenital hydrocephalus. It is quite clear that defects in cartilage formation in these mutants are due to pre-existing defects at pre-cartilaginous stages of development and we need not discuss them further here.

Amongst the other, probably cartilage-specific, mutants discrete biochemical lesions can be seen. We have good evidence that Mov-13 affects the collagen alpha chain, although the effects of such a drastic loss on cartilage have not yet been described. Specific antibody studies show that the core protein of proteoglycans is absent in both the cmd mouse and the nanomelic chick. Sulphation of proteoglycans is subnormal in the brachymorphic mouse and glycolysation deficient in the achondroplastic turkey. Defective oxidative phosphorylation is probably at the root of the defect in the achondroplastic rabbit. We must count these genes as successes for the biochemical approach.

But cartilage evidently comprises more than biochemistry. Rittenhouse *et al.* (1978) did not report a great similarity between the ultrastructure of cmd and nanomelia, for the simple reason that no great similarity exists at this level, despite almost identical biochemical lesions. We might suppose that, even allowing for the species difference, cartilage devoid of proteoglycan core protein would give a very similar phenotype in any species, but evidently this is not necessarily so.

In other mutants, as exemplified by the cho and cn mice, very clear ultrastructural abnormality cannot be tied to a specific biochemical lesion, although many detailed investigations have been made on these mutants.

In yet other examples the biochemical lesions are less clear cut and results tend to vary with the age of material used. It seems that a biochemical snapshot taken at a particular age may give a misleading impression, since cartilage appears to have some flexibility and ability to compensate for its own deficiencies.

There is also the question of cell kinetics. Where these have been investigated (smc, cn, stm mice, ac rabbit) it is clear that they are abnormal. This might be secondary to a biochemical lesion, but in cn there is no clear evidence to suggest a biochemical lesion, and in stm it seems that the process of cell division is primarily affected. Reduced or altered matrix formation may well affect the cell kinetics, but equally defective cell kinetics might lead to reduced synthesis of matrix.

Analysis of the chondrodystrophies discussed here leads us to the following conclusions. First, diversity of mutation at the molecular level, defects in core protein, sulphation, glycolization, oxidative phosphorylation, or disturbance of cell division or kinetics and many other as yet unidentified lesions all lead to cartilage with a reduced matrix component, which produce the chondrodystrophic phenotype.

Second, the time of action of these first causes varies from the fetal period (presumably from the first formation of cartilage matrix) to some time between birth and weaning. There is no good correlation between measurable upsets in histological appearance, affected growth, and abnormal biochemical parameters. Mutants affected early tend to have cleft palates and to die perinatally; in many cases secondary centres of ossification are also affected.

Third, there is no good correlation between biochemistry and ultra-structure or at least we do not yet have the expertise to see it.

HUMAN CHONDRODYSTROPHIES AND THE ANIMAL MODELS

From the above discussion it should be clear that direct comparison of human chondrodystrophies and animal models is not easy.

First, the human chondrodystrophies are a disparate group (as Rimoin (1975) pointed out in a classic review) and suffered from portmanteau diagnosis as 'achondroplasia' if the limbs were shortened or 'Morquio's disease' if the trunk were reduced. Rimoin was able to make a very useful classification of human chondrodystrophies and to illustrate the ultrastructural appearance of many of them. Beyond this point, however, lie several difficulties. First we must resolve the problem of mode of inheritance. There seems to be a strongly held, and sometimes stated, view that a dominant mutation in man, such as achondroplasia, must be modelled by a dominant condition in mouse, rat, or chick. This is not necessarily true, although in many cases it holds. For some reason, nearly all chondrodystrophies in laboratory animals are recessive. Dominance can be modified and no undue stress should be placed upon mode of in-heritance.

Second, we have the problem of non-equivalence of appearance under the electron or light microscope of human and animal material. This is likely to hinder comparison. We can attempt links, as Rimoin does, but the process may lead to frustration. The Malamute dwarf dog and stm mouse have similar ultrastructural appearances but 'this particular defect has not been observed in any of the human chondrodystrophies' (Rimoin 1975). The growth plates in rabbit and man differ, making comparison difficult and the possible relationship between the ac rabbit and thanatotrophic dwarfism difficult to assess. Seegmiller *et al.* (1971) likened cho to diastrophic dwarfism in man 'but endochondral histopathology in this mouse mutant differs markedly from that observed in human diastrophic dwarfism' and 'the chondrodystrophic mouse cannot be linkened to any particular human disease at this time' (Rimoin 1975).

Rimoin suggested that animal models may form valuable models of human chondrodystrophies but pointed out that there is often no similarity between animal mutant and human disorder bearing the same name and that before an animal model can be said to represent a particular human condition a tight correlation must be demonstrated in terms of clinical disease, extraskeletal abnormalities, mode of inheritance (but see above), radiographic skeletal abnormalities, and the histological and ultrastructural appearance of cartilage. These are noble sentiments but rather far from being realized.

It may seem strange that not one good model of a human cartilage defect has been found. Are we to suppose that mouse, chicken, turkey, dog, etc. all have their own specific set of unique cartilage abnormalities? Or could it be that defects with the same biochemistry exhibit different ultrastructural pictures in different species? We know that many different defects of biosynthesis, extrinsic or intrinsic, result in dwarfism. We might also suspect (Brown and Harne 1982) that the clinical manifestation of dwarfing may depend on the action of a specific defect on cartilage rather than on the defect itself. They suggest that a type II collagen defect (as in Dmm) may produce a rhizomelic defect because of its rather drastic effect on the quantity of matrix, whilst a cartilage-specific proteoglycan defect (bm) might result in a mesomelic growth disturbance based upon a lesser loss of matrix. The time of onset of a defect must also be taken into account here. Is all mutant cartilage abnormal from the time the first matrix is laid down? Or might a defect affect only a critical stage, either specifically or by being present at a period of intense growth? Disruption of growth plate will always produce similar clinical defects.

It seems possible that when and if biochemical assays on human dystrophic cartilage are published (the ethical aspects of this have to be carefully considered) we may find greater correlation between cartilage diseases of man and animals than seems possible at this stage of our knowledge.

REFERENCES

Abbot, J. and Holtzer, H. (1966). The loss of phenotypic traits by differentiated cells. III. Reversible behaviour of chondrocytes in primary culture. *Journal of Cell Biology* **28**, 473–87.

Argraves, W. S., McKeown-Longo, P. J., and Goetinck, P. F. (1981). Absence of proteoglycan core protein in the cartilage mutant nanomelia. *FEBS Letters* **132**, 265–8.

Ahrens, P. B., Solursh, M. and Reiter, R. S. (1977). Stage related capacity for limb chondrogenesis in cell culture. *Developmental Biology* **60**, 69–82.

——, ——, —— (1979) Position related capacity for differentiation of limb mesenchyme in cell culture. *Developmental Biology* **69**, 436–50.

Athenstaedt, H. (1968). Permanent electric polarisation and pyroelectric behaviour of the vertebrate skeleton. I. The axial skeleton of vertebrates (excluding Mammalia). *Zeitschrift für Zellforschung und microscopische Anatomie* **91**, 135–52.

—— (1970). Permanent longitudinal electric polarisation and pyroelectric behaviour of collagenous structures and nervous tissue in man and other vertebrates. *Nature* **228**, 830–4.

—— (1974). Pyroelectric and piezoelectric properties of vertebrates. *Annals of the New York Academy of Science* **238**, 68–93.

Bargman, G. J., Mackler, B., and Shepard, T. H. (1972). Studies of oxidative energy deficiency. I. Achondroplasia in the rabbit. *Archives of Biochemistry and Biophysics* **150**, 137–46.

Beresford, W. A. (1981). *Chondroid bone, secondary cartilage and metaplasia.* Urban & Schwarzenberg, Baltimore/Munich.

Biggars, J. D. (1957). The growth of cartilaginous embryonic chick bones after freezing. *Experientia* **13**, 483–4.

—— and Gwatkin, R. B. L. (1964). Effects of X rays on the morphogenesis of the embryonic chick tibiotarsus. *Nature* **202**, 152–4.

Blackwood, H. J. J. (1966). Growth of the mandibular condyle of the rat studied with tritiated thymidine. *Archives of Oral Biology* **11**, 493–500.

Bonucci, E. (1969). Further investigations on the organic/inorganic relationships in calcifying cartilage. *Calcified Tissue Research* **3**, 38–54.

——, Del Marco, A., Nicoletti, B., Petrinell, P., and Pozz, L. (1976). Histological and histochemical investigations of achondroplastic mice: a possible model of human achondroplasia. *Growth* **40**, 241–51.

——, Gherardi, G., Del Marco, A., Nicoletti, B., and Petronelli, P. (1977). An electron microscopic investigation of cartilage and bone in achondroplastic (cn) mice. *Journal of Submicroscopial Cytology* **9**, 299–306.

Bourett, L. A., Goetinck, P. F., Hintz, R., and Rodan, G. A. (1979). Cyclic $3',5'$-AMP changes in chondrocytes of the proteoglycan deficient chick mutant nanomelia. *FEBS Letters* **108**, 353.

Breen, M., Richardson, R., Bondareff, W., and Weinstein, H. G. (1973). Acid glycosaminoglycans in developing sternocostal cartilage of the hydrocephalic ch/ch mouse. *Biochimica et Biophysica Acta* **304**, 828–36.

Brewer, A. K., Johnson, D. R., and Moore, W. J. (1977). Further studies on skull growth in achondroplastic mice. *Journal of Embyrology and experimental Morphology* **39**, 59–70.

Brighton, C. T. (1978). Structure and function of the growth plate. *Clinical Orthopaedics and Related Research* **136**, 22–32.

—— and Hunt, R. M. (1974). Mitochondrial calcium and its role in calcification. *Clinical Orthopaedics and Related Research* **100**, 406–16.

——, —— (1976). Histochemical localisation of calcium in the growth plate mitochondria and matrix vesicles. *Federation Proceedings, Federation of American Societies for Experimental Biology* **35**, 143–7.

Bringel, S. A., Sande, R. D., and Newbrey, J. (1983). Dwarfism in the Alaskan Malamute: ultrastructural features of dwarf growth plate chondrocytes. *Calcified Tissue International* **35**, 216–24.

Brown, K. S. and Harne, L. (1982). Brachymorphism, cartilage matrix deficiency and disproportionate micromelia: three inborn errors of cartilage biosynthesis in mice. In *Animal models of inherited metabolic diseases*, pp. 245–9. Alan Liss, New York.

——, Cranley, R. E., Greene, R., Kleinman, H. K., and Pennypacker, J. P. (1981). Disproportionate micromelia (Dmm) an incomplete dominant mouse dwarfism with abnormal cartilage matrix. *Journal of Embyrology and experimental Morphology* **62**, 165–82.

Brown, W. H. and Pearce, L. (1945). Hereditary achondroplasia in the rabbit. I. Physical appearance and general features. *Journal of experimental Medicine* **82**, 241–61.

Caplan, A. I. (1970). Effects of the nicotinamide sensitive teratogen 3-acetylpiridine on chick limb cells in culture. *Developmental Biology* **69**, 436–50.

—— and Hascall, V. C. (1980). Structure and developmental changes in proteoglycans. In *Dilation of the uterine cervix* (ed. F. Nafalin and P. G. Stubblefield). Raven Press, New York.

Caterson, B. and Baker, R. J. (1978). The interaction of link proteins with proteoglycan monomers in the absence of hyaluronic acid. *Biochemical and Biophysical Research Communications* **80**, 496–503.

Chiakulas, J. J. (1957). The specificity and differential fusion of cartilage derived from mesoderm and mesectoderm. *Journal of experimental Zoology* **136**, 287–99.

Cleall, J. F., Wilson, G. N., and Garnett, D. S. (1968). Normal craniofacial skeletal growth of the rat. *American Journal of Physical Anthropology* **29**, 225–42.

De Luca, S., Caplan, A. I., and Hascall, V. C. (1978). Biosynthesis of proteoglycans by chick limb bud chondrocytes. *Journal of Biological Chemistry* **253**, 4713–20.

——, Heinegard, D., Hascall, V. C., Kimura, J., and Caplan, A. I. (1977). Chemical and physical changes in proteoglycans during development of chick limb bud chondrocytes grown in vitro. *Journal of Biological Chemistry* **252**, 6600–8.

——, Lohmander, L. S., Nilsson, B., Hascall, V. C., and Caplan, A. I. (1980). Proteoglycans from chick bud chondrocyte cultures: keratan sulphate and oligosaccharides which contain mannose and sialic acid. *Journal of Biological Chemistry* **255**, 6077–83.

Diegelmann, R. and Petrofsky, B. (1972). Collagen biosynthesis during connective tissue development in the chick embryo. *Developmental Biology* **28**, 443–53.

Diewert, V. (1980). Differential changes in cartilage cell proliferation and cell density in the rat craniofacial complex during secondary palate development. *Anatomical Record* **198**, 219–28.

Dodds, G. S. (1930). Row formation and other types of arrangement of cartilage cells in endochondral ossification. *Anatomical Record* **46**, 385–99.

Dr Brul, E. L. and Laskin, D. M. (1961). Preadaptive potentialities of the mammalian skull: an experiment in growth and form. *American Journal of Anatomy* **109**, 117–32.

Duke, J. and Elmer, W. A. (1977). Effect of the brachypod mutation on cell adhesion in aggregates of mouse limb mesenchyme. *Journal of Embyrology and experimental Morphology* **42**, 209–17.

——, —— (1978). Cell adhesion and chondrogenesis in brachypod mouse limb mesenchyme: fragment fusion studies. *Journal of Embyrology and experimental Morphology* **48**, 161–8.

——, —— (1979). Effect of the brachypod mutation on the early stages of chondrogenesis in mouse embryo hind limbs: an ultrastructural analysis. *Teratology* **19**, 367–76.

Durkin, J. F. (1972). Secondary cartilage a misnomer? *American Journal of Orthodontics* **62**, 15–41.

Ede, D. A. and Flint, O. P. (1972). Patterns of cell division, cell death and chondrogenesis in cultured aggregates of normal and talpid[3] mutant embryos. *Journal of Embryology and experimental Morphology* **27**, 245–60.

——, Flint, O. P., Wilby, O. K., and Colquhoun, P. (1977). The development of precartilaginous condensations in limb bud mesenchyme in vitro and in vivo. In *Vertebrate limb and somite morphogenesis* (ed. D. A. Ede, J. R. Hinchliffe, and M. Balls), pp. 161–180. Cambridge University Press, London/New York.

Elmer, W. A. and Selleck, D. K. (1975). In vitro chondrogenesis of limb mesoderm from normal and brachypod mouse embryos. *Journal of Embryology and experimental Morphology* **33**, 371–86.

Epperlein, H. H. and Lehmann, R. (1975). Ectomesenchymal–ectodermal interaction system (EEIS) of *Triturus alpestris* in tissue culture. 2. Observations on differentiation of visceral cartilage. *Differentiation* **4**, 159–74.

Fell, H. B. (1925). The histogenesis of cartilage and bone in the long bones of the embryonic fowl. *Journal of Morphology and Physiology* **40**, 417–58.

—— and Grüneberg, H. (1939). The histology and self differentiating capacity of the abnormal cartilage in a new lethal mutation in the rat (Rattus norvegicus). *Proceedings of the Royal Society* **B127**, 257–77.

Ferguson, J. M., Wallace, M. E., and Johnson, D. R. (1978). A new type of chondrodystrophic mutation in the mouse. *Journal of Medical Genetics* **15**, 128–31.

Fitzgerald, M. J. T. and Shtieh, M. M. (1977). Interstitial versus appositional growth in developing non-articular cartilage. *Journal of Anatomy* **124**, 503–4.

Fraser, R. A. and Goetinck, P. F. (1971). Reduced synthesis of chondroitin sulfate by cartilage from the mutant nanomelia. *Biochemical and Biophysical Research Communications* **43**, 494–503.

Gaffney, L. J. (1975). Chondrodystrophy, an inherited lethal condition in turkey embyros. *Journal of Heredity* **66**, 339–43.

Goetinck, P. F. (1983). Mutations affecting limb cartilage. In *Cartilage* (ed. B. K. Hall), Vol. II, pp. 165–89. Academic Press, New York.

——, Pennypacker, J. P., and Rogal, P. D. (1974). Proteochondroitin sulphate synthesis and chondrogenic expression. *Experimental Cell Research* **87**, 241–8.

Gould, R. P., Selwood, L., Day, A., and Wolpert, L. (1974). The mechanism of cellular orientation during early cartilage formation in the chick limb and regenerating amphibian limb. *Experimental Cell Research* **83**, 283–96.

Greene, R. M., Brown, K. S., and Pratt, R. MN. (1978). Autoradiographic analysis of altered glycosaminoglycan synthesis in the epiphysial cartilage of neonatal brachymorphic mice. *Anatomical Record* **191**, 19–30.

Grüneberg, H. (1963). *The pathology of development.* Blackwell, Oxford.

—— and Lee, A. J. (1973). The anatomy and development of brachypodism in the mouse. *Journal of Embryology and experimental Morphology* **30**, 119–41.

Hains, R. W. (1975). The histology of epiphysial union in mammals. *Journal of Anatomy* **120**, 1–25.

——, Mohiuddin, A., Okpa, F. J., and Viega-Pires, J. A. (1967). The sites of early epiphysial union in the limb girdles and major long bones of man. *Journal of Anatomy* **101**, 823–31.

Hall-Craggs, E. C. B. (1969). Influence of epiphysis on regulation of bone growth. *Nature* **221**, 1245.

Happle, R. (1977). Dermatologische Leitsymptome einer Sonderform der Chondrodysplasia punctata. *Hautarzt* **28** (Suppl. 2), 260–3.

—— (1981). Cateracts as a marker of genetic heterogeneity in chondrodysplasia punctata. *Clinical Genetics* **19**, 64–6.

——, Phillips, R. J. S., Roessner, A., and Junemann, G. (1983). Homologous genes for X-listed chondrodysplasia punctata in man and mouse. *Human Genetics* **63**, 24–7.

Hardinham, T. E. (1979). The role of link protein in the structure of cartilage proteoglycan aggregates. *Biochemical Journal* **177**, 237–47.

Hascall, V. C. (1977). Interaction of cartilage proteoglycans with hyaluronic acid. *Journal of Supramolecular Structure* **7**, 101–20.

—— and Heinegard, D. (1972a). Aggregation of cartilage proteoglycans. I. The role of hyaluronic acid. *Journal of Biological Chemistry* **249**, 4232–41.

——, —— (1974b). Aggregation of cartilage proteoglycans. II. Oligoasaccharide competitors of the proteoglycan–hyaluronic acid interaction. *Journal of Biological Chemistry* **249**, 4242–9.

——, —— (1979). Structure of cartilage proteoglycans. In *Glycoconjugate research* (ed. J. D. Gregory and R. Jeanloz), Vol. 1. Academic Press, New York. York.

—— and Sajdera, S. (1970). Physical properties and polydisperity of proteoglycan from bovine nasal cartilage. *Journal of Biological Chemistry* **245**, 4920–30.

——, Oegama, T. R., Brown, M., and Caplan, A. I. (1976). Isolation and characterisation of proteoglycans from chick embryo limb bud chondrocytes grown in vitro. *Journal of Biological Chemistry* **251**, 3511–19.

Heinegard, D. and Hascall, V. C. (1974). Aggregation of cartilage proteoglycans. III. Characterisation of the proteins isolated from trypsin digests of aggregates. *Journal of Biological Chemistry* **249**, 4250–6.

Hewitt, A. T., Varner, H. H., and Martin, G. R. (1980). The role of chondronectin and proteoglycan in chondrocyte attachment to collagen. *Journal of Cell Biology* **91**, 154a.

Hicks, M. J. (1982). Analysis of the differential growth of chondrogenic elements in the embryo chick limb. PhD thesis. University College of Wales, Aberystwyth.

Hinchliff, J. R. and Ede, D. A. (1968). Abnormalities in bone and cartilage development in the talpid[3] mutant of the fowl. *Journal of Embryology and experimental Morphology* **19**, 327–39.

—— and Johnson, D. R. (1982). Growth of cartilage. In *Cartilage* (ed. B. K. Hall), Vol. 2, pp. 255–95. Academic Press, New York.

Hindin, D. and Erickson, R. P. (1981). The brachymorphic mutation of mice and altered developmental patterns of limb bud 3':5' cyclic adenosine monophosphate. *Experientia* **37**, 1149–50.

Holmes, L. B. and Trelstad, R. L. (1980). Cell polarity in precartilage mouse limb mesenchyme cells. *Developmental Biology* **78**, 511–20.

Holtfreter, J. (1968). Mesenchyme and epithelia in inductive and morphogenic processes. In *Epithelial–mesenchymal interactions* (ed. R. Fleischmajer and R. E. Billingham), pp. 1–30. Williams & Wilkins, Baltimore.

Holtzer, H. and Abbott, J. (1968). Oscillations of the chondrogenic phenotype in vitro. In *The stability of the differentiated state* (ed. H. Ursprung). Springer Verlag, New York/Berlin.

Howell, D. S. (1971). Current concepts of calcification. *Journal of Bone and Joint Surgery* **53A**, 250–8.

——, Pita, J. C., Marquez, J. F., and Gatter, R. A. (1969). Demonstration of macromolecular inhibitors of calcification and nucleation factors in fluid from calcifying sites in cartilage. *Journal of Clinical Investigation* **48**, 630–41.

Hoyte, D. A. N. (1971). Mechanisms of growth in the cranial vault and base. *Journal of Dental Research* **50**, 1447–61.

Hunter, S. J. and Caplan, A. I. (1982). Control of cartilage differentiation. In *Cartilage* (ed.. B. K. Hall), Vol. 2, pp. 87–113. Academic Press, New York.

Janners, M. T. and Searls, R. L. (1970). Changes in rate of cellular proliferation during the differentiation of cartilage and muscle in the mesenchyme of the embryonic chick wing. *Developmental Biology* **23**, 136–65.

Johnson, D. R. (1974). The in vivo behaviour of achondroplastic cartilage from the cartilage anomaly (can/can) mouse. *Journal of Embryology and experimental Morphology* **31**, 313–18.

—— (1977). Ultrastructural observations on stumpy (stm) a new chondrodystrophic mutant in the mouse. *Journal of Embryology and experimental Morphology* **39**, 279–84.

—— (1978). The growth of the femur and tibia in three genetically distinct chondrodystrophic mutants of the house mouse. *Journal of Anatomy* **125**, 267–75.

—— (1983). Abnormal cartilage from the mandibular condyle of stumpy (stm) mutant mice. *Journal of Anatomy* **137**, 715–28.

—— (1984). The ultrastructure of the manibular condylar cartilage from mice carrying the spondylo-metaphyseal chondrodysplasia (smc) gene. *Journal of Anatomy* **138**, 463–70.

—— and Hunt, D. M. (1974). Biochemical observations on the cartilage of achondroplastic (can) mice. *Journal of Embyrology and experimental Morphology* **31**, 319–28.

—— and Wise, J. M. (1971). Cartilage anomaly (can) a new mutant gene in the mouse. *Journal of Embryology and experimental Morphology* **25**, 21–31.

Johnston, L. and Nash, D. J. (1982). Saggital growth trends of the development of cleft palate in mice homozygous for the 'paddle' gene. *Journal of Craniofacial Genetics and Developmental Biology* **2**, 265–76.

Jolly, R. J. and Moore, W. J. (1975). Skull growth in achondroplastic (cn) mice: a craniometric study. *Journal of Embryology and experimental Morphology* **33**, 1013–22.

Karasawa, K., Kimata, K., Ito, K., Kato, Y., and Suzuki, S. (1979). Morphological and biochemical differentiation of limb bud cells in chemically defined medium. *Developmental Biology* **70**, 287–305.

Kember, N. F. (1960). Cell division in endochondral ossification. A study in cell proliferation in rat bones by the method of tritiated thymidine autoradiography. *Journal of Bone and Joint Surgery* **42B**, 824–39.

—— (1972). Comparative patterns of cell division in epiphysial cartilage plates in the rat. *Journal of Anatomy* **111**, 137–42.

—— (1982). Cell kinetics of cartilage. In *Cartilage* (ed. B. K. Hall), Vol. 1, pp. 149–80. Academic Press, New York.

Kimata, K., Barrach, H. J., Brown, K. S., and Pennypacker, J. P. (1981). Absence of proteoglycan core protein in cartilage from cmd/cmd (cartilage matrix deficiency) mouse. *Journal of Biological Chemistry* **256**, 6961–8.

——, Oike, Y., Ito, K., Karasawaka, K., and Suzuki, S. (1978). The occurrence of low bouyant density proteoglycans in embyronic chick cartilage. *Biochemical and Biophysical Research Communications* **85**, 1431–9.

Kitamura, K. and Yamagota, T. (1976). The occurrence of a new type of proteochondroitin sulphate in the developing chick embryo. *FEBS Letters* **71**, 337–40.

Kleinman, H. K., Klebe, R. J., and Martin, G. R. (1981). Role of collagenous matrices in the adhesion and growth of cells. *Journal of Cell Biology* **88**, 473–85.

——, Pennypacker, J. P., and Brown, K. S. (1977). Proteoglycan and collagen of achondroplastic (cn/cn) neonatal mouse cartilage. *Growth* **41**, 171–7.

Kochhar, D. M., Aydelotte, M. B., and Vest, T. K. (1976). Altered collagen fibrillogenesis in embryonic mouse limb cartilage deficient in matrix granules. *Experimental Cell Research* **102**, 213–22.

Konyukhov, B. V. and Ginter, E. (1966). A study of the action of the brachypodism-H gene on development of the long bones of the hind limbs in the mouse. *Folia Biologica, Praha* **12**, 199–206.

—— and Bugrilova, R. S. (1968). The growth inhibiting factor in embryos of mutant stock brachypodism-H mice. *Folia Biologica, Praha* **13**, 65–9.

—— and Paschin, Y. V. (1967). Experimental study of the achondroplasia gene effects in the mouse. *Acta biologica Academiae scientarum hungaricae* **18**, 285–94.

——, —— (1970). Abnormal growth of the body, internal organs and skeleton in the achondroplastic mice. *Acta biologica Academiae scientarum hungaricae* **21**, 347–54.

Kosher, R. A., Savage, M. P., and Chan, S. C. (1979). Cyclic AMP derivatives stimulate the chondrogenic differentiation of the mesoderm subjacent to the apical ectodermal ridge of the chick limb bud. *Journal of experimental Zoology* **209**, 221–8.

Krotoski, D. M. and Elmer, W. A. (1973). Alkaline phosphatase activity in fetal hind limbs of the mouse mutation brachypodism. *Teratology* **7**, 99–106.

Kuettner, R. E. and Pauli, B. K. (1978). Resistance of cartilage to neoplastic and normal invasion. In *Mechanisms of localised bone loss* (ed. J. E. Horton, T. M. Tarpley, and W. F. Davis). Information Retrieval Inc, Washington.

Lacroix, P. (1951). *The organisation of bones* (translated S. Gilder). Churchill, London.

Landauer, W. (1965). Nanomelia, a lethal mutation of the fowl. *Journal of Heredity* **56**, 131–8.

—— (1968). A bibliography of micromelia. In *Limb development and deformity*, (Ed. Chester A. Swingard), pp. 540–620. Chas T. Thomas, Springfield, Illinois.

Lane, P. W. and Dickie, M. M. (1968). Three recessive mutations producing disproportionate dwarfing in mice. *Journal of Heredity* **59**, 300–8.

Lash, J. W. and Vasan, N. S. (1982). Glycosaminoglycans of cartilage. In *Cartilage* (ed. B. K. Hall), Vol. 1, pp. 101–13. Academic Press, New York.

Leach, R. M. and Bass, E. G. (1977). The effect of inherited chondrodystrophy on the hexosamine content of cartilage from turkey embryos. *Poultry Science* **56**, 1043–5.

Lehniger, A. L. (1970). *The molecular basis of cell structure and function*. Worth, New York.

Lennon, D. P. In Hunter, S. J. and Caplan, A. J. (1982). Control of cartilage differentiation. *Cartilage* (ed. B. J. Hall), Vol. II, pp. 87–117. Academic Press, New York.

Levitt, D. and Dorfman, A. (1973). Control of chondrogenesis in limb bud cell cultures by bromodeoxyuridine. *Proceedings of the National Academy of Sciences, USA* **70**, 2201–5.

——, Ho, L., and Dorfman, A. (1975). Effect of 5-bromodeoxyuridine on ultrastructure of developing limb bud cells in vitro. *Developmental Biology* **43**, 75–90.

Lohmander, L. S., Mudsen, K., and Hinek, A. (1979a). Secretion of proteoglycans by chondrocytes. Influence of colchicine, cytochalasin B and beta D xyloside. *Archives of Biochemistry and Biophysics* **192**, 148–57.

——, Hascall, V. C., Caplan, A. I. (1979b). Effects of 4-methyl 6-umbelliferyl-beta-D-xylopyranoside on chondrogenesis and proteoglycan synthesis in chick limb bud mesenchymal cell cultures. *Journal of Biological Chemistry* **254**, 100551–61.

Mathews, M. B. (1965). The interaction of collagen and acid mucopolysaccharides. *Biochemical Journal* **96**, 710–16.

—— (1967). Chondroitin sulphate and collagen in inherited skeletal defects of chickens. *Nature, London* **213**, 1255–6.

—— and Decker, L. (1967). The effect of acid mucopolysaccharides and acid mucopolysaccharide-proteins on fibril formation from collagen solutions. *Biochemical Journal* **109**, 517–26.

Matsutani, E. and Kuroda, Y. (1978). Enhancement of chondrogenesis of cultured quail limb bud mesenchymal cells by cellophane films. *Cell Structure and Function* **3**, 237–48.

—— and Kuroda, Y. (1980). Effect of cell association on the in vitro chondrogenesis of mesenchyme cells from quail limb buds. *Cell Structure and Function* **5**, 239–46.

Mayne, R. and Von der Mark, K. (1982). Collagens of cartilage. In *Cartilage* (ed. B. K. Hall), Vol. 1, pp. 181–214. Academic Press, New York.

McKeown, P. J. and Goetinck, P. F. (1979). A comparison of the proteoglycans synthesised in Meckel's and sternal cartilage from normal and nanomelic chick embryos. *Developmental Biology* **71**, 203–15.

McKeown-Longo, P. F. (1981). Proteoglycan link protein synthesis by cartilage from normal and nanomelic chick embryos. *Federated Proceedings of the American Society of Experimental Biologists* **40**, 1840.

——, Sparks, K. J., and Goetnick, P. F. (1982). Preparation and characterisation of an antiserum against purified proteoglycan link proteins from avian cartilage. *Collagen and Related Research* **2**, 231–44.

Messier, B. and Leblond, C. P. (1960). Cell proliferation and migration as revealed by radioautography after injection of thymidine H^3 into male rats and mice. *American Journal of Anatomy* **106**, 1075–7.

Milaire, J. (1965). Étude morphogénétique de trois malformations congénitales de l'atopode chez la souris (syndactylisme—brachypodisme—hémimélie dominante) par des méthodes cytochimiques. *Mémoires de l'Academie r. de Belgique*. Class des sciences Collection in 40° 2me serie **16**.

Miller, E. J. (1976). Biochemical characteristics and biological significance of the genetically distinct collagen. *Molecular and Cellular Biochemistry* **13**, 165–92.

—— (1977). The collagens of the extracellular matrix. In *Cell and tissue interactions* (ed. J. W. Walsh and M. M. Burger), pp. 71–86. Raven Press, New York.

—— and Matukas, V. A. (1969). Chick cartilage collagen: a new type of $alpha_1$ chain not present in bone or skin of the species. *Proceedings of the National Academy of Sciences, USA* **69**, 2069–72.

Miller, W. A. and Flynn-Miller, K. L. (1976). Achondroplastic, brachymorphic and stubby chondrodystrophies in mice. *Journal of Comparative Pathology* **86**, 349–64.

——, Flynn, K. L., and Drinnan, A. J. (1974). Dental and histological changes associated with chondrodystrophies in mice. *International Association of Dental Research* p. 189.

Minor, R. R. (1980). Collagen metabolism. *American Journal of Pathology* **98**, 228–80.

Monson, C. B. and Seegmiller, R. E. (1981). Ultrastructural studies of cartilage matrix in mice homozygous for chondrodysplasia. *Journal of Bone and Joint Surgery* **63A**, 637–44.

Moscona, A. (1965). Development of heterotypic combinations of dissociated embryonic chick cells. *Proceedings of the Society for Experimental Biology and Medicine* **92**, 410–16.

Nash, D. J. (1969). *Mouse News Letter* **40**, 20.

Nero, Z. and Dorfman, A. (1972). Stimulation of chondromucoprotein synthesis in chondrocytes by extracellular chondromucoprotein. *Proceedings of the National Academy of Sciences, USA* **69**, 2064–72.

Okayama, M., Pacifici, M., and Holtzer, H. (1976). Differences amongst sulphated proteoglycans synthesised in non-chondrogenic cells, presumptive chondroblasts and chondroblasts. *Proceedings of the National Academy of Sciences, USA* **73**, 3224–8.

Olson, B. R. (1981). Collagen biosynthesis. In *Cell biology of extracellular matrix* (ed. E. D. Hay), pp. 139–77. Plenum, New York.

Orkin, R. W., Pratt, R. M., and Martin, G. M. (1976). Undersulphated chondroitin sulphate in cartilage matrix of brachymorphic mice. *Developmental Biology* **50**, 82–94.

——, Williams, B. R., Cranley, R. E., Poppke, D. C., and Brown, K. S. (1977).

Defects in the cartilaginous growth plate of brachymorphic mice. *Journal of Cell Biology* **73**, 287–99.

Ovadia, M., Parker, C. H., and Lash, J. W. (1980). Changing patterns of proteoglycan synthesis during chondrogenic differentiation. *Journal of Embyrology and experimental Morphology* **56**, 59–70.

Owens, E. M. and Solursh, M. (1981). In vitro histogenic capacities of limb mesenchyme from various stage mouse embryos. *Developmental Biology* **88**, 297–311.

——, —— (1982). Cell–cell interaction by mouse limb cells during in vitro chondrogenesis: analysis of the brachypod mutation. *Developmental Biology* **91**, 376–88.

Palmoski, M. J. and Goetinck, P. (1972). Synthesis of proteochondroitin sulphate by normal, nanomelic and 5-bromodeoxyuridine treated chondrocytes in cell culture. *Proceedings of the National Academy of Sciences, USA* **69**, 3385–8.

Pearce, L. and Brown, W. H. (1945*a*). Hereditary achondroplasia in the rabbit. II. Pathological aspects. *Journal of experimental Medicine* **82**, 261–80.

——, —— (1945*b*). Hereditary achondroplasia in the rabbit. III. Genetic aspects; general conclusions. *Journal of experimental Medicine* **82**, 281–95.

Pennypacker, J. P.and Goetinck, P. F. (1976). Biochemical and ultrastructural studies of collagen and proteochondroitin sulphate in normal and nanomelic cartilage. *Developmental Biology* **50**, 35–47.

——, Kimata, K., and Brown, K. S. (1981). Brachymorphic mice (bm/bm): a generalised biochemical defect expressed primarily in cartilage. *Developmental Biology* **81**, 280–7.

Phillips, R. J. S., Hasker, S. H., and Moseley, H. J. (1973). Bare patches, a new sex linked gene in the mouse associated with a high production of XO factors. A preliminary report of breeding experiments. *Genetical Research* **22**, 91–9.

Pita, J. C., Muller, E., and Howell, D. S. (1975). Disaggregation of proteoglycan aggregate during endochondral ossification: physiological role of lysosome. In *Dynamics of connective tissue macromolecules* (ed. P. M. C. Burleigh and A. R. Poole), pp. 247–58. North Holland, Amsterdam.

Pleskova, M. V., Rodinov, V. M., Bugrilova, R. S., and Konyukhov, B. V. (1974). The partial purification of growth inhibiting factor of the brachypodism-H mouse embryos. *Developmental Biology* **37**, 417–21.

Pratt, R. M., Solomon, D. S., Diewert, U. M., Erickson, R. P., Burns, R., and Brown, K. S. (1980). Cortisone induced cleft palate in the brachymorphic mouse. *Teratogenesis, Carcinogenesis, Mutagenesis* **1**, 15–23.

Rhodes, R. K. and Elmer, W. A. (1972). Aberrant collagen biosynthesis in fibulae of the mouse mutant brachypodism. *American Zoologist* **12**, 709.

——, —— (1975). Aberrant metabolism of matrix components in neonatal fibular cartilage of brachypod (bp^H) mice. *Developmental Biology* **46**, 14–27.

Rigal, W. M. (1962). The use of tritiated thymidine in studies of chondrogenesis. In *Radioisotopes* (ed. P. Lacroix and A. M. Brady), pp. 197–225. Blackwell, Oxford.

Rimoin, D. L. (1975). The chondrodystrophies. *Advances in Human Genetics* **5**, 1–118.

Rittenhouse, E., Dunn, L. C., Cookingham, J., Calo, C., Spiegelman, M., Dooher, G. B., and Bennett, D. (1978). Cartilage matrix deficiency (cmd) a new autosomal recessive lethal mutation in the mouse. *Journal of Embyrology and experimental Morphology* **43**, 71–84.

Schenk, R. K., Spiro, D., and Weiner, J. (1967). Cartilage resorption in the tibial epiphyseal plate of growing rats. *Journal of Cell Biology* **34**, 275–91.

Schnieke, A., Harbers, K., and Jaenisch, R. (1983). Embryonic lethal mutation in mice induced by retrovirus insertion into the alpha 1 (I) collagen gene. *Nature* **304**, 315–20.

Schwartz, N. B., Belch, J., Henry, J., Hupert, J., and Sughara, K. (1982). Enzyme defect in PAPS synthesis of brachymorphic mice. *Federated Proceedings of the American Society for Experimental Biology* **41**, 852.

Searls, R. L. (1973). Newer knowledge of chondrogenesis. *Clinical Orthopaedics and Related Research* **96**, 327–44.

Seegmiller, R., Fraser, F. C., and Sheldon, H. (1971). A new chondrodystrophic mutant in mice. Electron microscopy of normal and abnormal chondrogenesis. *Journal of Cell Biology* **48**, 580–93.

——, Ferguson, C. C., and Sheldon, H. (1972). Studies on cartilage. VI. A genetically determined defect in tracheal cartilage. *Journal of Ultrastructural Research* **38**, 288–301.

——, Myers, R. A., Dorfman, A., and Horowitz, A. L. (1981). Structural and associative properties of cartilage matrix constituents in mice with hereditary chondrodysplasia (cho). *Connective Tissue Research* **9**, 69–77.

Serafini-Fracassini, A. and Smith, J. W. (1974). *The structure and biochemistry of cartilage.* Churchill-Livingstone, Edinburgh/London.

Shambaugh, J. and Elmer, W. A. (1980). Analysis of glycosaminoglycans during chondrogenesis of normal and brachypod mouse limb mesenchyme. *Journal of Embryology and experimental Morphology* **56**, 225–38.

Shapiro, F., Holtrop, M. E., and Glimcher, M. J. (1973). Organisation and cellular biology of the perichondral ossification groove of Ranvier. A morphological study in rabbits. *Journal of Bone and Joint Surgery* **59A**, 703–723.

Shepard, T. H. and Bass, G. L. (1971). Organ culture studies of achondroplastic rabbit cartilage: evidence for a metabolic defect in glucose utilisation. *Journal of Embryology and experimental Morphology* **25**, 347–63.

——, Fry, L. R., and Moffett, B. C. (1969). Microscopic studies of achondroplastic rabbit cartilage. *Teratology* **2**, 13–22.

Shimomura, Y., Wezeman, F. H., and Ray, R. D. (1973). The growth cartilage plate of the rat rib: cellular proliferation. *Clinical Orthopaedics and related Research* **90**, 246–54.

Silberberg, R. and Lesker, P. (1975). Skeletal growth and development of achondroplastic mice. *Growth* **39**, 17–33.

——, Hasler, M., and Lesker, P. (1976). Ultrastructure of articular cartilage of achondroplastic mice. *Acta anatomica* **96**, 162–75.

Silbermann, M. and Frommer, J. (1972*a*). Vitality of chondrocytes in the mandibular condyle as revealed by collagen formation: an autoradiographic study with ³H proline. *American Journal of Anatomy* **135**, 359–63.

——, —— (1972*b*). The nature of endochondral ossification in the mandibular condyle of the mouse. *Anatomical Record* **172**, 659–67.

——, —— (1974). Ultrastructure of developing cartilage in the mandibular condyle of the mouse. *Acta anatomica* **90**, 330–46.

Sissons, H. A. (1956). Experimental study of the effect of local irradiation on bone growth. In *Progress in Radiobiology* (ed. J. S. Michel, B. C. Homes, and C. L. Smith), pp. 436–48. Oliver & Boyd, Edinburgh.

Solursh, M. and Meier, S. (1973). A conditioned medium factor produced by chondrocytes that promotes their own differentiation. *Developmental Biology* **30**, 279–89.

——, —— (1974). Effects of cell density on the expression of differentiation by cultured chick embryo chondrocytes. *Journal of experimental Biology* **187**, 311–22.

—— and Rieter, R. S. (1975*a*). The enhancement of in vitro survival and chondrogenesis of limb bud cells by cartilage conditioned medium. *Developmental Biology* **44**, 278–87.

——, —— (1975*b*). Determination of limb bud chondrocytes during a transient block of the cell cycle. *Cell. Differentiation* **4**, 131–7.

——, Ahrens, P. B., and Rieter, R. S. (1978). Tissue culture analysis of steps in limb chondrogenesis. *In Vitro* **14**, 51–61.

——, Rieter, R. S., Ahrens, P. B., and Pratt, R. M. (1979). Increase in levels of cyclic AMP during avian limb chondrogenesis in vitro. *Differentiation* **15**, 184–6.

——, ——, ——, and Vertel, B. M. (1981). Stage and position related changes in chondrogenic response of chick embryonic wing mesenchyme to treatment with dibutyl cyclic AMP. *Development Biology* **83**, 9–19.

——, Jensen, K. L., Singley, C. T., Linsenmayer, T. F., and Rieter, R. S. (1983). Evidence of the two distinct regulating steps in cartilage differentiation. *Developmental Biology* (in press).

Stearns, K. and Goetinck, P. F. (1979). Stimulation of chondroitin sulphate synthesis by beta D xyloside in chondrocytes of the proteoglycan deficient mutant nanomelia. *Journal of Cell Physiology* **100**, 33–8.

Stephens, T. D. and Seegmiller, R. E. (1976). Normal production of cartilage glycosaminoglycan in mice homozygous for the chondrodysplasia gene. *Teratology* **133**, 317–26.

Stockwell, R. A. (1979). *Biology of cartilage cells.* Cambridge University Press, London/New York.

Stocum, D. L., Davis, R. M., Leger, M., and Condrad, H. E. (1979). Development of tibiotarsus in the chick embryo: biosynthesis activities of histologically distinct regions. *Journal of Embryology and experimental Morphology* **54**, 155–70.

Sugahara, K. and Schwartz, N. B. (1979). Defect in 3′ phosphadenosine 5′ phosphosulphate formation in brachymorphic mice. *Proceedings of the National Academy of Sciences, USA* **76**, 6615–81.

Tang, L., Rosenberg, L., Reiner, A., and Poole, R. A. (1979). Proteoglycans from bovine nasal cartilage. Properties of a soluble factor of link protein. *Journal of Biological Chemistry* **254**, 10523–31.

Tanner, J. M. (1962). *Growth in adolescence.* Blackwell, Oxford.

Thorngren, K. G. and Hansson, L. J. (1973). Cell kinetics and morphology of the growth plate in normal and hypophysectomised rat. *Calcified Tissue Research* **13**, 113–29.

Thorogood, P. B. (1982). Morphogenesis of cartilage. In *Cartilage* (ed. B. K. Hall), Vol. II, pp. 223–50. Academic Press, New York.

—— and Hinchliffe, J. R. (1975). An analysis of the condensation process during chondrogenesis in the embryonic chick hind limb bud. *Journal of Embryology and experimental Morphology* **33**, 582–606.

Thurston, M. N., Johnson, D. R., Kember, N. F., and Moore, W. J. (1983). Cell kinetics of growth cartilage in stumpy: a new chondrodystrophic mutant in the mouse. *Journal of Anatomy* **136**, 407–15.

——, ——, —— (1985*a*). Cell kinetics of growth cartilage of achondroplastic (cn) mice. *Journal of Anatomy* **140**, 425–34.

——, ——, —— (1985*b*). Cell kinetics of growth cartilage in spondylo-metaphyseal chondrodysplasia (smc) mice. *Journal of Anatomy* **140**, 435–45.

Tonna, E. A. (1961). The cellular component of the skeletal system studied autoradiographically with tritiated thymidine during growth and ageing. *Journal of Biophysical and Biochemical Cytology* **9**, 813–24.

Toole, B. P. (1972). Hyaluronate turnover during chondrogenesis in the developing chick limb axial skeleton. *Developmental Biology* **29**, 321–9.

—— and Gross, J. (1971). The extracellular matrix in the regenerating newt limb: synthesis and removal of hyaluronate prior to differentiation. *Development Biology* **25**, 57–77.

—— and Lowther, D. A. (1968). The effect of chondroitin sulphate-protein on the formation of collagen fibrils in vitro. *Biochemical Journal* **109**, 857–66.

——, Jackson, G., and Gross, J. (1972). Hyaluronate in morphogenesis: inhibition of chondrogenesis in vitro. *Proceedings of the National Academy of Science, USA* **63**, 1384–6.

Trelstad, R. L. (1977). Mesenchymal cell polarity and morphogenesis of chick embryo. *Developmental Biology* **59**, 153–63.

——, King, A. H., Igarashi, S., and Gross, J. (1970). Isolation of two distinct collagens from chick cartilage. *Biochemistry* **9**, 4493–8.

Umansky, R. (1966). The effect of cell population density on the developmental fate of reaggregating mouse limb mesenchyme. *Developmental Biology* **13**, 31–56.

Upholt, W. B., Vertel, B. M., and Dorfman, A. (1979). Translation and characterisation of messenger RNAs in differentiating chicken cartilage. *Proceedings of the National Academy of Sciences, USA* **76**, 4847–51.

Vilmann, H. and Moss, M. L. (1980). Studies on orthocephalisation. V. Peripheral positional stability of the rat cranial frame in the period between 14 and 150 days. *Acta anatomica* **107**, 330–5.

Von der Mark, K. (1980). Immunological studies on collagen type transition in chondrogenesis. *Current Topics in Developmental Biology* **14**, 199–225.

——, Osdoby, P., and Caplan, A. I. (1982). Effect of 4-methylumbelliferyl-beta-D-xyloside on collagen synthesis in chick limb bud mesenchymal cell cultures. *Developmental Biology* **90**, 24–30.

——, Von der Mark, K., and Gay, S. (1976). Studies of differential collagen synthesis during development of the chick embryo by immunofluorescence. 1. Preparation of collagen type I and type II specific antibodies and their application to early studies of the chick embryo. *Developmental Biology* **48**, 237–49.

Walker, K. V. R. and Kember, N. F. (1972a). Cell kinetics of growth cartilage in the rat tibia. I. Measurements in young male rats. *Cell and Tissue Kinetics* **5**, 401–8.

——, —— (1972b). Cell kinetics of growth cartilage in the rat tibia: II. Measurements during ageing. *Cell and Tissue Kinetics* **5**, 409–19.

Webber, R. J., Fox, R. R., and Sokoloff, L. (1981). In vitro cultures of rabbit growth plate chondrocytes. 2. Chondrodystrophic mutants. *Growth* **45**, 269–78.

Wiebtain, O. and Muir, H. (1973). The inhibition of sulphate incorporation in isolated adult chondrocytes by hyaluronic acid. *FEBS Letters* **37**, 42–6.

Wolpert, L. (1981). The cellular basis of skeletal growth during development. *British Medical Bulletin* **37**, 215–19.

Wood, G. C. (1960). The formation of fibrils from collagen solutions. 3. Effects of chondroitin sulphate and some other naturally occurring polyanions on the rate of formation. *Biochemical Journal* **75**, 605–12.

4. The bony skeleton

BONE

Bone, like cartilage, can be considered as being made up of its constituent cells set in a matrix. In bone, however, the organic matrix has another dimension, since it is mineralized. Defects in bony tissue might therefore occur in the cellular phase of bone, in the deposition of matrix, or its mineralization. Calcified bone is also a hindrance to growth and is continually removed and remodelled. We shall see that many inherited defects in bone concern this remodelling system.

CELLS OF BONE

Although the periosteum covering bone contains fibroblasts and the bone marrow contained within many bones has representatives of the haemopoietic series of cells, we shall be mainly concerned with the cells to be found in bone proper, the osteoblasts, osteocytes, and osteoclasts.

Origin of bone cells

The source of bone cells has been a subject of much debate. Owen (1977) supported the view that osteoblasts and osteoclasts have separate lines of cellular differentiation. Because of the interest in fracture repair and degenerative diseases much of the work on the origin of bone cells has been done in adults on the nature of the small stem cell population which is called into action after trauma. The stem cells here, maintained in the adult against need, are not necessarily the same as those which produce bone in the fetus, but may give valuable insight into the origin of the latter. We shall therefore consider first the cells which produce bone *de novo* in the adult.

Origin of osteoblasts and osteocytes

In the postnatal animal the osteogenic (and chondrogenic) stem cell is thought to be a component of bone marrow (Friedenstein 1976; Owen 1980). The stromal part of bone marrow consists of a network of reticular fibres and the cells which secrete them together with adventitial cells (perivascular cells) and endothelial cells which line the walls of blood vessels. Stromal cells are fibroblasts, which form the supporting network of the haemopoietic cell system; in the adult these two cells are histologically distinct, and there is no evidence that there is interchange of cells between the two systems (Friedenstein 1976). In the fetal long bone the condensation of mesenchymal cells and the differentiation of the bony collar (primary centre of ossification) occurs before colonization by haemopoietic stem cells from the liver and the formation of haemopoietic marrow (Moore and Metcalf 1970; Owen 1978) or in its absence in organ culture (Johnson 1980). In postnatal tissue following transplantation of a marrow plug to an ectopic site or removal of marrow from a long bone (Tavassoli and Crosby 1968; Patt and Maloney 1975), stromal cells proliferate, differentiate into osteoblasts, and form bone, as well as reticular cells and lining tissue of the sinusoidal microcirculation before the arrival of fresh haemopoietic stem cells. Friedenstein (1976) showed that bone marrow stromal cells are capable of generating bone spontaneously when implanted in diffusion chambers *in vivo*—other similar tissue derived from other organs did so only in the presence of an inducer.

If marrow cells are cultured *in vitro*, the haemopoietic element decreases within the first week and fibroblastic colonies are left. Several studies (e.g. Friedenstein *et al.* 1970; Friedenstein 1976) suggest that each fibrocyte colony is derived from a single cell. In rabbit marrow cultures implanted in diffusion chambers two main types of colony can be distinguished (Ashton *et al.* 1980), fast-growing tight colonies and slow-growing loose colonies. Some of the former stain intensely for alkaline phosphatase; Ashton suggested that these may be colonies with osteogenic potential, although this remains unproven. Certainly either marrow or fibroblast cultures implanted *in vivo* will produce calcified tissue (Friedenstein 1973; Ashton *et al.* 1980) and the first sign of such a development is alkaline phosphatase activity appearing in the cytoplasmic membrane of a few cells. Later most cells become alkaline phosphatase positive but this reaction decreases again as mineralization proceeds, so that after several weeks only a surface layer of cells adjacent to the mineralized tissue stain for alkaline phosophatase (a situation analagous to that seen in native bone). Ashton *et al.* (1980) showed that alkaline-phosphatase-positive cells appeared 7–14 days after implantation, and by implication that these cells were unlikely to be preosteoblasts—relatively mature cells of the osteoblast line accidentally transferred with the bone marrow implant.

MORPHOLOGY

Osteoblasts

Osteoblasts, plump cells with basophilic cytoplasm, are seen as a single layer of cells applied to the unmineralized osteoid of forming bone, so providing a barrier between the bone and the overlying connective tissue of the periosteum or the connective tissue and blood vessels of the marrow.

The osteoid seam is separated from fully mineralized bone beneath by the calcification front. Fine osteoblast processes penetrate the osteoid to make contact with processes of the deeper lying osteocytes. Osteoblasts are often classified as active or resting according to their morphology—active cells being taller.

Osteoblasts secrete collagen, the main constituent of bone matrix, as well as the glycoprotein-based ground substance. Mineral is deposited within the matrix as amorphous tricalcium phosphate which is converted to hydroxyapatite. Calcospheroids or matrix vesicles, apparently the sloughed-off tips of osteoblast cytoplasmic processes, appear to act as centres for mineralization. Seventy-five per cent of the mineral in bone matrix is deposited over the course of a few days, the balance over the next few months. This short initial phase of intense mineral deposition is thought to be under the control of osteoblasts; the later, longer phase not.

Osteocytes

Osteoblasts eventually become surrounded by mineralized bone, when they are spoken of as osteocytes. Osteocytes are in contact with one another, and with a blood supply via thin cytoplasmic processes which run through canaliculi in the bone matrix. Their morphology is variable, resembling that of osteoblasts if near the surface, but more flattened in deeper locations. Aaron (1976) pointed out that variable appearance is probably an indication of varying function. The role of the osteocyte is not fully understood. In some locations osteocytes fail to fill the lacuna in which they sit, and this has given rise to suggestion that they may have osteolytic properties. Boyde (1972), however, found no supporting evidence of this in rats treated with parathormone. The importance of osteocytes, in fact, becomes clear only when they die; the surrounding bone then becomes structurally inadequate and is replaced by living bone.

OSTEOCLASTS

Osteoclasts are syncytia formed by the fusion of precursor cells. Definitive evidence is now available that at least the majority of osteoclast precursors are circulatory cells of haematogeneous origin. The nucleolar differences between chick and quail cells enabled chimera experiments (Kahn and Simmons 1975) to be performed and parabiosis studies (Gothlin and Ericsson 1973) confirm these findings. The precursor cells are now thought to be of the monocytic–macrophage series in common with foreign-body and inflammatory giant cells (Murch *et al.* 1982; Papadimitriou 1978; Sutton and Weiss 1966). Chambers (1978) found that osteoclasts, when not in contact with bone, are ultrastructurally identical to foreign-body giant cells, whereas the latter, grown in contact with bone, develop the characteristic morphology of the osteoclast.

Osteoclasts share many properties with the mononuclear phagocytic cell series including migration towards resorbing bone, the ability to resorb bone *in vitro* (Chambers 1981*a,b*; Kahn *et al.* 1981), and similar ultrastructure. Experiments using tritiated thymidine (Fischman and Hay 1962) and charcoal (Jee and Nolan 1963) support this view.

Thesingh (1983*a*) co-cultured the stripped metatarsal bone of 17-day-old mouse embryos from which the periosteum had been removed with other tissues; bone resorption was taken to indicate that the latter produced osteoclasts. Thesingh found that pieces of 9- and 10-day-old mouse embryo generated osteoclasts (presumably from the liver which has precursors of the macrophage–monocyte cell line from nine days onwards). So did liver, lung, heart, kidney, kidney capsule, bladder, and periosteum from 12–17-day-old fetuses. Adult organs or tissues never produced osteoclasts, and more osteoclasts were produced by younger tissue. These could, of course, all be circulating cells originating in the embryonic liver.

Function of osteoclasts—modelling and remodelling of bone

Bone cells are collectively responsible for both the formation of new bone and the resorption of pre-existing matrix. During the process of growth, during the continual turnover of healthy bone, and after trauma old bone must be removed before new bone can be deposited. Parfitt (1976) described the presence of a well-defined cutting cone of osteoclasts advancing longitudinally through the cortex of cortical bone, and reformation by a closing cone of osteoblasts. In less dense trabecular bone the process is less dramatic and deposition and resorption are often seen on either side of a single narrow spicule of bone. The nature of the evident coupling between deposition and resorption is still a matter of conjecture.

The osteoclast is typically a multinucleated cell; the number of nuclei may vary from two to several hundred, and appears to vary with species. The nuclear number is typically small in rodents, larger in cats, and intermediate in man (Hancox 1972*a*). The cytoplasm appears foamy under the light microscope and may be acidophilic or slightly basophilic. Typically one side of the cell is applied to bone matrix and the osteoclast sits in a slight depression (Howship's lacuna) which its efforts have generated in the bony surface. The remainder of the cell usually lies in a vascular channel. Accounts of the ultrastructure of the osteoclast are given by Hancox (1972*a,b*), Gothlin and Ericsson (1976), Holtrop and King (1977), and Bonucci (1981).

The ruffled border

The area of the osteoclast cell membrane in contact with bone is known as the ruffled border, a complex system of cytoplasmic folds which greatly increases the surface area of this part of the cell. It may be localized or spread out along the bone surface: a section through an active osteoclast will not necessarily pass through a ruffled border.

The ruffled border is thought to serve as a site at which bone is resorbed. The bone beneath a ruffled border presents a frayed appearance with no lamina limitans, and bone matrix crystals and unmineralized collagen fibres can often be seen amongst the folds of the border (Cameron 1972; Dudley and Spiro 1961). Lysosomal enzymes involved in the breakdown of osteoid can be located histochemically in the ruffled border (Doty and Schofield 1972). The necessary acidic environment for rapid demineralization is provided by accumulation of lactic and citric acids between the folds of the border; this is also optimal for acid hydrolases (Marks and Walker 1976).

The clear zone

Encircling the ruffled border is the clear zone. Within this zone the cell membrane follows the contour of the bone and the lamina limitans is intact. The clear zone is devoid of cell organelles such as mitochondria, vacuoles, or ribosomes and is thought to be concerned with the adhesion of the osteoclast to the bone, thus providing support for the active, motile ruffled border (Lucht 1972).

Vacuoles and other organelles

Vacuoles, vesicles, and cytoplasmic bodies are abundant in the osteoclast cytoplasm outside the clear zone (Scott 1967). The largest (phagosomes) are found close to the ruffled border (Lucht 1972) and bone salt crystals and matrix material can be identified in them. Others contain lysosomal enzymes (Doty and Schofield 1972). These vesicles are thought both to fuse with and discharge their contents into phagosomes and to secrete enzymes into the space between ruffled border and bone matrix. Bone breakdown is thus both extra- and intercellular.

Mitochondria are present in large numbers, as would be expected in a large active cell. Rough endoplasmic reticulum is variable in extent and Golgi apparatus is also present.

Multinucleate osteoclasts are thought not to divide. Mitoses are occasionally seen amongst their nuclei, but these are never synchronous and only concern a small number of the total nuclear number at any one time. The consensus of opinion is that some cells at the time of incorporation into osteoclasts are already in an advanced stage of the mitotic cycle, and that this division is completed.

SEPARATION OF CELL TYPES

Attempts have been made (Wong and Cohn 1974; Luben *et al.* 1976; Wong *et al.* 1977; Chen and Feldman 1979) to separate osteoblasts and osteoclasts by sequential digestion of bone by clostidial peptidase. Those authors did not achieve a clean separation but recently Braidman *et al.* (1983) have used equilibrium density centrifugation of enzyme-released cells from rat fetal calvaria followed by culture in differential conditions. Braidman was able to demonstrate two distinct cell populations, one of multinucleate cells with up to seven nuclei which did not proliferate in culture, had low alkaline and high acid phosphatase, high aryl sulphatase activity, and synthesized little collagen. In contrast the second population consisted of smaller cells with irregular cytoplasmic projections, low aryl sulphatase and acid phosphatase, high alkaline phosphatase, and which synthesized collagen actively. The former type responded to calcitonin but not to parathyroid hormone (see below) while the converse was true of the latter. It seems that this method may prove important for the isolation of osteoblasts and osteoclasts.

BONE MATRIX

The principal constituent of bone matrix is collagen but it also contains other proteins and carbohydrates, glycoproteins, or proteoglycans whose functions are not well understood, as well as mineral salts.

Collagen synthesis

Bone collagen is type I with little carbohydrate and less than 10 hydroxylysine residues per chain and is a heteropolymer with two different genetically determined chains in each molecule. The overlapping molecules in the usual banded native fibre leaves 'hole zones' in which the first mineral crystals may be laid down. The three-dimensional array of collagen fibres is described by Boyde (1972); in lamellar bone the fibres are regular but, in rapidly formed woven bone, collagen fibres run haphazardly in all directions.

Each collagen polypeptide (alpha) chain is synthesized within a fibroblast, chondroblast, or osteoblast as a precursor or pro-alpha chain with long extensions on each end of the molecule, subsequently removed. These extensions assist in the assembly of a three-chain helical precursor, procollagen or tropocollagen, which is exported from the cell. Once in the matrix the extensions are removed by specific peptidases and fibril formation begins. Cross-linking to form fibrils gives mature collagen its specific properties, but the exact linkages are not known.

Breakdown of collagen

The breakdown processes of collagen have not been easy to establish (Harris and Krane 1974) but seem to devolve around collagenases similar to those first described in the anuran tadpole. These enzymes split collagen molecules transversely into two fragments which denature spontaneously above 32°C and are then degraded further by extracellular neutral proteases or by lysosomal enzymes after phagocytosis. Hydroxyproline and hydroxylysine formed by this means are oxidized or excreted in urine as peptides. Hydroxyproline in urine thus gives an estimate of collagen turnover.

Non-collagen bone protein

Ten to 12 per cent of the organic matrix of bone is in the form of proteins other than collagen. These are chiefly glycoproteins and proteoglycans. Glycoproteins are short branched molecules with sugars which may include galactose, glucose, galactosamine, glucosamine, mannose, and sialic acid; proteoglycans have long side arms with regularly repeating disaccharide units which may be sulphated and thus render the proteoglycan acidic. In bone these include chondroitin-4 and -6 sulphate and keratan sulphate. Glycoproteins may be synthesized in bone (e.g. sialoprotein) or resident in bone although synthesized elsewhere (e.g. alpha$_2$HS glycoprotein); plasma proteins will also be present in bone (Herring 1968; Leaver *et al.* 1975; Vaughan 1975; Triffit *et al.* 1976).

Sialoprotein has a molecular weight of 23 000, a single polypeptide chain, large amounts of aspartic, glutamic, and sialic acid, and a high metal binding capacity. Alpha$_2$HS glycoprotein seems to be synthesized in the liver and concentrated in bone. Albumin seems likely to leak into bone compartments from blood, although some appears to be incorporated into bone matrix (Owen and Triffit 1976). Another potentially important bone protein is one containing delta-carboxygluconic acid (Price *et al.* 1976). Delta carboxylation of glutamic acid gives it a high affinity for calcium. The bone protein differs from those of similar composition involved in blood clotting in containing hydroxyproline, suggesting a link with collagen-synthesizing cells. A similar or identical protein is found in tissues undergoing ectopic calcification (Liam *et al.* 1976) This gluconic-acid-containing protein binds strongly to hydroxyapatite crystals, but not to amorphous calcium phosphate and strongly inhibits the formation of hydroxyapatite crystal nuclei.

Bone mineral

The mineral content of bone matrix may be crystalline or amorphous. Bone from young animals and new bone has a higher proportion of amorphous mineral. The crystals are those of calcium hydroxyapatite $[Ca_{10}(PO_4)_6(OH)_2]$ but fluoride, strontium, sodium, magnesium, copper, zinc, lead, and radium can fit into the crystal lattice or be absorbed on to its surface, often with clinical significance. This may be of positive benefit— fluoride, in small quantities, improves crystal stability—or negative: lead is a cumulative poison and absorbed radium or strontium may be a health hazard. The amorphous phase is not of constant composition and its equilibrium with the crystalline phase may be altered by many factors (such as renal failure). Bones may contain 99 per cent of the calcium of the body, 85 per cent of the phosphorus, and 66 per cent of the magnesium (Potts and Deftos 1974).

Mineralization

Smith (1979) reviewed the main theories of bone mineralization. He pointed out that studies by Vaughan (1975) and Rasmussen and Bordier (1974) placed very different emphasis on the respective importance of various bone fractions on mineralization: the former favoured protein/ carbohydrate components and the latter collagen/mineral interactions. In the absence of any conclusive evidence we may list, along with Smith, the most likely factors causing mineralization.

1. Bone fluid. The extracellular fluid of bone, protected by its layer of osteoblasts from other fluid compartments seems to differ from them in containing more potassium and less calcium, magnesium, and sodium (Vaughan 1975). The concentration of Ca and P in this fluid must, presumably, be high enough to maintain mineralization, and may rise sufficiently high to initiate it. At the time of writing, however, bone fluid has not been directly obtained or analysed.

2. Bone matrix. We have already noted that mineralization of collagen may start in the hole-zones of the Petruska and Hodge (1964) model of collagen structure. Fifty per cent of bone mineral may be accommodated in these sites, with a continuous distribution of mineral from one microfibril to the next. It has been suggested that phosphate is important in the transition from non-mineralized to mineralized collagen at the calcification front. Sialoprotein, $alpha_2HS$ glycoprotein, and, of course, the delta carboxyglutamic-acid-containing protein may also be important because of their metal-binding affinities.

3. Bone cells. Osteoblasts and calcifying vesicles demonstrate that cellular activity is necessary for mineralization. There is abundant evidence to show the importance of the latter (Anderson 1976a,b). Electron microscopic studies (Mutsuzana and Anderson 1971; Peress *et al.* 1974) showed that they contain lipids and enzymes associated with calcification. The ultrastructure of calcifying vesicles often reveals dense apotitic crystals

(Anderson 1969) and less dense material which may represent amorphous calcium phosphate (Thybery 1974). Recent scanning electron microscope studies of epiphysial cartilage (Ornoy and Langer 1978) have shown that matrix vesicles in cartilage originate by budding from chondrocytes and are deposited in the matrix prior to mineralization. Bonucci (1971) and Bernard and Pense (1969) described small globular structures in membranous bone matrix which are probably matrix vesicles similar to those seen in cartilage.

Ornoy *et al.* (1980) concluded on the basis of SEM and TEM studies that osteoblastic cell processes are covered with small, membrane-bounded, globular matrix vesicles. These seem to be budded off, and some act as initial loci of calcification. Once calcification has been initiated the vesicles lose their membrane and aggregate to form larger spherules (calcospherites). Matrix vesicles contain alkaline phosphatase, ATPase, and pyrophosphatase (Anderson 1976a,b).

Inhibitors

Calcification is not a universal property of collagen-based matrices, despite the fact that extracellular fluids contain enough calcium and potassium to initiate and maintain mineralization. It has thus been postulated that inhibitors of mineralization are present in non-mineralizing connective tissue. Russel and Fleisch (1976) suggested pyrophosphate as a likely candidate since it inhibits calcium phosphate crystal formation *in vivo* and acts to prevent ectopic calcification when given parenterally. Pyrophosphate is a normal constituent of body fluids, and Russel and Fleisch suggested that it is locally removed at sites of calcification. This would require a pyrophosphatase (such as alkaline phosphatase). A change in alkaline phosphatase levels might thus initiate a defect in mineralization (as in hypophosphataemia, Thompson *et al.* 1969).

CONTROL OF BONE MATRIX COMPOSITION

The control of bone matrix composition has latterly become synonymous with mineral homeostasis, for much of the work done in the last 30 years has concentrated on minerals to the exclusion of the non-mineralized matrix. Much of our knowledge is thus based on the regulation of calcium and phosphorus.

Calcium and phosphorus balance

There is considerable exchange of calcium between bone matrix, bone tissue fluid, and extracellular fluid. Paterson (1974) and Smith (1979) gave diagrams of calcium exchange. Only a small fraction of skeletal calcium is in the pool of exchangeable material but Reeve (1977) suggested that the exchange between bone fluid and extracellular fluid may be as high as 300 mmol per day in man. Much circulating plasma calcium is protein bound, most of the remainder ionized.

Phosphate homeostasis is much more of a grey area; we should remember that, like calcium, plasma phosphate may also be protein bound (Paterson 1974).

Vitamin D

Vitamin D is converted to active metabolites while in the body, one of which, 1,25-dihydroxycholecalciferol (1,25 $(OH)_2D$), produced by the kidney, acts as a hormone rather than a vitamin. Naturally occurring vitamin D (synthesized in the skin) and D_2, ergocalciferol (a plant sterol), are metabolized by an hydroxylation at position 25, then a position 1 alpha hydroxylation at the other end of the molecule. The 25 hydroxylation occurs in the liver, the 1 hydroxylation exclusively in kidney mitochondria. These two steps convert D_2 or D_3 into 1,25 $(OH_2)D$. The alpha hydroxylation is stimulated by parathormone and hypophosphataemia. Increases in circulating calcium and phosphate will suppress this step and allow the production of a less active $24,25(OH_2)D$ alternative. A third possible form, $1,24,25(OH_3)D$, is present in trace amounts.

Vitamin D increases calcium transport across the gut wall, via the effect of $1,25(OH_2)D$ on calcium-binding protein synthesis. $1,25(OH_2)D$ acts directly on bone to cause an increase in resorption. The similar effect of parathormone is thought to be vitamin-D-mediated, and does not occur in the absence of the latter.

Parathyroid hormone (parathormone, PTH)

This is an 84-chain amino acid polypeptide produced by the parathyroid gland but circulating as fragments, only those containing an N terminal being biologically active. The fragmentation is due to activity in the kidney and liver (Tomlinson and O'Riordon 1978). PTH secretion increases as ionized calcium in the plasma falls. Increased PTH increases bone resorption in the presence of vitamin D. PTH acts on bone cells via cyclic AMP (Peacock 1976). PTH is bound on specific osteoclast receptors and releases adenyl cyclase as second messenger. The effect of PTH on osteocytes is debated (Parfitt 1976); there may be stimulation of osteocytes to produce osteolysis. When PTH is added to cultured bone (Lucht and Maunsbach 1973), osteoclastic activity increases with an increase in size of the whole cell. The number of osteoclasts also increases (Bingham *et al.* 1969; Tetevessian 1973) as does the number of osteoclasts in contact with bone (Wezeman *et al.* 1979). The number of nuclei and vacuoles in each cell is also increased (Addison 1980). Cameron (1972) noted the swelling of mitochondria in osteoclasts treated with PTH, perhaps damage caused by higher than physiological concentrations of the hormone. Lucht and Maunsbach (1973) demonstrated the appearance of many coated lysosomes and large pinocytotic vesicles containing cell debris resembling osteoblasts and osteocytes. The ruffled border does not seem to be increased in size by PTH.

Calcitonin (CT)

Calcitonin is a smaller peptide (32 amino acids) with a terminal disulphide bridge. Only nine of the 32 amino acids are common to porcine, salmon, and human calcitonin. Calcitonin secretion, from the C cells of the thyroid, is stimulated by hypercalcaemia.

When calcitonin is added to bone cultures, bone resorption is reduced and many osteoclasts become detached from the bone matrix. The cells themselves become smaller, ruffled borders are reduced or disappear, and vacuoles are smaller and fewer in number and contain no bone crystals (Holtrop *et al.* 1974). Osteoclasts incubated on glass or plastic in the presence of calcitonin lose their motility (Chambers and Magnus 1982) and also the ruffled edge of their lamellipod. The quiescent state brought about by calcitonin is reversible, being abolished by the removal of calcitonin from the culture medium. CT does not affect osteoblasts, peritoneal macrophages, or inflammatory giant cells in this way. The CT seems to interact with a trypsin-sensitive receptor, as prior treatment with trypsin removes the response of the osteoclasts to CT. Quiescence is produced by CT concentration within physiological limits. Chambers and Magnus suggested that motility of osteoclasts in culture may be equated with resorptive activity *in vivo* and that the ruffled border is equivalent to the lamellipod. CT might therefore control the resorptive activity of osteoclasts by 'immobilizing' the ruffled border.

The relationship between parathyroid hormone, calcitonin, osteoblasts, and osteoclasts seems to be a complicated one. Chambers and Magnus (1982) suggested that the system is controlled by the osteoblast. Osteoclasts, although stimulated by parathormone, do not possess parathormone receptors, while osteoblasts probably do (Chambers 1980; Rodan and Martin 1981). The osteoclast response may therefore be indirect, following an initial response by the osteoblast. However, Warshawski *et al.* (1980) and Martin and Partridge (1981) suggested that osteoclasts do have calcitonin receptors. The CT response, direct or indirect, seems usually to be limited to multinucleated osteoclasts, as neither mononuclear phagocytes nor inflammatory giant cells respond to CT and may lack a specific receptor. Chambers and Magnus (1982) noted, however, that a small proportion of the mononuclear cells in their culture showed a CT response, suggesting that they represented mononuclear osteoclasts, perhaps a pre-fusion precursor.

Other hormonal effects

Other hormones also effect skeletal structure (Jowsey 1977). Growth hormone, or, more accurately, the group of growth-hormone-mediated somatemedin peptides manufactured by the liver, seem to act mainly on cartilage but can also stimulate collagen synthesis. Excess growth hormone and, hence, excess somatemedin produce gigantism and, after epiphyses have fused, acromegaly. The bones in gigantism may appear radiologically osteopetrotic. Deficiency produces proportional dwarfism.

Thyroxine is also implicated; in thyrotoxicosis there is an increased turnover of bone with resorption in excess of formation. Oestrogen deficiency can lead to post-menopausal osteoporosis, as does testosterone deficiency. Excess corticosteroids can lead to skeletal stunting; renal androgens are probably responsible for the adolescent growth spurt. 10^{-6}M hydrocortisone inhibits the growth of cat osteoclasts in culture (Suda *et al.* 1983). At lower concentrations (10^{-7}–10^{-9}M) the cells round up and detach themselves from the substrate and cell fusion is inhibited. Morphological studies of hydrocortisone-modified cells by light and electron microscopy show them to resemble inactive osteoclasts. Glucocorticoids are used clinically to control bone resorption in some haematological neoplasms; it has been suggested that they block the effect of osteoclast activating factor (OAF, see below) on osteoclasts (Strumpf *et al.* 1978).

LOCAL FACTORS

Classical investigations of metabolic bone disease have concerned them-selves with circulating hormone levels: however it is now clear that other substances, which are essentially local in action, have important effects on bone metabolism. Perhaps we should turn to these for an explanation of the situation often seen in bone of an osteoblast depositing bone on one side of a narrow bony lamina whilst an osteoclast nibbles away at its other face a few μm distant.

Prostaglandins

These are a family of 20 carbon fatty acids of widespread origin and effect (Samuelsson *et al.* 1975). The prostaglandin E group are potent resorbers of bone *in vitro* (Klein and Raisz 1970; Tashjian 1975) and may be involved in bone neoplasia.

Klein and Raisz demonstrated that PGE_1 and PGE_2 were effective in very low concentrations (10^{-8}M) whilst PGA_1 and PGF_1 were less effective, showing small effects at 10^{-6}M. The effect seemed to parallel calcium loss due to parathyroid hormone. The resorptive properties of prostaglandins were ascribed to the known ability of these tissues to increase the cAMP content of bone, although norepinephrine, which shares the latter property, did not increase resorption.

Dietrich *et al.* (1975) confirmed in similar conditions that the E series prostaglandins were most potent with PGE_2 causing a significant Ca release at $10^{-9}M$. PGE, although known to stimulate cAMP production, was not shown to affect adenyl cyclase or phosphodiesterase (Chase and Aurbach 1970). Raisz *et al.* (1977*a*) demonstrated longer-term activity of more stable metabolites of PGE_2 and PGE (the half-life of PGE itself is probably less than one minute) which were almost as active as their basic forebears (13,14 dihydro derivatives) whilst others ($15k-H_2-PGE_2$) were ineffective. Endoperoxides derived from PGE_2 gave a rapid transient response quite unlike that of the base prostaglandin. These compounds (PGG_2, PGH_2) were produced via the cyclo-oxygenase pathway, known to be elevated in malignancy and other conditions resulting in bone loss. Raisz *et al.* speculated that the endoperoxidases are normally denatured to thromboxane A_2 or prostaglandin enzymatically within bone. Tashjian *et al.* (1977) were thinking along the same lines but with apparent differences in results. They found that 13,14 dihydro-PGE_2 had only 6 per cent of the activity of PGE_2 and other analogues were less active. The endoperoxidases tested were less than 1/1000 as effective as PGE_2 itself.

Recently, Davidovich *et al.* (1983) used monoclonal antibodies to identify the local distribution of another prostaglandin (PGF_2»ga) on bone cells. Although PGF_2»ga may not be directly involved in resorption (Atkins *et al.* 1979), it probably is involved in bone cell metabolism, and its distribution may reflect that of PGE_2. Davidovich was able to localize the prostaglandin in unfixed undermineralized frozen bone sections placed on adhesive tape, a modification of a technique developed for cyclic nucleotides. The PGF_2»ga proved to be located over the plasma membrane of cat maxilla osteoblasts and osteoclasts. The staining distribution was patchy: was this an artefact, or are the prostaglandins really located only on selected cells?

Osteoclast activating factor (OAF)

Raisz *et al.* (1975) cultured normal human leucocytes with phytohaemagglutinin. The supernatant from these cultures contained a substance, OAF, which caused a rapid release of previously incorporated ^{45}Ca from bone, thus resembling PTH. The response was only transitory, inhibited by calcitonin, and partially inhibited by increasing phosphate concentration; cortisol was a more effective inhibitor. Maximally effective doses of OAF were not enhanced by PTH or PGE_2, but submaximal doses were additive. Collagen synthesis in rat calvaria was inhibited by doses of both PTH and OAF stimulating bone resorption.

Mundy and Raisz (1977) were able to purify OAF and found that it exists in two forms. Big OAF had a mw. of between 25 000 and 12 500, and further treatment degraded this to little OAF with a mw. close to 140. Big OAF was converted to little OAF by equilibration with 1M NaCl or 2M urea; little OAF reaggregated in buffers of low ionic strength. Mundy and Raisz thought that little OAF was unlikely to be a prostaglandin, and suggested that either little OAF is an active subunit of big OAF or that the latter is protein bound.

Epidermal growth factor

Epidermal growth factor is also known to stimulate bone resorption. It may act via prostaglandin production (Tashjian and Levine 1978), but there is some evidence of direct action (Raisz *et al.* 1980). Tashjian *et al.* (1982) reported that platelet-derived growth factor (PDGF) stimulates bone resorption via a prostaglandin-mediated mechanism. This work was done on mouse calvaria in culture, which obviously contain many cell types. Key *et al.* (1983) were able to demonstrate the effect of PDGF on monocyte cultures: monocytes are capable of resorbing bone in their own right, and are, of course, thought to be osteoclast precursors.

THE MUTANTS

Having discussed briefly the basics of bone anatomy and physiology we may now turn to the mutants which produce abnormal bone. Most emphasis has been placed on a large group of osteopetroses, but defects in bone synthesis have also been recorded. Once again we find hangover effects, in talpid and shaker with syndactylism, and add to these certain conditions which seem to affect only the bony phase of the skeleton. In other cases (e.g. cn, Bonucci *et al.* 1976, 1977) abnormal cartilage produces normal bone.

The talpid[3] chick

As we noted in Chapter 3, Hinchliffe and Ede (1968) noticed a failure of cartilage replacement bone in talpid[3] embryos. Grafts of the shoulder girdle region, containing both membrane and cartilage replacement bone rudiments and grown on the chorioallantoic membrane, showed good development of membrane bone, but none in cartilage replacement bone, even when maintained for a week past the stage when ossification should have appeared (Fig. 4.1). This parallels the situation in the creeper chick (Landauer 1927) where cartilage replacement bone was delayed, and such mutants as spondylo-metaphysial chondrodysplasia (Johnson 1984) where secondary centres of ossification are abnormal and grossly delayed.

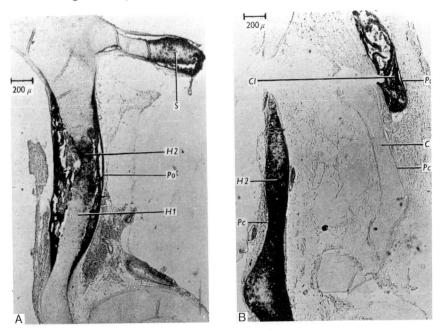

Fig. 4.1. (A) Shoulder girdle grafted from a normal chick into a normal host for 14 days and stained alkaline phosphatase; (B) a similar graft from a talpid[3] chick: C: cartilage; Cl: clavicle; H1, H2: alkaline phosphatase negative and positive respectively; Pc: perichondrium; Pe: periostium; S: scapula. (From Hinchliffe and Ede 1968.)

Hinchliffe and Ede noted that the perichondrium of the coracoid and scapula (cartilage replacement bones in chick) grafts remained fibroblastic, with no differentiation of osteoblasts and consequently no bone formation. Alkaline phosphatase is not produced by the talpid perichondrium; Fleisch and Bisaz (1962*a,b*) felt that alkaline phosphatase removes a phosphate inhibitor which prevents apatite crystal growth. Hinchliffe and Ede suggested that periosteal ossification commences at the same stage as chondroblast expansion, a phase of cartilage development known to be abnormal in talpid[3] and that this early disorganization is expressed as a failure of the cartilage to 'induce' (in whatever way) osteogenic tissue in the surrounding perichondrium.

Shaker with syndactylism (sy, Chr 18)

In shaker with syndactylism the feet are syndactylous, the skeleton is less densely constructed than normal, and there are defects of the inner ear (Grüneberg 1956, 1962). Deol (1963) identified a defect of the membranous skeleton, and speculated on how this might affect bone. He pointed out that the abnormal mesenchyme of the otic capsule has an abnormal periosteum. The floccular fossa lacks a periosteal lining and the ossification of the semicircular canals is so abnormal that they sometimes pass through bone marrow (Fig. 4.2). Deol argued that the same conditions may occur elsewhere in the skeleton and account for the widespread abnormalities found there, both of bone shape and bone consistency.

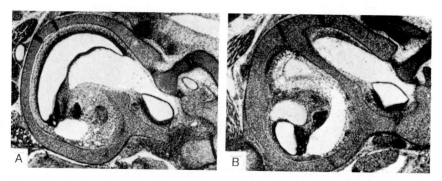

Fig. 4.2.   Left cochlear region of (A) normal and (B) sy/sy 17-day embryos. Note the thicker cartilaginous capsule and the shelf between cochlear coils in (B). (From Deol 1963.)

Tiptoe walking Yoshimura (twy, Chr ?)

Hosoda *et al*. (1981) described a recessive mutant (twy) with multiple osteochondral lesions which seem to simulate osteoarthritis or ankylosing spondylitis. Homozygotes (twy/twy), which walk on tiptoe, show an exaggerated vertebral curve and later develop a motor disturbance in the limbs; the cartilage matrix of the intervertebral discs is degenerate, and irregular calcification is seen (Fig. 4.3). Calcification bridges adjacent vertebrae in some animals. Calcification of the Achilles tendon, hair follicles, and whiskers (Fig. 4.4) was also noted. This is obviously a case of ectopic calcification, but Ca levels in serum were only a little raised above those of controls.

Gyro (Gy, Chr X)

Sela *et al*. (1982) looked at Gyro using light and electron microscopy. Gy/– males (Gyro is sex-linked) showed osteomalacia of the maxilla, tibia, and inner ear, with broad bands of osteoid showing partial or complete failure of mineralization. Unmineralized foci were also present in mineralized areas. Sections from epiphysial cartilage revealed an irregular transition

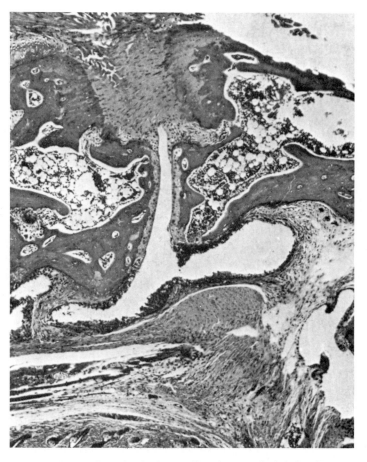

Fig. 4.3.   Focal destruction of articular cartilage in a twy limb joint. (From Hosoda *et al.* 1981.)

between calcified cartilage and diaphysial bone, and impaired cartilage calcification. Electron micrographs showed an abundance of bone cells in a collagen-rich matrix, with many longitudinal calcifying fronts composed of hydroxyapatite crystals (Fig. 4.5). Matrix vesicles were abundant, all surrounded by normal-looking trilaminar membranes and many contained hydroxyapatite crystals (Fig. 4.6). These were also seen, abnormally, associated with the cell membranes of bone-forming cells. Sela *et al.* considered that this, in association with the absence of calcospheroids, could indicate early vesicular maturation, with a shortening of the normal lag period between the production of matrix vesicles and the precipitation of crystals within them (Anderson 1969; Mulhrad *et al.* 1981).

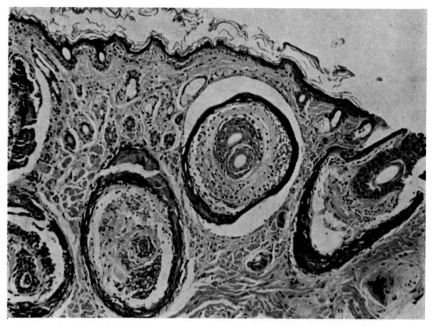

Fig. 4.4. Parafollicular calcification of the whiskers of a twy mouse. (From Hosoda *et al.* 1981.)

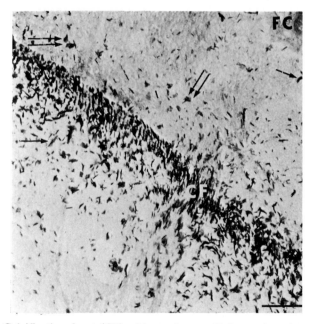

Fig. 4.5. Calcification front (CF) of bone from a Gy/− mouse: (single arrows) apatite crystals; (double arrows) matrix vesicles. (From Sela *et al.* 1982.)

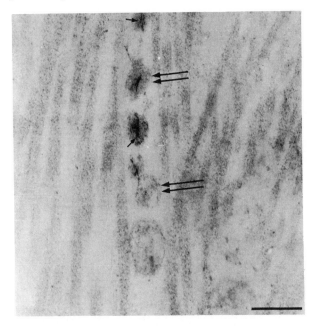

Fig. 4.6. Electron micrograph of (single arrows) intravesicular crystals and (double arrows) matrix vesicles from a Gy/− mouse. (From Sela *et al.* 1982.)

Hypophosphataemia (Hyp, Chr X)

Meyer *et al.* (1979) looked at another sex-linked mouse mutant, hypophosphataemia. This is dominant and produces a phenotype which includes reduced renal absorption of phosphate, hypophosphataemia, and dwarfing. Hyp/− males compared to normal male littermates were hypocalcaemic, hypophosphataemic, hypermagnesaemic, and had elevated plasma alkaline phosphatase. Femur ash weighed less than half the value obtained for normal controls, but the Ca:P ratio was normal. The ash was high in sodium and potassium but low in magnesium. Histologically the width of osteoid in the femur was increased by a factor of 5 and the videodensitometry measurement (obtained by digitizing spot densities at many points on enlarged photographs of radiographs) reduced to two-thirds normal (Fig. 4.7). Osteoid covered most endosteal and periosteal surfaces and lined Haversian canals. Osteoblasts were flattened and fibroblastic in appearance. They were confined to lacunae not lined with osteoid, i.e. where a mineralized bone surface was exposed. In midshaft microradiographs osteocyte lacunae were enlarged and had indistinct outlines. Poorly and well mineralized lamellae tended to alternate in Hyp bone. Sections through the epiphysial plate (Fig. 4.8) showed areas of mineralized cartilage and abnormal metaphysial trabecular bone. The cartilage was distorted, and large areas of hypertrophic cells were present.

Iorio *et al.* (1979) noted that alveolar bone from Hyp individuals showed an overall whorled arrangement. The lacunae are larger than in normal sibs and osteoid gaps are frequently seen. A streaky appearance of haematoxylin staining in calcified areas of mutant alveolar bone also suggests hypocalcification. Alveolar bone also appeared sparser than normal.

Meyer *et al.* suggested that the spectrum of abnormalities seen in Hyp resembles human osteomalacia, and there are indeed many points of comparison. They suggest that, because osteomalacia in man is a consequence of low plasma phosphate and because Hyp mice have low plasma phosphate and osteomalacia, the latter is a secondary consequence of the former and due to phosphate loss secondary to low renal resorption. Phosphate supplements will increase the growth of Hyp mice (Eicher *et al.*

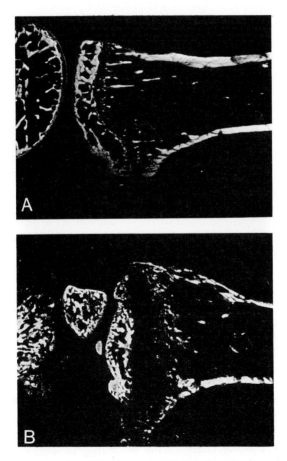

Fig. 4.7. Microradiographs of the distal epiphysis of the femur and adjacent tibia of (A) Hpy/− and (B) +/− male mice. (From Meyer *et al.* 1979.)

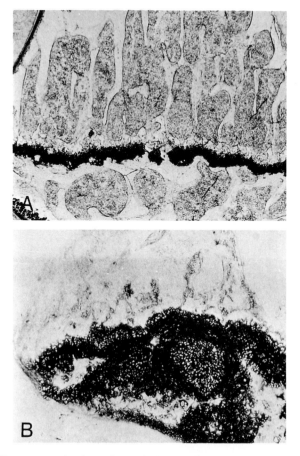

Fig. 4.8. Paragon stained section of the epiphysis of (A) normal and (B) hypophosphataemic mice. (From Meyer *et al.* 1979.)

1976). The lack of response to vitamin D or 1,25 dihydroxy vitamin D (O'Doherty *et al.* 1977) suggests that the disease is vitamin-D-resistant. In fact vitamin D levels are normal in both Hyp mice (Meyer *et al.* 1978) and vitamin-D-resistant rickets in man. Hyp differs, however, in having elevated plasma magnesium but low bone magnesium. In vitamin-D-resistant rickets in man, plasma magnesium is normal or low (Williams and Winters 1972; Anast 1967). Hyp mice are obviously unable to transport Mg into the bones from a more than adequate supply in the plasma. A more trivial difference is the status of the heterozygous female; in mouse the gene is almost completely dominant but in women the expression is low, with abnormal dentin being the most characteristic feature.

Marie *et al.* (1982) reported that phosphate salts (Pi) heals the rickets but does not correct defective mineralization in Hyp mice. Combination of Pi with $1,25(OH_2)D_3$ produced a dose-dependent elevation in serum Ca and P and improved bone mineralization. However the persistence of osteomalacia in the face of this regime suggests a specific bone-cell resistance to mineral and/or hormonal influences in Hyp; the symptoms are not fully cured even by excessive extracellular levels of Ca and P.

Fragilitas ossium (fro, Chr ?)

Guenet *et al.* (1981) described a condition which may form a model for osteogenesis imperfecta. fro is an autosomal recessive gene; homozygotes are small with abnormalities of all limbs consisting of curving of the long bones. Tail, spine, and ribs appear grossly normal. Ninety per cent of homozygotes die within the first three days of life; occasional survivors are small, but breed normally in both sexes. Radiological study showed shortened long bones with diaphysial curving, the diaphysis being wider and more translucent than normal. Cortices were thin and irregular and several individuals had discontinuities of the cortex. In sections prepared for light microscopy the diaphyses were confirmed to be abnormally wide and curved and callosities associated with bone remodelling were found. Osteoblasts were not reduced in number, but were flattened. Growth cartilage looked normal with regular columns and vascular invasion. Biochemical data for this mutant is eagerly awaited.

Tight skin (Tsk, Chr 2)

Green *et al.* (1976) described a mouse mutation, tight skin, where skeletal growth is enhanced. Heterozygotes look normal up to two months of age and can be recognized by the difficulty experienced when trying to pick them up by pinching the back skin. Although of normal weight their skeleton is larger than normal. Loose connective tissue is hyperplastic. The size of all bones and cartilages measured was increased, although the histology of bone and cartilage appeared normal. The condition thus has a similarity to acromegaly, and growth hormone levels were accordingly measured. Pituitary size was normal and growth hormone levels normal by acrylamide gel electrophoresis and radioimmune assay, although the latter were very variable. Blood sugar levels were also normal. The Tsk homozygote seems to die at about eight days of gestation, but no distinctive phenotype was recognized. The authors suggested that Tsk may represent a defect in one of the somatamedins, or more probably is a deficiency of a receptor protein binding somatomedins to connective tissue and/or embryonic cells.

Hair-loss (hl, Chr 15)

Although primarily a hair structure gene, hair-loss (Hollander and Gowen 1959) has a skeletal effect. +/hl offspring of hl/hl mothers have an extremely fragile skeleton which fractures easily and leads to excessive mortality until the age of 2 weeks. Hollander (1960) suggested that hl/hl mice maintain a smaller pool of calcium than normal and that their +/hl offspring are unable to fulfil their requirements from hl milk.

Progressive ankylosis (ank, Chr ?)

Sweet and Green (1981) described a recessive gene causing ankylosis. In mice aged 4–5 weeks the first abnormality noted is the inability to curl the toes. Progressive loss of joint mobility then occurs with increasing age. Histological examination of the feet of 31-week-old mice showed non-inflammatory joint problems. The toe bones had cartilaginous and bony fusions across joints and much proliferation of fibrous tissue, cartilage, and bone around joints. Joint cavities were often obliterated. These massive changes were first hinted at in 33-day-old mice where joint cavities are reduced due to proliferation of cells resembling osteoblasts from the synovial membranes. In the elbow and knee at 31 weeks a jumbled mass of cartilage, bone, tendon, and ligament was seen with areas of necrosis. The joint cavities, however, were little affected although the synovial membrane of the knee is reported to be thickened, with a surface layer several cells deep; these cells again resembled osteoblasts. Other joints were similarly affected. The authors suggest that at least three processes are involved, increased calcification (in the tracheal and laryngeal cartilages, as well as around the joints), increased cellular proliferation around joints, and degeneration, the latter almost certainly secondary. These changes are unlike the partially genetic degenerative changes which accompany ageing in some strains of mice (Sokoloff 1967; Sokoloff *et al.* 1962) and do not resemble any human disease closely: the nearest approach is ankylosing spondylitis, but this is confined in man largely to the vertebral column, sacroiliac, and hip joints and is often associated with considerable inflammation.

Microphthalmia (mi, Chr 6)

Microphthalmia is usually considered as an osteopetrosis and is dealt with as such below. However, it serves here as an example of the finding that in man (Zawisch 1947; Englefeldt *et al.* 1954; Hohling and Czitober 1971; Bonucci *et al.* 1975) and in animals (Doykos *et al.* 1967; Walker 1975*b*,*c*) calcified tissue in osteopetrotic animals is often abnormal.

Al-Douri and Johnson (1983) studied the bone of the maxilla of microphthalmic mice under the electron microscope. Osteoblasts at 5 days old were plentiful with much endoplasmic reticulum whose cisternae were often dilated by aggregates of electron-dense material. In the matrix collagen fibres were sparse and calcospherites abundant (Fig. 4.9); the latter were of varying size and often showed a dark central locus of calcification. The osmiophilic lamina (which normally surrounds the osteoblast) was poorly developed or absent in mi/mi mice. At 15 days the mi matrix contained only isolated bundles of collagen fibres between areas thought to represent persistent calcospherites (Fig. 4.10). Osteocyte lacunae were irregular and pericellular space reduced, so that many osteocytes seemed to be in intimate contact with bone matrix. Many osteocytes were dead, dying, or represented by lacunae containing only cellular debris. Dentin in the incisors was poorly calcified.

It seems that many mi osteoblasts do not survive long as osteocytes, and that those that do are surrounded by abnormal bone. The calcospherites

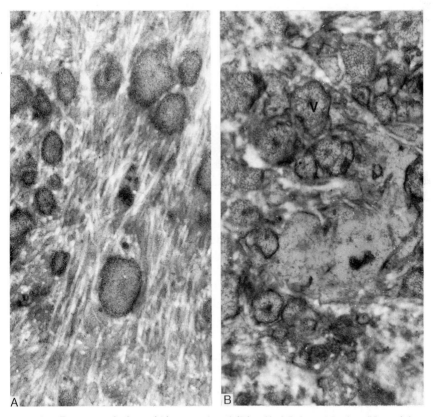

Fig. 4.9. Bone matrix from (A) normal and (B) mi/mi 5-day-old mice. V, vesicles = calcospherites.

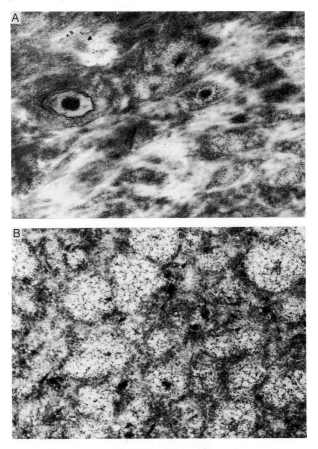

Fig. 4.10.   Bone matrix from (A) normal and (B) mi/mi 15-day-old mice.

are irregular and perhaps calcify precociously and the collagen content is reduced. Jaffe (1972) reported a lack of active osteoclasts, empty bone lacunae, and septic necrosis in human osteopetrosis and Zawisch (1947) a scarcity of collagen fibres, while Bonucci *et al.* (1975) noted abnormal collagen in osteopetrosis fetalis. Incorporation of ^{3}H-thymidine into collagen is abnormal in grey lethal and microphthalmic mice (Walker 1966*b*), incisors-absent rat (Marks 1973), and the osteopetrotic rabbit (Marks and Walker 1976).

These findings *in toto* suggest that osteopetrosis may involve all skeletogenic cells, a concept further discussed below.

THE OSTEOPETROSES

The osteopetroses have provided a large group of mutations with similar phenotypic effects, but which one suspects are based upon very diverse bases, as indeed are the osteopetroses in man.

Grey lethal (gl, Chr 10)

Pride of place amongst osteopetrotic mutants must go to grey lethal, on grounds of seniority alone; like the other osteopetroses, however, we are still far from understanding the defect behind this gene. Grey lethal was first described by Grüneberg before World War II (Grüneberg 1935, 1936, 1938) and his description of the skeleton will serve as a model for all the osteopetroses. The skeleton of grey lethal is smaller than that of normal mice, but denser (Fig. 4.11). Numerous bone spicules encroach on the

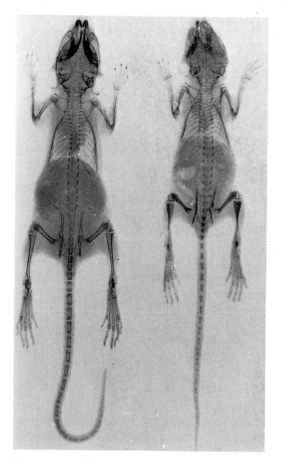

Fig. 4.11. Radiographs of (A) normal and (B) grey lethal littermates aged 21 days. (From Grüneberg 1963.)

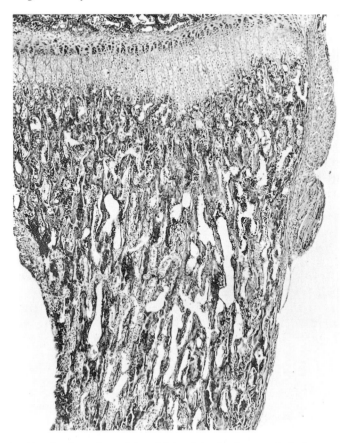

Fig. 4.12.   Proximal portion of the tibial epiphysis of a grey lethal mouse aged 31 days. (From Walker 1972.)

marrow cavity as a result of the absence of secondary bone resorption (Fig. 4.12). These are often poorly mineralized. The failure of bone resorption also leads to a characteristic misshaping of all the bones in the body (Bateman 1954). The teeth fail to erupt, according to Grüneberg (1963) due to failure of resorption of bone; part of the dentine also remains uncalcified and soft, and bone spicules invade the tooth germs. At the base of the continuously growing rodent incisor (which fails to erupt) a large odontoma forms.

An addition to the grey lethal syndrome, which may be due to the workings of spurious pleiotropism or to a very closely linked gene, is the grey coat colour due to absence of phaeomelanin. The connection (if any) between this and osteopetrosis has never been satisfactorily elucidated. Nor has the cause of death; Grüneberg (1963) stated that the immediate cause of death is starvation secondary to inadequate feeding and unerupted dentition. The animals certainly fail to gain weight in their third week and usually die soon after weaning. However a diet of mush made from rat cake and water does not increase their lifespan (although successful in other animals with dental problems secondary to osteopetrosis) and grey lethal mice often die suddenly in the hands of the observer, with no signs of fitting or convulsions.

The osteopetrosis is another facet of the syndrome not fully explained. The obvious focus for a study of osteopetrosis is the osteoclast, but, amazingly, little seems to have been published on this subject in grey lethal. Barnicot (1947) vitally stained calvaria from normal and grey lethal mice with neutral red and recorded the presence of some very large osteoclasts of irregular shape (Fig. 4.13). These giant osteoclasts often lay against ridges of bone. A count of osteoclasts showed their number to be reduced in grey lethal aged 0–18 days and an attempt to ascertain the amount of 'osteoclast substance' by cutting outlines of individual cells from photomicrographs and weighing them showed an overall defect in gl, despite the few giant cells. No more recent account of grey lethal osteoclasts is to be found: no one seems to have subjected them to electron microscopy or to more modern stereology. Walker (1966b) described the ultrastructure of the gl osteoclast after intense treatment with para-thormone, but failed to include any description of the osteoclast before

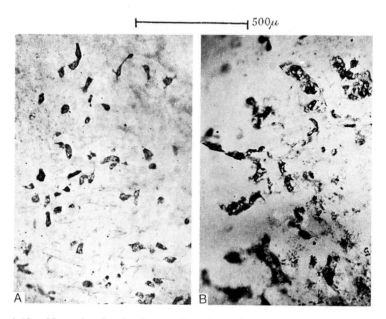

Fig. 4.13. Neutral red stained osteoclasts from the parietal bone of (A) normal and (B) grey lethal mice. (From Barnicot 1947.)

treatment. He did, however (Walker 1973*a*), note the presence of cytosomes containing nuclei in various stages of disintegration in untreated gl osteoclasts. Johnson and Laing (unpublished observations) noted that osteoclasts in 12-day-old gl mice appear to be normal and active, with extensive ruffled border and release of collagen fibres from underlying bone (Fig. 4.14). They did not see degenerating nuclei as described by Walker.

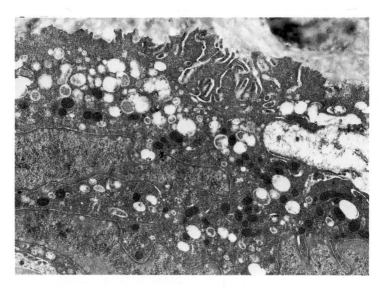

Fig. 4.14.   Osteoclast with ruffled border from the maxilla of a 14-day-old grey lethal mouse.

Is the defect in the bone or in its resorption? Three types of experiment suggest the latter. Barnicot (1941) transplanted grey lethal bone into normal littermates and found that it was resorbed; normal bone transplanted into grey lethals was unchanged. He also showed (Barnicot 1945) that grey lethals can be induced to resorb bone in the presence of normally fatal doses of parathormone, to which they are immune. After a short period of time, however, the mice become refractory to the hormone. On a metabolic level the response is less striking. Serum calcium is only slightly elevated (Walker 1966*b*) and ability to incorporate tritiated proline (several times higher than normal in untreated gl mice) is reduced by only 15 per cent. Collagen degradation is increased in treated gl (Walker 1966*a*) where osteoclast, hyperplastic chondrocyte, and osteocyte can be seen to elaborate collagenase.

Third, Barnicot (1948) showed that, if parietal bone plus parathyroid gland were transplanted into the meninges of a normal mouse, resorption occurred; this was so if the transplant was of normal bone and normal parathyroid or grey lethal bone and grey lethal parathyroid.

Grüneberg (1963) hypothesized as follows. Grey lethal bone is resorbable. In transplants it is resorbed by physiological levels of parathormone, although *in situ* much higher levels are required to produce a response. The bone plus parathyroid transplant suggests that the gl parathyroid produces adequate parathormone; perhaps the grey lethal mouse inactivates or destroys parathormone at a greatly increased rate.

Walker (1972) showed that grey lethal osteopetrosis could be cured by parabiosis to normal sibs. Walker's (1966b) experiments are also interesting. Although designed to show the effect of parathormone on gl mice, and so not always informative on normal and grey lethal metabolism in the untreated state, Walker did measure the collagenolytic activity of gl bone by its ability to release radioactivity from ^{14}C collagen gels. After three days culture his untreated normals released a mean of 1120 counts over blank and his gl controls 3730 indicating high collagenase activity. This could originate from hypertrophic chondrocyte, osteocyte, or osteoclast. After parathormone treatment (to which normal and gl responded more or less equally) evidence was seen of histolytic activity by osteocytes, and a tendency to osteocyte fusion noted.

Murphy (1968, 1969) made a detailed metabolic study of grey lethal mice. She confirmed the presence of hypophosphataemia first demonstrated by Watchorn (1938) and also demonstrated slight hypocalcaemia, very high serum alkaline phosphatase (around twice the normal level), low total blood phosphorus, and normal blood glucose. gl homozygotes have a high circulating citrate and bone citrate, the latter attributed by Murphy to a reduction in available pyridine nucleotides for further oxidative decarboxylation. Bone ash was slightly elevated due to increased calcium. Incubation of bone slivers *in vitro* showed increased passive ion solubilities of calcium and citrate.

The conjunction of hypophosphataemia, hypocalcaemia, and the lack of bone resorption suggests hyperthyrocalitoninism. This is supported by Walker's (1966b) finding that, in parathormone-treated gl mice, large parafollicular light cells can be seen. Marks and Walker (1969) were able to demonstrate an excess of these cells, which secrete thyrocalcitonin (Bussoloti and Pearse 1967) in untreated gl mice.

Murphy (1972) collected crude gl plasma and injected it into other mice. BALB/c mice starved overnight showed a significant hypocalcaemic response 60 minutes after injection; normal plasma produced no such response. The hypocalcaemic factor in gl blood was stable after storage at −30°C for many weeks, could be partially purified and absorbed on to silica gel, and was released into the supernatant after subsequent high-speed centrifugation for 24 h. The factor thus behaved like a small peptide and

Murphy suggested that it was calcitonin. The calcitonin levels of gl blood (if indeed this is calcitonin) are very high. Oddly the calcitonin content of gl thyroid glands (measured by bioassay) was not significantly increased. Patients with high levels of circulating calcitonin due to medullary thyroid tumours also have high levels of calcitonin in the thyroid (Tashjian *et al.* 1970). Murphy suggested that the gl mouse may not store calcitonin in the thyroid gland, but may follow a regime of rapid synthesis and secretion.

Both Murphy and Tashjian and Bottomly (in Murphy 1972) failed to elicit hypocalcaemic response to gl plasma from rats. This, and the variable response obtained when CBA or C57BL mice were used in place of BALB/c, is a little surprising, since porcine and salmon calcitonin are both effective in man.

Bioassay of calcitonin in man does not suggest its involvement in osteopetrosis, nor are the hypocalcaemia, hyperphosphataemia, increased alkaline phosphatase, and increase in C cell numbers characteristic of gl present. Gudmundsson *et al.* (1970) and Verdy *et al.* (1971) found circulatory calcitonin levels lower than normal in five osteopetrotic patients.

There is also the problem that grey lethal osteopetrosis, in common with most murine osteopetroses, may be cured by parabiosis or when irradiated mice are infused with cells from the marrow or spleens of normal littermates (Walker 1975a,b). If the disease is due to high calcitonin levels, should these not affect the normal osteoclasts derived from infused cells? If this is not the case, then perhaps gl osteoclasts are unreceptive to calcitonin, and the high levels represent an oversecretion constantly attempting to activate 'deaf' osteoclasts.

Micropthalmia (mi, Chr 6)

Microphthalmia (Hertwig 1942) is morphologically very similar to grey lethal. Bateman's (1954) careful comparison shows that the mi skeleton is almost indistinguishable from gl. Coincidentally coat colour is also affected, so that mi/mi is a black-eyed mock albino: this is often not easy to see as the eyes themselves are reduced, as the name suggests. The mi heterozygote has a head spot and reduced hair pigmentation.

The osteopetrosis of mi is also due to a failure to resorb bone, although not quite a complete one. mi mice live much longer than gl, especially on a suitable background. Hertwig (1942) observed a single individual which reached maturity and bred; on a suitable background mi mice aged 150 days are not uncommon. Although mi teeth show exactly the same abnormalities as those in gl they sometimes erupt; delayed eruption of molars in older mice is common and even incisors are sometimes seen. In this context our finding (Johnson and Al-Douri 1982) of normal looking and apparently active osteoclasts over the tip of mi teeth is relevant. Freye

(1956) studied the eruption of mi teeth in some detail. Konyukhov and Osipov (1966) reported erupted teeth and noted that parathormone stimulated tooth eruption.

Although the emphasis in research on mi has been mainly cellular rather than biochemical, it is clear that, despite the skeletal similarity, the aetiology of mi and gl is quite distinct.

Murphy (1973) gave an account of the biochemistry of the mi mouse. On the genetic background she used (A/A, b/b), normal mice were cinnamon in coat colour; +/mi had a white flash on the head and belly spots. No pathology was seen in +/mi mice.

mi homozygotes are smaller than normal littermates from birth, this difference increasing progressively after five days. Tooth eruption was poor, but 28 per cent of homozygotes on soft food survived and often lived for a year or more. Histological examination showed a typical osteopetrosis with calcified cartilage and bone spicules in the marrow cavities. Reynolds and Murphy (in Murphy 1973) found from studies with ^{45}Ca that mi, like gl resorbs bone at one-third or less of the normal rate. Adams and Carr (1965) and Marks and Walker (1976) suggested a reduction of 30–50 per cent *in vivo*. Wong *et al.* (1983) suggested that bone resorption is almost completely inhibited. They found little, if any, loss of ^{3}H-tetracycline or ^{3}H collagen from mi bone *in vitro*, compared to marked continuous loss in normal littermates, suggesting lack of bone remodelling. Blood/bone calcium level was 42–70 per cent of normal, suggesting that blood calcium is maintained by diet. The blood/bone ratio is decreased strikingly and steadily from 7–14 to 35–42 days and at birth the incorporation of ^{3}H-tetracycline and ^{3}H collagen given previously *in utero* was normal.

Murphy (1973) found a slightly decreased mean serum calcium level in mice aged 9–66 days and a markedly decreased serum phosphorus; many litters showed no hypocalcaemia. Blood acid phosphatase was normal, but alkaline phosphatase was increased with a regression in this enzyme level between 10 and 40 days in both normal and mi mice. Citrate levels are significantly increased in mi mice, by 140 per cent in femur and 53 per cent in blood. Ash weights of bone were normal. No hypocalcaemic activity of mi serum was seen. Parafollicular C cells are hyperplastic (Marks 1977*a*). Osteopetrosis in mi is not cured by neonatal removal of the parathyroids (Walker 1973*a*).

It is clear that biochemically mi differs from gl with mild hypocalcaemia, hypophosphataemia, increased citrate and serum alkaline phosphatase, and without the excess of Ca bone ash seen in gl. There is no indication of an excess of thyrocalcitonin.

The osteoclasts of mi mice are clearly implicated since bone resorption is abnormal. Packer (1967) reported that mi osteoclasts seen under the light microscope were almost universally small and rounded and had significantly fewer nuclei than normal. Johnson and Laing (unpublished) confirmed this; the osteoclasts released from macerated 10–16 day-old mi bones only once had more than two nuclei: those from normal littermates had up to 10. Osteoclast counts performed by Packer, using Barnicott's (1947) method showed a normal number of cells, or even a slight increase. mi osteoclasts were rich in glycogen.

Holtrop *et al.* (1981) looked at the ultrastructure of mi osteoclasts and reported that they were smaller than those of normal littermates; ruffled borders were absent and clear zones were significantly smaller than normal (Fig. 4.15). The amount of osteoclast profile not in contact with bone was increased (29 per cent vs. 12–16 per cent in normals). The cytosol was poorly differentiated with fewer vacuoles and less ribosomes than normal and a lower frequency of mitochondria. No difference was found in the number of osteoclast profiles seen. The lamina limitans of bone from mi was pronounced and wider than normal.

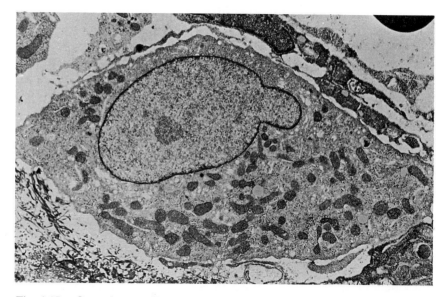

Fig. 4.15. Osteoclast profile from a mi/mi mouse. Ruffled border and clear zone are absent. (From Holtrop *et al.* 1981.)

Johnson and Al-Douri (1982) were able to confirm these findings on 3-day-old mice and extend them to younger and older material (0–15 days). Their impression was that by 15 days the number of osteoclasts was somewhat diminished; otherwise, with the important exception of the tooth tips (Fig 4.16 and see above) they agreed with Holtrop's description.

How does this picture of impaired osteoclast activity fit the disease?

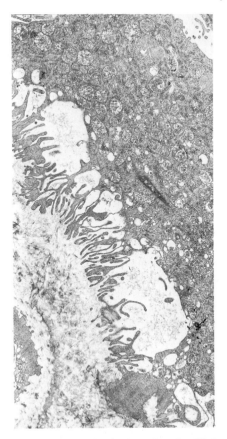

Fig. 4.16. Osteoclast from above the incisor tip of a 10-day-old mi/mi mouse. Note the well-developed ruffled border.

Osteopetrosis in mi can be cured by parabiosis with normal littermates (Walker 1973b; Barnes *et al*. 1975) or by injection of marrow or spleen cells from normal animals into irradiated mi homozygotes (Walker 1972, 1973b, 1975a,b). Conversely, normal irradiated littermates became osteopetrotic when infused with cells from mi. Appearance of normal osteoclasts with ruffled borders precedes the remission of osteopetrosis. These cells have been shown to be of donor origin by experiments using the beige strain of mice, which has giant lysosomes, as donor (Marks and Walker 1981; Loutit and Townsend 1982a,b). Marshall *et al*. (1983) showed that mi mice can be cured by intravenous injection of cultured bone marrow; this is indirect evidence that the active cell line or its progenitors are either derived from a haemopoietic stem cell or can be efficiently co-cultured with it. These findings clearly implicate the mi osteoclast and suggest that it might be functionally abnormal in line with the morphological observations. Chambers and Loutit (1979) assessed the capabilities of peritoneal macrophages (the precursors of osteoclasts and able to absorb bone in their own right) from mi mice. They injected ^{45}Ca into normal and osteopetrotic mice and, four days later, prepared sterilized bone powder from their long bones. The bone powder was added to cultures of peritoneal macrophages obtained by lavage. Normal macrophages caused more release of ^{45}Ca into the medium than mi. There was also a bone effect; mi bone, no matter what kind of macrophages were working on it, was more resistant to dissociation than normal bone. This is interesting in the light of our findings on abnormal bone in mi (Al-Douri and Johnson 1983). mi macrophages were equal to their normal peers in phagocytosis of carbon or latex particles.

Resorption of mi/mi bone was also studied by Raisz *et al.* (1977*b*). They cultured calvaria and long bones, these being removed shortly after an injection of ^{45}Ca. The rate of leakage of ^{45}Ca into the culture medium was lower in mi than in normal littermates. PTH, prostaglandin in E_2, vitamins D_3 and A, and OAF added to the medium failed to affect leaching from mi bones. *In vivo* (Konyukhov and Osipov 1966) the result was different, PTH increasing bone resorption as well as the number of erupted teeth. However the plasma calcium response to PTH was sluggish and insignificant (Marks 1977*a*).

Thesingh (1983*b*) showed that the first osteoclasts invading the calcified cartilage of developing fetal long bones are predominantly mononucleated in mi, but multinucleated in normal littermates. These cells resemble osteoclasts in light microscopic and electron microscopic appearance and can be morphologically distinguished from mononuclear macrophages. As soon as blood cell formation occurs in the marrow cavity, multinucleated osteoclasts appear, albeit of abnormal appearance. Thesingh suggested that the earliest osteoclasts in mi, derived from the liver, do not form syncytia, whereas those formed locally in bone marrow spaces do. Thesingh also showed that newborn mi macrophages were unable to fuse to form foreign-body giant cells.

Johnson and Laing (unpublished) showed that the apparent inability of precursors to form syncytial osteoclasts in mi is not shared by the foreign-body cells of the mi spleen. mi spleen cultured in the presence of cellophane (as an irritant) produced foreign-body giant cells with nuclear number equal to that of normal littermates. These experiments, however, were performed on 10-day-old mice and might reflect the different generations of osteoclasts seen by Thesingh.

It is clear that the defect in mi resides in the osteoclasts, or in some of them. Cultured calvaria from mi mice fail to respond to a number of potent stimulators of resorption including PTH, PGE_2, $1,25(OH)_2D_3$, Vitamin A, and OAF from both human peripheral leucocytes and mouse spleen cultures (Raisz *et al.* 1977*a*). If there is subnormal bone resorption (as some groups of workers suggest), then the local control of osteoclasts may still operate (as in the bone above erupting teeth) but over a much reduced level of activity, so that only a few cells respond and so that a threshold compatible with normal stimulation is achieved only in areas where osteoclast activity is normally intense.

The presence of abnormal bone in mi (Johnson and Al-Douri 1983) and the observation that bone formation is increased in mi (Marks and Walker 1969), a condition alleviated by radical thyroidectomy (Walker 1973*b*), suggest that more than just the osteoclast is involved.

Osteosclerosis (oc, Chr 19)

This mutation in the C57BL strain was first described by Dickie (1967). Homozygous oc/oc mice are recognized at 10 days by the failure of eruption of their incisors. Affected mice are small and may have a kinked tail and clubbed hindfeet: they also show circling behaviour which increases in severity with age. Most affected mice die between 30 and 40 days. Long bones have a typical osteopetrotic appearance. Marks and Walker (1969) noted that bone deposition rate is abnormally high and that the number of C cells in the parathyroid is elevated; serum calcium was subnormal. There is little or no response to PTH and acid phosphatase activity of osteoclasts is greatly reduced. Marks and Walker (1976) speculated that impaired action of osteoclasts results in low serum calcium which in turn elevates the number of calcitonin-secreting cells.

Osteopetrosis (op, Chr 12)

Marks and Lane (1976) described another osteopetrotic mutation in the mouse. op mice are smaller than normal littermates, but the condition is not lethal if soft food is provided. However, breeding performance is poor and lifespan somewhat reduced.

Histology shows the usual spongiosa in the long bones with a marrow cavity developing belatedly in mice aged 8–10 months. Osteoclasts are small and few in number and acid phosphatase activity not localized next to the bone.

Marks (1982) found the number of osteoclasts to be around 10 per cent of that seen in normal littermates, and that they are small in size. There is greatly exaggerated development of the clear zone and ruffled border (Fig. 4.17) but the absence of frayed bone beneath them and the absence of matrix crystals in and around the cells suggests that they are immature or non-functional. The distribution of acid phosphatase within the osteoclasts is also abnormal and osteoclasts contain toluidine-blue-positive and electron-dense lipid vacuoles. Marks fortuitously saw one of these lipid

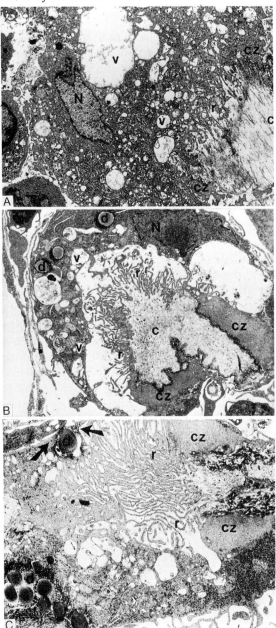

Fig. 4.17.   TEM of osteoclasts from 16-day-old (A) normal and (B) op/op mice. Note the excessive development of the clear zone (cz) and ruffled border (r) in (B). Also the dense cytoplasmic vesicles (d) away from the mineralized area (c) in addition to the usual clear vacuoles (v). (C) In a 20-day-old op/op mouse one of these dense vacuoles (arrowed) is seen entering or leaving the osteoclast cytoplasm. (From Marks 1982.)

bodies entering or leaving an osteoclast. He suggested that op mice have a deficiency of osteoclast precursors, and that the large ruffled borders and clear zones represent an attempt at compensation.

Lipid masses were seen in vascular and non-vascular sites within the skeleton (Fig. 4.18), occupying up to 25 per cent of the non-vascular soft tissue space, but were not seen in any non-skeletal tissue examined. There is an unexplained excess of megakaryocytes.

The number of monocytes in the blood is greatly decreased (Wiktor-Jedrzejezack *et al.* 1982) and the composition of the bone marrow more like that of peripheral blood. Blood pictures showed reduced leucocyte counts and lymphocyte counts, but normal haematocrit and granulocyte counts. Cells obtained from long-bone marrow cavities were reduced 10-fold compared to normal littermates. These cells, and those of spleen and thymus were low in erythroblasts and (proliferating) granulocytes. Fewer cells were also produced by peritoneal lavage. Monoblast number in spleen was increased nearly 15-fold, but monocytes in blood reduced by 20 times and macrophages by around 50 times. Marrow and spleen colonies from op were nearly always fibroblastoid and macrophages were scarce.

Wiktor-Jedtizejezack *et al.* concluded that the osteopetrosis in op is due to a deficiency in monocyte and macrophage populations and hence osteoclasts. The deficiency apparently resides in the absence of a specific CSA, a glycoprotein stimulating macrophage development. *In vitro* op haemopoietic cells will differentiate if provided with normal CSA supplied

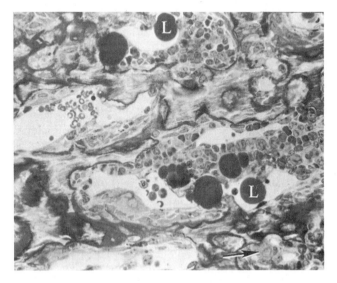

Fig. 4.18. Tibial metaphysis from 19-day-old op/op mouse. Note lipid masses (L) and single small osteoclast (arrow). (From Marks and Lane 1976.)

as supernatant from normal spleen colonies. The inability to transfer the disease by transplantation of haemopoietic tissue also suggests that op cells are unstimulated rather than unresponsive to CSA. These mice, unlike gl and mi (but like the tl rat—see below) are not cured by spleen transfusion. Nisbet *et al.* (1982) noted that the ectopic bone found when normal bone marrow was transplanted into op hosts was not osteopetrotic, and was thus apparently unaffected by the disease of the host; presumably the transplanted marrow was producing normal osteoclasts. Nisbet *et al.* doubted that these cells would survive for 84 days, the period of the transplant, however.

Wiktor-Jedrzejczack *et al.* were struck by the similarity of this phenomenon to the anaemia of the Sl/Sl^D mouse (McCulloch *et al.* 1965), which is also neither cured nor transmitted by marrow transfusion and is thought to be a consequence of abnormal haematopoietic microenvironment, governed by the stromal tissue of bones and spleen, which, of course, is of different embryological origin to the haematopoietic cells.

Bone matrix formation in op (as measured by the ability to incorporate ^{3}H proline) is significantly elevated before the fortieth day, but significantly reduced in animals aged 81 days to 10 months. The point of reversal appears to be at around 45 days. Serum calcium levels are normal at all ages measured but serum phosopherus levels significantly reduced, averaging only 70 per cent of the normal value. op/op mice are unable to respond to PTH by raising serum calcium levels.

Marks and Lane (1976) noted that op differs in several respects from other osteopetrotic mouse mutants, being longest lived and undergoing a remission (like the ia rat—see below). The remission is preceded by a decrease in bone deposition and the concentration of parafollicular cells. During the remission, blood Ca remains normal. They pointed out that parathyroid hyperplasia has been noted in all mouse osteopetroses so far described, but that parathyroidectomy does not cure the osteopetrosis (Walker 1973a) although it does decrease the excessive bone formation. They suspected that this hyperplasia is therefore secondary, a compensatory reaction to a reduction in osteoclast function.

The incisors-absent (ia) rat

The incisors-absent rat was first described by Greep (1941). The condition is not lethal: the rats have a normal lifespan, albeit with delayed sexual maturity, especially in females. Between 30 and 50 days after birth the osteopetrosis, until then producing a picture very similar to that seen in the mouse mutants described above, enters a phase of spontaneous remission. One hundred days later ia/ia rats are no longer osteopetrotic and indistinguishable from normal littermates, except for reduced bodyweight (Bhasker *et al.* 1950). Skeletal effects before remission are classical, but the dentition later erupts only partially due to ankyloses and odontoma formation (Schour *et al.* 1944, 1949).

Experimental procedures used in the study of the ia rat resemble those used in the gl and mi mice. Gilette *et al.* (1956) performed transplantation experiments essentially similar to those of Barnicott (1941). They transplanted ia ribs and found that after transplantation these resembled the host tissue; they concluded that ia osteoclasts were more resistant than normal to stimuli for resorption. Marks (1973, 1974) determined that the rate of release of ^{3}H from bone after incorporation of tritiated proline, stimulated by parathyroid extract, was reduced by 20 per cent at two weeks and increased by 20 per cent at three weeks.

ia osteoclasts are classically described as being unable to resorb bone because they lack a ruffled border in young individuals, a feature which develops between 25 and 50 days (Fig. 4.19, Marks 1978*a*). However, Johnson and Al-Douri (1982) showed that ruffled borders are present over the erupting incisor as early as 10 days, although this is very localized (Fig. 4.20). Marks (1981) also reviewed the link between tooth eruption and alveolar bone resorption in ia.

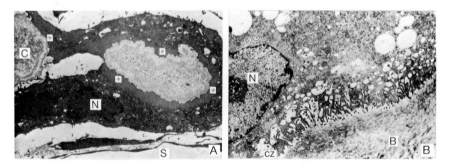

Fig. 4.19. (A) The osteoclasts of 25-day-old ia/ia rats have an extensive clear zone (asterisks) but no ruffled border. (B) By 51 days a poorly developed ruffled border can be seen (arrowed). (From Marks 1978*a*.)

The ia osteoclast characteristically accumulates lysosomes in the region adjacent to the clear zone or elsewhere beneath the cell membrane. With advancing age osteoclasts accumulate large numbers of these structures, which can easily be seen by the electron microscope. They differ in shape and size but are characteristically reactive for acid phosphatase, trimeta-phosphatase, and aryl sulphatase (Schofield *et al.* 1974; Fig. 4.21). Marks (1973) suggested that ia osteopetrosis may be a lysosome storage disease.

There is also excessive acid phosphatase activity (Handelman *et al.* 1964) brought about presumably by the cells' ability to synthesize but not release

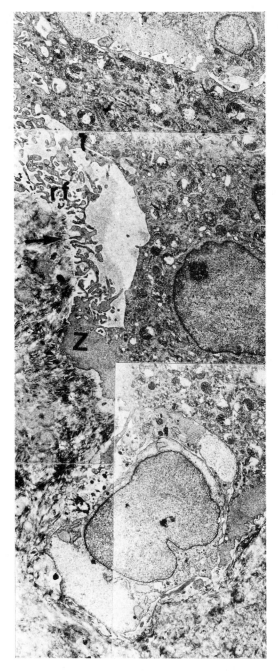

Fig. 4.20.   Montage of an osteoclast with a poorly developed ruffled border (rf) and clear zone (Z) from above the erupting incisor of a 10-day-old ia/ia rat.

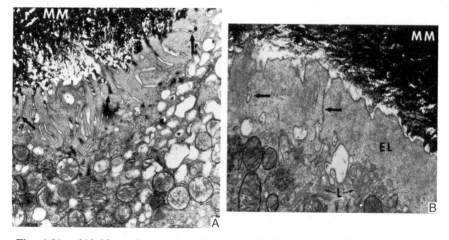

Fig. 4.21. (A) Normal osteoclast showing ruffled border and large number of dense bodies and vacuoles. Mineral is present in the infoldings (arrow). (B) Rarely observed infoldings (arrowed) in ia/ia 24-day osteoclast. Note the presence of L, lysosomes; MM, mineralized matrix; EL, ectoplasmic layer. (From Schofield *et al.* 1974.)

the enzyme: this is exaggerated by exogenous parathyroid extract. At 23 days after birth, when osteoclasts are generally becoming active, Marks (1973) noted a decrease in acid phosphatase levels in these cells, but an increase in the region adjacent to the bone surface.

Biochemical studies show normal serum calcium, low phosphate, low serum alkaline phosphatase activity, and elevated serum citrate concentration (Vaes and Nichole 1963). There is no significant difference in calcium or phosphorus contents of bone ash (Kienny *et al.* 1958) except that bone Ca was transiently elevated at 10–19 and 30–9 days. The latter possibly represents an increase in net bone resorption following remission of the disease. Bhasker *et al.* (1952) reported the beneficial results on tooth eruption of parathormone, but Marks (1976*a*) found no beneficial effect, either of PTH alone or supplimented with vitamin A. Parathormone treatment, putting the animals on a calcium-free diet, or fasting overnight activates osteoclasts, with development of large cytoplasmic vacuoles and ruffled borders (Schofield *et al.* 1978).

ia osteopetrosis can be cured by temporary parabiotic union (Marks 1976*b*) although this cannot be performed on animals young enough to allow the eruption of incisor or first molar. Whole-body irradiation followed by infusion of normal spleen cells (Marks 1976*a*, Miller and Marks 1982) liver, thymus, or bone marrow (Marks 1978*b*) also resolves bone resorption.

It seems as if the fault in ia may lie in osteoclasts not recognizing local or general hormonal stimuli before 50 days; just what changes at this stage is quite unknown.

The toothless (tl) rat

Cotton and Gains (1974) described a second osteopetrotic mutant of the rat—toothless. Toothless rats exhibit a typical osteopetrosis with failed tooth eruption. Serum alkaline phosphatase is elevated but blood levels of calcium and phosphorus are within normal limits. Morphological (Cotton and Gains 1974) and histological studies (Leonard and Cotton 1974) initially failed to demonstrate the presence of any osteoclasts.

Marks (1977*b*) showed that osteoclasts are in fact present, although not numerous or easy to find, They are always located peripherally in femur and tibia, smaller than in normal littermates, and do not possess extensive vacuolar systems adjacent to bone; nor is acid phosphatase activity localized close to bone surfaces. The excessive bone was characterized by the presence of many lipid droplets (cf. the op mouse, Fig. 4.22). tl rats are resistant to parathyroid extract, showing no increase in serum calcium concentration.

Whole-body irradiation followed by injection of spleen cells alone, or spleen cells plus bone marrow cells from normal littermates, failed to decrease the degree of osteopetrosis of tl animals within the 30-day experimental period. The treatment did not increase the number of osteoclasts present in tl rats.

Once again the osteoclast is implicated. It is odd, however, that infusion of normal osteoclasts is unable to ameliorate the effects of the disease, as it does in most other osteopetrotic mutations. Either, as Marks suggested the tl disease is fundamentally different (but closely allied to the op mouse) or the extent of the deficiency of working osteoclasts is so great that it is beyond redemption.

The osteopetrotic (op) rat

A third osteopetrotic rat mutant was described by Moutier *et al.* (1974). The toothless osteopetrotic individuals can be maintained on a suitable diet of crushed food pellets. Op osteopetrosis can be cured by temporary parabiosis with normal littermates (Togama *et al.* 1974). Milhaud *et al.* (1976) claimed that the condition can be ameliorated by a thymus graft, a result which has been contested by Nisbet *et al.* (1983). The latter authors found that op rats which had been thymectomized at weaning were not stimulated to resorb bone by intraperitoneal thymus grafts during a 50-day experimental period. Thymectomized op rats were, however, cured in four weeks by intraperitoneal or intravenously administered bone marrow grafts. This result was similar to the recovery shown by unthymectomized op animals. Nisbet *et al.* cited the results of Moutier *et al.* (1979) who used

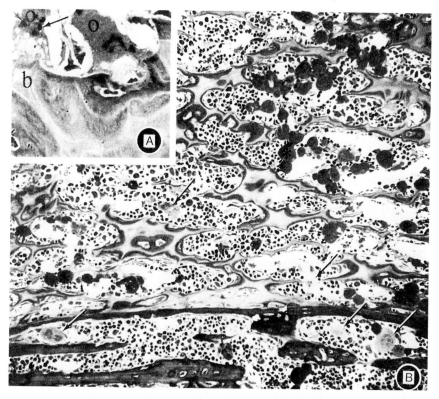

Fig. 4.22. (A) Osteoclast from tl/tl rat stained for acid phosphatase. The enzyme activity (arrow) is away from the bone (b). (B) Trabecular bone from tl/tl rat with numerous lipid droplets (*) and megakaryocytes (arrow). (From Marks 1977*b*.)

mutant rats carrying both op and a gene which led to absence of the thymus; here again athymic individuals responded to marrow grafts. Athymic rats not carrying op do not show osteopetrosis. Thymic cells do not stimulate resorption in thymectomized or unthymectomized mi mice (Loutit and Sansom 1976; Nisbet 1981). Nisbet argued that the thymus is probably not involved primarily in osteopetrosis in either mi or op. The ia rat can apparently be cured by dissociated thymic cells (Marks 1978*b*) but in these experiments a lethal dose of radiation was used to suppress the reaction to the graft because ia was on a mixed background. Following such irradiation haemopoietic stem cells are necessary for survival and it may have been these, rather than specific thymus cells that led to the subsequent cure.

Nisbet therefore suggested that op resembles mi in being based on an inadequacy of osteoclasts derived from myeloid tissue, and that the involvement of the thymus in these events is not fundamental.

The osteopetrotic (os) rabbit

Pearce (1984*a,b*) described an osteopetrotic condition in the rabbit with obliteration of marrow cavity, deficient calcification of bony tissue, and malformed, partially erupted teeth. The animals are anaemic and have abnormal white blood cells and die at 4–5 weeks. There is hypophosphataemia and hypocalcaemia and increased serum alkaline phosphatase level. Osteoblasts are increased in number and osteoclasts present in normal numbers. (Walker 1973*a*) noted that these cells lacked an erosion border. He also assayed isocitric, malic, and lactic dehydrogenases and acid phosphatase, which were all markedly reduced. Lactic dehydrogenase levels were normal.

OSTEOPETROSES IN RAT AND MOUSE —AN OVERVIEW

Several possible alternatives for the osteopetrotic mouse and rat may be written. Walker (1973*a*) summarized these as follows. After parabiosis mi and gl mice (and most other osteopetoses) are cured. Does the normal mouse provide blood-borne stem cells for competent osteolytic cells or does the normal mouse correct an endocrine disorder in the mutant? Walker tested these two hypotheses by making a temporary parabiosis for 2–3 weeks; at the end of this period little, if any, excees bone had been mobilized. But in a subsequent observation period, after the mice had been separated, recovery occurred. Walker therefore favoured the former, stem cell, hypothesis.

Stem cell hypothesis

The reversal of osteopetrosis two weeks after parabiosis suggests that the therapeutic factor is long-lived. Parathormone, calcitonin, vitamin D, and oestrogen (which have all been implicated in osteopetrosis) are short-lived and would have been metabolized rapidly after the division of the parabiotic pair. Progenitors of osteoclasts could have passed over and seeded the mutant marrow in sufficient numbers to account for the change. We now have evidence (Marks and Walker 1981; Loutit and Townsend 1982*a,b*) that recognizable osteoclast precursors transferred from the beige strain of mice to mi become osteoclasts. The op mouse and tl rat clearly do not belong with the majority in this respect, not being cured by transfusion of precursors. We have seen that the fault here lies not in the osteoclast but in its activation; although this work was done on the op mouse we may extend it to the tl rat by implication.

Involvement of thyrocalcitonin

Laboratory animal osteopetroses are resistant to parathormone and vitamin D (as is Albers–Schoenberg disease in man). Sluggish or absent calcaemic response to large doses of PTH and failure of a chronic parathormonal regime to remove excess bone and cartilage is typical. Barnicot (1945) suspected that the gl mouse suffered from hypoparathyroidism conditioned by an excessive rate of destruction of parathormone. He cited the following evidence in favour of this view. Lethal doses of PTE are tolerated by gl, parathyroid bodies are enlarged and produce excess hormone, and osteopetrotic bone is readily absorbed by normal littermates. Barnicot exonerated both osteoclasts and parathyroid chief cells and attributed the disease to an 'antiparathyroid factor' which he did not identify. Walker (1966a, 1971) suggested the excessive action of thyrocalcitonin and Murphy (1972) detected high levels of a calcitonin-like substance in gl plasma. Overproduction of calcitonin would also account for hypocalcaemia, hyperphosphataemia, resistance to PTH and PTE; in fact C cells become very productive after repeated injections of PTE. There is, however, no evidence that increased levels of calcitonin lead to excessive bone formation or accumulation and attempts to show that elevated calcitonin activity is responsible for osteopetrosis consistently fail. Removal of the source of calcitonin does not alter the state of bone in gl or mi mice. Surgical thymectomy with or without parathyroidectomy followed by radioiodine to destroy regenerating follicles failed to reverse osteopetrosis (Walker, unpublished, in Walker 1973b). Administration of calcitonin to intact or thyroidless mice or rats (Marks 1972; Walker 1971) does not induce osteopetrosis. Medullary carcinoma of the thyroid in man is associated with high blood calcitonin, but not osteopetrosis. Calcitonin levels in Albers–Schoenberg disease are normal, or only slightly elevated (Solcia *et al.* 1968; Verdy *et al.* 1971). The serum of mi mice does not contain the same active principle as gl (Murphy 1973).

Hyperparathyroid hypothesis

PTE and PTH both stimulate osteoblasts. Osteopetrosis associated with increased deposition of bone is accompanied by enlarged, hyperactive parathyroid glands. Could increased parathormone levels lie behind osteopetroses? Walker (1973b) found that if one member of a pair of parabiotic mice is given daily subcutaneous injections of 0.5 USP units/g PTE only the mouse receiving the hormone will show osteosclerosis, presumably because of the rapidity with which the hormone is metabolized. If osteopetrosis were due to hyperparathyroidism, then parabiosis should not cure the disease, Walker argued. Also enhanced rate of bone deposition is not universal; in some mutants bone formation is decreased. Transplantation of parathyroids and thyroids from normal rabbits to osteopetrotic sibs (where bone deposition is markedly reduced) has no corrective effect. Administration of thyroxine and PTE are similarly without effect. Walker considered that hyperparathyroidism here is acquired as a secondary consequence of failure of bone cells to respond to parathormone, itself consequent upon abnormal osteoclast morphology and physiology.

Osteoblast deficiencies

It is clear (Tables 4.1 and 4.2) that, although the various osteopetroses have little in common on a hormonal basis, osteoclasts are almost

TABLE 4.1. *Comparison of various osteopetroses*

	Mouse					Rat		Rabbit	Man	
	mi	gl	oc	op	ia	op	tl	os	Homozygous	Heterozygous
General										
Lifespan	months	< 30d	30–40d	N	N	N	N	< 35d	< 10y	N
Spontaneous remission	No	No	—	Yes	Yes	—	No	No	No	No
Hepatosplenomegaly/anaemia	—	No	—	No	—	—	Yes	—	Yes	Yes/No
Cure by parabiosis/transfusion	Yes	yes	—	No	Yes	Yes	No	—	Yes	—
Parafollicular cells in thyroid	U	U	U	U	—	—	—	—	—	—
Bone										
Osteoblasts	A, U	—	—	A	A	A	—	Up	—	—
Bone citrate	U	U	—	—	—	—	—	—	—	—
Matrix	A	—	—	—	N	—	—	calc.D	A,calc Up	—
Serum										
[Ca]	D	D	D	N	N	—	N	D	N	N
[P]	D	D	—	D	D	—	N	D	N	N
[Citrate	U	U	—	—	U	—	—	—	—	—
Alkaline	U	U	—	—	—	—	U	U	N	N
PTH										
Bone resorption	U	U	—	—	U/Nc	—	—	—	—	—
Tooth eruption	U	U	—	—	U/Nc	—	—	—	—	—
Serum [Ca]	Nc	Nc	—	Nc	U	—	Nc	—	—	—

N, Normal. A, abnormal. D, down. Nc, no change. U, up.

TABLE 4.2. *Comparison of osteoclasts in various osteopetroses*

	Mouse			Rat				Rabbit	Man	
	mi	gl	oc	op	ia	op	tl	os	Homo-zygous	Hetero-zygous
Numbers	N/D	D	—	D	—	—	D	N	U/D	U
Size/nuclear number	D	U	—	D	—	—	D	—	U	U
Ruffled border	A*	—	—	U	A*	—	—	A	A	A
Clear zone	A*	—	—	U	A*	—	—	A	A	A
Lysosomes/vacuoles	D	—	—	—	U	—	D	—	Large	—
Resorption	local	—	—	N	local	—	—	—	—	—
Acid phosphatase	D	D	—	A	U/A	—	A	D	—	—
PTH *in vitro*	+	+	—	—	+	—	—	—	—	—

* Except over developing incisors, N, normal. A, abnormal. D, down. U, up. †, positive effect.

invariably abnormal. This abnormality takes different forms in different species. In part this may be due to the investigations performed, but it is clear, for instance, that in ia rats acid phosphatase is concentrated in vacuoles and not released, in op mice and tl rats it is abnormally distributed, and in mi, gl, and the os rabbit levels are reduced. Size of the osteoclast and nuclear number are other variables. Nuclear number is reported as decreased in most cases, as is size, but the gl osteoclast population may include giant individuals. Ruffled border development was studied by Marks (1973) in the ia rat; he showed the clear link between development of a ruffled border and the resumption of bone resorption in the remission seen in this mutant. But again copious (though apparently inactive) ruffled borders have been described in the op mouse and apparently normal osteoclasts in gl. The animal models of osteopetrosis thus present a fascinating series of inherited but distinct conditions. The permutations of different symptoms seem to be endless. The ia rat and the op mouse undergo remission of symptoms; the rest do not. op mice and tl rats contain lipid bodies in bone and excess megakaryocytes and cannot be cured by a transplant, but the others can. gl has raised thyrocalcitonin-like activity in the blood; the others do not. Some deposit excessive amounts of bone; others do not.

The common denominator seems to be the osteoclast, but it is clearly affected in a series of different ways. The ia osteoclast retains acid phosphatase, which may be fundamental. The maldistribution of acid phosphatase in many osteopetroses are clearly due to an osteoclast stem cell defect. There is evidence that bone formation may be quantitatively raised or lowered or qualitatively abnormal, either as a symptom of reduced or increased calcium, or in its own right.

Then we have the problem of remission. Osteopetrosis is clearly a symptom of imbalance between bone deposition and resorption. In ia rats and op mice the balance appears to become more normal in later life—is this chicken or egg? Can the osteoclastic equipment of these mutants cope with the lessened amount of bone turnover in the adult, but not with the excessive demands of bone growth in the young? Or is there a qualitative change in osteoclast function with maturity, just as Thesingh (1983*a*,*b*) suggested a change at around the time of birth? We know that in ia and mi, osteoclasts over the erupting teeth are normal in function at a time when those elsewhere are not; how much is local resorption of bone dependant on local demand and local control, rather than hormone level? Frost (1966) suggested that bone remodelling is linked to bone formation.

HUMAN OSTEOPETROSES

How does the range of animal osteopetroses link to the human condition? The classical osteopetrosis in man is Albers–Schoenberg disease, first described in 1904. The condition has adopted various synonyms, marble bone disease, osteopetrosis generalista, and osteopetrosis. Carter and Fairbank (1974) recognized two types—autosomal recessive (malignant) and dominant (benign).

Malignant type

This is often well developed at birth, and death is usually in the first decade, from haemorrhage or infection. The bones are dense, sclerotic, and the metaphyses filled with trabeculae formed of calcified cartilage with a covering of woven bone (Rubin 1964). Bone resorption is assumed to be decreased (Schapiro 1980). Fractures are common; the bone is brittle, perhaps due to abnormal orientation of collagen fibres (Schapiro *et al.* 1980) and/or excessive mineralization (Rubin 1964). Tooth eruption is delayed, and teeth are often carious, leading to osteomalacia of mandible and maxilla. Anaemia is associated with loss of bone marrow space, and compensatory hepatosplenamegaly is present.

Osteoclast numbers have been reported as both high (Jowsey 1977) and low (Dent *et al.* 1965) and Schapiro *et al.* (1980) noted the presence of large conspicuous osteoclasts with more nuclei than normal. Under the electron microscope (Bonucci 1975; Schapiro *et al.* 1982) osteoclasts are seen to lack ruffled borders and clear zones and to contain large lysosomal vesicles. Osteoblasts appear normal (Schapiro *et al.* 1982) but are reduced in number (Bonucci 1981).

There are no consistent biochemical abnormalities; serum Ca and P are usually normal, as are acid and alkaline phosphatase levels (Rubin 1964). Bone marrow transplantation (Coccia *et al.* 1980) was successful in curing a 5-year-old girl of malignant osteopetrosis.

Benign type

This type is less severe, and often not noted until adulthood (Jaffe 1972). Radiographic findings are similar to the malignant type, but the disease is often asymptomatic, although fractures and anaemia have been described. Osteoclasts resemble those seen in the malignant disease (Jowsey 1977).

It is clear that at least two human conditions mirror the double handful of those seen in laboratory animals. The finding that malignant osteopetrosis is susceptible to transplantation is welcome. Otherwise we can do no more than suggest that Albers–Schoenberg disease is a typical reflection of the animal models and, like them, based upon a stem cell deficiency.

REFERENCES

Aaron, J. E. (1976). Histology and microanatomy of bone. In *Calcium, phosphate and magnesium metabolism* (ed. B. E. C. Nordin), pp. 298–356. Churchill Livingstone, London.

Adams, P. J. V. and Carr, T. E. F. (1965). Some observations on the calcium and strontium metabolism in the microphthalmic mutant of the mouse. In *Calcified tissue* (ed. R. J. Richelle and M. S. Dallemague), pp. 145–155. Gerster, Liège.

Addison, W. C. (1980). The effect of parathyroid hormone on the numbers of nuclei in feline osteoclasts in vivo. *Journal of Anatomy* **130**, 479–86.

Al-Douri, S. M. J. and Johnson, D. R. (1983). Ultrastructurally abnormal bone and dentin produced by microphthalmic mice. *Journal of Anatomy* **136**, 715–22.

Anast, C. S. (1967). Magnesium studies in relation to vitamin D resistant rickets. *Paediatrics* **40**, 425–35.

Anderson, H. C. (1969) Vesicles associated with calcification in the matrix of epiphyseal cartilage. *Journal of Cell Biology* **41**, 59–72.

—— (1976*a*). Matrix vesicles and and calcification. *Federation Proceedings* **35**, 105–7

—— (1976*b*). Matrix vesicles of cartilage and bone. In *The biochemistry and physiology of bone* IV, 2nd edn. (ed. G. H. Bourne), pp. 135–55. Academic Press, New York, London.

Ashton, B., Allen, T. D., Howlett, C. R., Eagleton, C. C., Hattori, A., and Owen, M. (1980). Formation of bone and cartilage by bone marrow stromal cells in diffusion chambers in vivo. *Clinical Orthopaedics and Related Research* **151**, 294–307.

Atkins, D., Greaves, M., Ibbotson, K. J., and Martin, T. J. (1979). Role of prostaglandins in bone metabolism, a review. *Journal of the Royal Society of Medicine* **72**, 27–34.

Barnes, D. W. H., Loutit, J. F., and Sansom, J. M. (1975). Histocompatible cells for the resolution of osteopetrosis in microphthalmic mice. *Proceedings of the Royal Society* **B188**, 501–05.

Barnicot, N. A. (1941). Studies on the factors involved in bone absorption. I. The effect of subcutaneous transplantation of bone of the grey lethal house mouse into normal hosts and normal bones into grey lethal hosts. *American Journal of Anatomy* **68**, 497–531.

—— (1945). Some data on the effect of parathormone on the grey lethal mouse. *Journal of Anatomy* **79**, 83–91.

—— (1947). The supravital staining of osteoclasts with neutral red: their distribution on the parietal bone of normal growing mice and a comparison with the mutants grey lethal and hydrocephalus. 3. *Proceedings of the Royal Society* **B.134**, 467–85.

—— (1948). The local action of parathyroid and other tissues on bone in intracerebral grafts. *Journal of Anatomy* **82**, 233–48.

Bateman, N. (1954). Bone growth, a study of the grey lethal and microphthalmic mutants in the mouse. *Journal of Anatomy* **88**, 212–62.

Bernard, G. W. and Pense, D. C. (1969). An electron microscopic study of initial membranous osteogenesis. *American Journal of Anatomy* **125**, 271–90.

Bhasker, S. N., Schour, I., Greep, R. O., and Weinmann, J. P. (1952). The corrective effects of parathyroid hormone on genetic anomalies in the dentition and tibia of the ia rat. *Journal of Dental Research* **31**, 257–70.

—— Weinmann, J. P., Schour, I., and Greep, R. O. (1950). The growth pattern of the tibia in normal and ia rats. *American Journal of Anatomy* **86**, 439–77.

Bingham, P. J., Brazell, L. A., and Owen, M. (1969). The effect of parathyroid extract on cellular activity and plasma calcium levels in vivo. *Journal of Endocrinology* **45**, 387–400.

Bonucci, E. (1971). The locus of initial calcification in cartilage and bone. *Clinical Orthopaedics* **78**, 108–39.

—— (1981). New knowledge of the origin, function and fate of osteoclasts. *Clinical Orthopaedics* **158**, 252–69.

—— Sartori, E., and Spina, M. (1975). Osteopetrosis fetalis. *Virchow's Archiv (Pathology and Anatomy)* **368**, 109–21.

——, Del Marco, A., Nicoletti, B., Petrinelli, P., and Pozz, L. (1976). Histological and histochemical investigations of achondoplastic mice: a possible model of human achondroplasia. *Growth* **40**, 241–51.

——, Gherardi, G., Del Marco, A., Nicoletti, B., and Petrinelli, P. (1977). An electron microscopic investigation of cartilage and bone in achondroplastic (cn) mice. *Journal of Submicroscopical Cytology* **9**, 299–306.

Boyde, A. (1972). Scanning electron microscopic studies of bone. In *The biochemistry and physiology of bone. I. Structure* (ed. G. H. Bourne, pp. 259–310. Academic Press, New York and London.

Braidman, I. P., Anderson, D. C., Jones, C. J. P., and Weiss, J. B. (1983). Separation of two bone cell populations from fetal rat calvaria and a study of their responses to parathyroid hormone and calcitonin. *Journal of Endocrinology* **99**, 387–99.

Bussolati, G. and Pearse, A. G. (1967). Immunofluorescent localisation of calcitonin in the C cells of pig and dog thyroid. *Journal of Endocrinology* **37**, 205–09.

Cameron, D. A. (1972). The ultrastructure of bone. In *The biochemistry and physiology of bone* Vol. I, pp. 191–231. (ed. G.H. Bourne), Academic Press, New York, London.

Carter, C. O. and Fairbank, T. J. (1974). *The genetics of locomotor disorders.* Oxford University Press, London.

Chambers, T. J. (1978). Multinucleate giant cells *Journal of Pathology* **126**, 125–48.

—— (1980). The cellular basis of resorption. *Clinical Orthopaedics* **151**, 283–93.

—— (1981*a*). Phagocytic recognition of bone by macrophages. *Journal of Pathology* **135**, 1–7.

—— (1981*b*). Resorption of bone by mouse peritoneal macrophages. *Journal of Pathology* **135**, 295–9.

—— and Loutit, J. F. (1979). A functional assessment of macrophages from osteopetrotic mice. *Journal of Pathology* **129**, 57–63.

—— and Magnus, C. J. (1982). Calcitonin alters behaviour of isolated osteoclasts. *Journal of Pathology* **136**, 27–39.

Chase, L. R. and Auerbach, G. D. (1970). The effect of parathyroid hormone on the concentration of adenosine 3',5'monophosphate in skeletal tissue in vitro. *Journal of Biological Chemistry* **245**, 1520–6.

Chen, T. L. and Feldman, D. (1979). Glucocorticoid receptors and actions on subpopulations of cultured bone cells. *Journal of Clinical Investigation* **63**, 750–8.

Coccia, P. F., Krivit, W., Cerventia, J., Clawson, C., Kersey, J. H., Kim, T. H., Nesbit, M. E., Ramsay, N. K. C., Warkentin, P. I., Teitelbaum, S. L., Kahn, A. J., and Brown, D. M. (1980). Successful bone marrow transplanation for infantile malignant osteopetrosis. *New England Journal of Medicine* **302**, 701–8.

Cotton, W. R. and Gains, J. F. (1974). Unerupted dentition secondary to congenital osteopetrosis in the Osborne–Mendel rat. *Proceedings of the Society for experimental Biology and Medicine* **146**, 554–61.

Davidovich, Z., Schonfeld, J. L., and Lally, E. (1983). Prostaglandin $F_{2\alpha}$ is associated with alveolar bone cells: immunohistochemical evidence using monoclonal antibodies. In *Factors and mechanisms influencing bone growth* (ed. A. D. Dixon and B. G. Sarnat), pp. 125–34. Alan Liss, New York.

Dent, C. E., Smellie, J. M. and Watson, L. (1965). Studies in osteopetrosis *Archives of Diseases of Childhood* **40**, 7–15.

Deol, M. S. (1963). The development of the inner ear in mice homozygous for shaker with syndactylism. *Journal of Embryology and experimental Morphology* **11**, 493–512.

Dickie, M. M. (1967). *Mouse News Letter* **36**, 39.

Dietrich, J. W., Goodson, J. M., and Raisz, A. L. G. (1975). Stimulation of bone resorption by various prostaglandins in organ culture. *Prostaglandins* **10**, 231–40.

Doty, S. B. and Schofield, B. H. (1972). Electron microscopic localisation of hydrolytic enzymes in osteoclasts. *Histochemical Journal* **4**, 245–58.

Doykos, J. D., Cohen, M. N., and Shklar, G. (1967). Physical, histological and roentgenographic characteristics of the grey lethal mouse. *American Journal of Anatomy* **121**, 29–40.

Dudley, H. R. and Spiro, D. (1961). The fine structure of bone cells. *Journal of biophysical and Biochemical Cytology* **11**, 627–49.

Eicher, E. M., Southard, J. L., Scriver, C. R., and Glorieux, F. H. (1976). Hypophosphataemia: a mouse model for human familiar hypophosphataemic (vitamin D resistant) rickets. *Proceedings of the National Academy of Sciences, USA* **73**, 4667–71.

Engelfeldt, B., Engstrom, A., and Zetterstrom, R. (1954). Biophysical studies on bone tissue. Osteopetrosis (marble bone disease). *Acta paediatrica scandanavica* **43**, 152–6.

Fischman, D. A. and Hay, E. D. (1962). Origin of osteoclasts from mononuclear leucocytes in regenerating newt limbs. *Anatomical Record* **143**, 329–37.

Fleisch, H. and Bisaz, S. (1962*a*). Mechanism of calcification: inhibitory role of pyrophosphate. *Nature* **195**, 911.

—— —— (1962*b*). Isolation from urine of pyrophosphate, a calcification inhibitor. *American Journal of Physiology* **203**, 671–5.

Freye, H. (1956). Untersuchungen über die Zahnanomalies des Mikropthalmus-syndromes der Hausmaus. *Zeischrift für menschliche Vererbungs-und Konstitutionslehre* **33**, 492–504.

Friedenstein, A. J. (1973). Determined and inducible osteogenic precursor cells. In *Hard tissue growth, repair and remineralisation,* Ciba Foundation Symposium no.**11**. Elsevier, New York.

—— (1976). Precursor cells of mechanocytes. *International Review of Cytology* **47**, 327–55.

——, Chailakhijan, R. K., and Lalykina, K. S. (1970). The development of fibroblast colonies in monolayer cultures of guinea pig bone marrow and spleen cells. *Cell and Tissue Kinetics* **3**, 393–402.

Frost, H. M. (1966). Bone dynamics and metabolic bone disease. *Journal of Bone and Joint Surgery* **48A**, 1192–203.

Gilette, R., Mardfin, F., and Schour, I. (1956). Osteogenesis in subcutaneous rib transplants between normal and ia rats. *American Journal of Anatomy* **99**, 447–71.

Gothlin, G. and Ericsson, J. L. E. (1973). On the histogenesis of the cells in fracture callus. Electron microscopic autoradiographic observations in parabiotic rats and studies on labelled monocytes. *Virchow's Archiv B, Cell Pathology/Zellpathologie* **12**, 318–29.

—— —— (1976). The osteoclast. *Clinical Orthopaedics and related Research* **120**, 201–31.

Green, M. C., Sweet, H. O., and Bunker, L. E. (1976). Tight-skin, a new mutation of the mouse causing excessive growth of connective tissues and skeleton. *American Journal of Pathology* **82**, 493–507.

Greep, R. O. (1941). An hereditary absence of the incisor teeth. *Journal of Heredity* **32**, 397–8.

Grüneberg, H. (1935). A new sub-lethal colour mutation in the house mouse. *Proceedings of the Royal Society* **B118**, 321–42.

—— (1936). Grey lethal, a new mutation in the house mouse. *Journal of Heredity* **27**, 105–9.

—— (1938). Some new data on the grey lethal mouse. *Journal of Genetics* **36**, 153–70.

—— (1956). Genetical studies on the skeleton of the mouse. XVIII. Three genes for syndactylism. *Journal of Genetics* **54**, 113–45.

—— (1962). Genetical studies on the skeleton of the mouse. XXXII. The development of shaker with syndactylism. *Genetical Research* **3**, 157–66.

—— (1963). *The pathology of development.* Blackwell, Oxford.

Gudmundsson, T. V., Galante, L., Horton, R., Matthews, E. W., Woodhouse, N. I. Y., MacIntyre, I. and Nagent de Deuxchaisnes, C. (1970). Human plasma calcitonin In *Calcitonin. Proceedings of the 2nd International Symposium* (ed. S. Taylor and G. Foster). Heinemann, London.

Guenet, J. L., Stanescu, R., Maroteaux, P., and Stanescu, V. (1981). Fragilitas ossium: a new autosomal recessive mutation in the mouse. *Journal of Heredity* **72**, 440–1.

Hancox, N. M. (1972a). The osteoclast. In *The biochemistry and physiology of bone,* 2nd edn. (ed. G. H. Bourne). pp. 45–66. Academic Press, New York/London.

—— (1972b). Osteoclastic bone resorption. In *The Biology of bone* (ed. N. M. Hancox) pp. 113–35. Cambridge University Press.

Handelman, C. S., Morse, A., and Irving, I. T. (1964). The enzyme histochemistry of the osteoclasts of normal and ia rats. *American Journal of Anatomy* **115**, 363–76.

Harris, E. D. and Krane, S. M. (1974). Collagenases. *New England Journal of Medicine* **291**, 557–63, 605–9, 652–61.

Herring, G. M. (1968). The chemical structure of tendon, cartilage, dentin and bone matrix. *Clinical Orthopaedics and Related Research* **60**, 261–99.

Hertwig, P. (1942). Neue Mutationen und Koppelungsgruppen bei der Hausmaus. *Zeischrift für Vererbungslehre* **80**, 220–46.

Hinchliffe, J. R. and Ede, D. A. (1968). Abnormalities in bone and cartilage development in the talpid³ mutant of the fowl. *Journal of Embryology and experimental Morphology* **19**, 327–39.

Hohling, H. J. and Czitober, H. (1971). Die Marmorknochenkrankheit des Erwaschsenen (marbus Albers–Schonber Osteopetrose-knochen. *Weiner Zeitschrift für innere Medzin und ihre Grenzgebiete* **53**, 305–11.

Hollander, W. F. (1960). Genetics in relation to reproductive physiology in mamals *Journal of Cellular and Comparative Physiology* **56** (Suppl. 1), 61–72.

Hollander, W. F. and Gowen, J. W. (1959). A single gene antagonism between mother and fetus in the mouse. *Proceedings of the Society for Experimental Biology and Medicine* **101**, 425–8.

Holtrop, M. E., Cox, K. A., Eilon, G., Simmons, H. A., and Raisz, L. G. (1981). The ultrastructure of osteoclasts in microphthalmic mice. *Metabolic Bone Disease and Related Research* **3**, 123–9.

—— and King, G. J. (1977). The ultrastructure of the osteoclast and its functional implications. *Clinical Orthopaedics and Related Research* **123**, 177–96.

—— Raisz, L. G. and Simmons, H. H. (1974). The effects of parathyroid hormone, colchicine and calcitonin on the ultrastructure and the activity of osteoclasts in organ culture. *Journal of Cell Biology* **60**, 346–55.

Hosoda, Y., Yoshimusa, Y. and Higaki, S. (1981). A new breed of mouse showing multiple osteochondral lesions—twy mouse. *The Ryumachi* **21**, (supp), 157–64.

Iorio, R. J., Bell, W. A., Meyer, M. H. and Meyer, R. A. (1979). Histologic evidence of calcification abnormalities in teeth and alveolar bone of mice with X-linked dominant hypophosphatemia (VDRR). *Annals of Dentistry* **38**, 38–44.

Jaffe, D. H. (1972). *Metabolic degenerative and inflammatory disease of bone and joints*. Lee & Febinger, Philadelphia.

Jee, W. .S. S. and Nolan, P. D. (1963). Origin of osteoclasts from fusion of phagocytes. *Nature* **200**, 225.

Johnson, D. R. (1980). Formation of marrow cavity and ossification in mouse limb bones grown in vitro. *Journal of Embryology and experimental Morphology* **56**, 301–7.

—— (1984). The ultrastructure of the mandibular condylar and rib cartilage from mice carrying the spondylo-metaphyseal chondrodysplasia (smc) gene. *Journal of Anatomy* **138**, 463–70.

—— and Al-Douri, S. (1982). Osteoclasts with ruffled borders from above the tip of the erupting incisors of osteopetrotic mice and rats. *Metabolic Bone Diseases and Related Research* **4**, 263–8.

Jowsey, J. (1977). *Metabolic diseases of bone*, Vol.1. W. B. Saunders, London.

Kahn, A. J., Malone, J. D., and Teitelbaum, S. L. (1981). Osteoclast precursors, mononuclear phagocytes and bone resorption. *Transactions of the Association of American Physicians* **94**, 267–78.

—— and Simmons, D. J. (1975). Investigations of cell lineage in bone using a chimera of chick and quail embryonic tissue. *Nature* **258**, 325–7.

Key, L. L., Carnes, D. L., Weichselbaum, R., and Anast, C. S (1982). Platelet derived growth factor stimulates bone resorption by monocyte monolayers. *Endocrinology* **112**, 761–2.

Kieny, A. D., Toepel, W. and Schour, I. (1958). Calcium and phosphorus metabolism in the ia rat. *Journal of Dental Research* **37**, 432–43.

Klein, D. C. and Raisz, L. G. (1970). Prostaglandins in stimulation of bone resorption in tissue culture. *Endocrinology* **85**, 1436–40.

Konyukhov, B. V. and Osipov, U. V. (1966). Study of maldevelopment of teeth in mice of the mutant strain microphthalmia. *Zhurnal obschei biologii* **27**, 620.

Landauer, W. (1927). Untersuchungen über Chondrodystrophie I. Allegmeine Erscheinungen und Skelett chondrodystrophischer Hühnerembryonen. *Wilhelm Roux Archiv für Entwicklungsmechanik der Organismen* **110**, 195–278.

Leaver, A. G., Triffit, J. T., and Holbrook, J. B. (1975). Newer knowledge of non-collagenous protein in dentin and cortical bone matrix. *Clinical Orthopaedics and Related Research* **110**, 269–92.

Leonard, E. P. and Cotton, W. R. (1974). Morphological and histochemical observations on the lack of osteoclasts in the tl strain of rat. *Proceedings of the Society for Experimental Biology and Medicine* **147**, 596–8.

Liam, J. B., Skinner, M., Glimcher, M. J., and Gallop, P. (1976). The presence of delta-carboxyglutamic acid in the proteins associated with ectopic calcification. *Biochemical and Biophysical Research Communications* **73**, 349–55.

Loutit, J. F. and Sansom, J. M. (1976). Osteopetrosis of micropthalmic mice—a defect of the haematopoietic stem cell? *Calcified Tissue Research* **20**, 251–9.

—— and Townsend, K. M. S. (1982*a*). Longevity of osteoclasts in radiation chimeras of beige and osteopetrotic microphthalmic mice. *British Journal of Experimental Pathology* **63**, 214–20.

—— —— (1982*b*). Longevity of osteoclasts in radiation chimeras of osteopetrotic beige and normal mice. *British Journal of Experimental Pathology* **63**. 221–3.

Luben, R. A., Wong, G. L. and Cohn, D. V. (1976) Biochemical characterisation of isolated bone cells: provisional identification of osteoclasts and osteoblasts. *Endocrinology* **99**, 526–34.

Lucht, U. (1972). Osteoclasts and their relationship to bone as studied by electron microscopy. *Zeitscrift für Zellforschung und Mikroskopische Anatomie* **135**, 211–28.

—— and Maunsbach, A. B. (1973). Effects of parathyroid hormone on osteoclasts in vivo. *Zeitschrift für Zellforschung und Mikroskopische Anatomie* **141**, 529–44.

Marie, P. J., Travers, R., and Glorieux, F. H. (1982). Bone response to phosphate and vitamin D metabolites in the hypophosphataemic mouse. *Calcified Tissue International* **34**, 158–64.

Marks, S. C. (1972). Lack of effect of thyrocalcitonin on formation of bone matrix in mice and rats. *Hormone and Metabolic Research* **4**, 296–300.

—— (1973). Pathogenesis of osteopetrosis in the ia rat: reduced bone resorption due to reduced osteoclast function. *American Journal of Anatomy* **138**, 165–89.

—— (1974). A discrepancy between measurement of bone resorption in vivo and in vitro in newborn osteopetrotic rats. *American Journal of Anatomy* **141**, 329–39.

—— (1976*a*). Tooth eruption and bone resorption: experimental investigation of the ia (osteopetrotic) rat as a model for studying their relationships. *Journal of Oral Pathology* **5**, 149–63.

—— (1976*b*). Osteopetrosis in the ia rat cured by spleen cells from a normal littermate. *American Journal of anatomy* **146**, 331–8.

—— (1977*a*). Pathogenesis of osteopetrosis in the microphthalmic mouse; reduced bone resorption. *American Journal of Anatomy* **149**, 276–96.

—— (1977*b*). Osteopetrosis in the toothless (tl) rat: presence of osteoclasts but failure to respond to parathyroid extract or to be cured by infusion of spleen or bone marrow cells from normal littermates. *American Journal of Anatomy* **149**, 289–97.

—— (1978*a*). Studies on the mechanism of spleen cell cure of osteopetrosis in the ia rat, appearance of osteoclasts with ruffled borders. *American Journal of Anatomy* **151**, 119–30.

—— (1978*b*). Studies of the cellular cure for osteopetrosis by transplanted cells: specificity of the cell type in ia rats. *American Journal of Anatomy* **151**, 131–7.

—— (1981). Tooth eruption depends on bone resorption: experimental evidence from osteopetrotic (ia) rats. *Metabolic Bone Disease and Related Research* **3**, 107–15.

—— (1982). Morphological evidence of reduced bone resorption in osteopetrotic (op) mice *American Journal of Anatomy* **163**, 157–67.

—— and Lane, P. W. (1976). Osteopetrosis, a new recessive skeletal mutation on chromosome 12 of the mouse. *Journal of Heredity* **67**, 11–18.

—— and Walker, D. G. (1969). The role of the parafollicular cells of the thyroid gland in the pathogenesis of congenital osteopetrosis in mice. *American Journal of Anatomy* **126**, 299–314.

—— —— (1976). Mammalian osteopetrosis: a model for studying cellular and hormonal factors in bone resorption. In *The biochemistry and physiology of bone* (ed. G. H. Bourne), pp. 227–95. Academic Press, New York/London.

—— —— (1981). The haematogeneous origin of osteoclasts: experimental evidence from osteopetrotic (microphthalmic) mice treated with spleen cells from beige mouse donors. *American Journal of Anatomy* **161**, 1–10.

Marshall, M. J., Nisbet, N. W., and Stockdale, M. (1983). Osteopetrosis in the Grüneberg (mi) mouse can be cured by cultured allogenic bone marrow. *Calcified Tissue International* **35**, 812–14.

Martin, T. J. and Partridge, N. C. (1981). Initial events in the activation of bone cells by parathyroid hormone, prostaglandins and calcitonin. In *Calcium regulating hormones* (ed. D. Cohn, R. Talmage, and L. Matthews), p. 147. Excerpta Medica, Amsterdam.

McCulloch, E. A., Siminovick, L., Till, J. L., Russel, E. S., and Bernstein, S. E. (1965). The cellular basis of the genetically determined haemopoietic defect in anaemic mice of the genotype Sl/Sl^D. *Blood* **26**, 339–410.

Meyer, R. A., Gray, R. W., and Meyer, M. H. (1978). Paradoxical normal levels of vitamin D metabolites in the plasma of X linked hypophosphataemic mouse. In *Program and Abstract, 60th annual meeting, The Endocrine Society*, p. 486.

—— Jowsey, J., and Meyer, M. H. (1979). Osteomalacia and altered magnesium metabolism in the X-linked hypophosphataemic mice *Calcified Tissue International* **27**, 19–26.

Milhaud, G., Labat, M-L., Graf, B., and Thillard, M. J. (1976). Guérison de l'ostéopetrose congénitale du rat (op) par greffe de thymus. *Comptes rendus de l'Académie de Sciences, Paris* **D283**, 531–3.

Miller, S. C. and Marks, S. C. (1982). Osteoclast kinetics in osteopetrotic (ia) rats cured by spleen cell transfers from normal littermates. *Calcified Tissue International* **34**, 442–27.

Moore, M. A. S. and Metcalf, D. (1970). Ontogeny of the haemopoietic system: yolk sac origin of in vivo and in vitro colony forming cells in the developing mouse embryo. *British Journal of Haematology* **18**, 279–90.

Moutier, R., Toyama, K., and Charrier, M.F. (1974). Genetic study of osteopetrosis in the Norway rat. *Journal of Heredity* **65**, 373–5.

—— —— Lamedin, H., and Ballet, J.-J. (1979). Guérison de l'ostéopetrose par injection de moelle osseuse allogènique chez le rat double-mutant ostéopetrotique-athymique. *Comptes rendez de l'Acàdemie des Sciences Paris* **289**, 919–21.

Muhlrad, A., Bab, I., and Sela, J. (1981). Dynamic changes in bone cells and extracellular matrix vesicles during healing of alveolar bone in rats: an ultrastructural and biochemical study. *Metabolic Bone Disease and Related Research* **2**, 234–6.

Mundy, G. R. and Raisz, L. G. (1977). Big and little forms of osteoclast activating factor. *Journal of Clinical Investigation* **60**, 122–8.

Murch, A. R., Grouds, M. D., Marshall, C. A., and Papadimitriou, J. M. (1982). Direct evidence that inflammatory MGCs form by fusion. *Journal of Pathology* **137**, 177–80.

Murphy, H. M. (1968). Calcium and phosphorus metabolism in the grey lethal mouse. *Genetical Research* **11**, 7–14.

—— (1969). Citrate metabolism in the osteopetrotic bone of the grey lethal mouse. *Calcified Tissue Research* **3**, 176–83.

—— (1972). Calcitonin-like activity in the circulation of osteopetrotic grey lethal mice. *Journal of Endocrinology* **53**, 139–50.

—— (1973). The osteopetrotic syndrome in the microphthalmic mutant mouse. *Calcified Tissue Research* **13**, 19–26.

Mutsuzana, I. and Anderson, H. C. (1971). Phosphates of epiphyseal cartilage. *Journal of Histochemistry and Cytochemistry* **19**, 801–8.

Nisbet, N. W. (1981). Bone absorption and the immune system. *Scandanavian Journal of Immunology* **14**, 599–605.

—— Waldron, S. F., and Marshall, M. J. (1983). Failure of thymic grafts to stimulate resorption of bone in the fatty/Orl-op rat. *Calcified Tissue International* **35**, 122–5.

—— Menage, J., and Loutit, J. F. (1982). Osteogenesis in osteopetrotic mice. *Calcified Tissue International* **34**, 37–42.

O'Doherty, P. J. A., De Luca, H. F. and Eicher, E. M. (1977). Lack of effect of vitamin D and its metabolites on intestinal phosphate transport in familial hypophosphatemaeia of mice *Endocrinology* **101**, 1325–30.

Ornoy, A., Atkin, I., and Levy, J. (1980). Ultrastructural studies on the origin and structure of matrix vesicles in bone of young rats. *Acta anatomica* **106**, 450–61.

—— and Langer, V. (1978). Scanning electron microscopy studies on the origin and structure of matrix vesicles in epiphyseal cartilage from young rats. *Israel Journal of Medical Sciences* **14**, 745–852.

Owen, M. (1977). Precursors of osteogenic cells. In 13th European symposium on calcified tissues. *Calcified Tissue Research (Suppl.)* **24**, R19.

—— (1978). Histogenesis of bone cells. *Calcified Tissue International* **25**, 205–7.

—— (1980). The origin of bone cells in the postnatal organism. *Arthritis and Rheumatism* **23**, 1073–80.

—— and Trifitt, J. T. (1976). Extravascular albumin in bone tissue. *Journal of Physiology* **257**, 293–307.

Packer, S. O. (1967). The eye and skeletal effects of two mutant alleles at the microphthalmia locus of *Mus musculus*. *Journal of experimental Zoology* **165**, 21–46.

Papadimitriou, J. M. (1978). Endocytosis and formation of macrophage polykaria: an ultrastructural study. *Journal of Pathology* **126**, 215–19.

Parfitt, A. M. (1976), The actions of parathyroid hormone on bone: relation to bone remodelling, turnover, calcium homeostasis and metabolic bone disease. *Metabolism* **25**, 809–44, 909–55, 1033–69, 1157–88.

Paterson, C. R. (1974). *Metabolic disorders of bone*. Blackwell Scientific, Oxford.

Patt, H. M. and Maloney, M. A. (1975). Bone marrow regeneration after local injury. A review. *Experimental Haematology, Copenhagen* **3**, 135–46.

Peacock, M. (1976). Parathyroid hormone and calcitonin. In *Calcium phosphate and magnesium metabolism* (ed. B. E. C. Nordin), pp. 405–44. Churchill Livingstone, London.

Pearce, L. (1948*a*). Hereditary osteopetrosis of the rabbit. I. General features and course of the disease: genetic aspects. *Journal of experimental Medicine* **88**, 579–95.

—— (1948*b*). Hereditary osteopetrosis of the rabbit. II. X ray, haematological and chemical observations. *Journal of experimental Medicine* **88**, 597–619.

Peress, N. S., Anderson, H. C., and Sajder, J. W. (1974). The lipids of matrix vesicles from bovine fetal epiphyseal cartilage. *Calcified Tissue International* **14**, 275–81.

Petruska, J. A. and Hodge, A. J. (1964). A subunit model for tropcollagen macromolecule. *Proceedings of the National Academy of Sciences, USA* **51**, 871–6.

Potts, J. T. and Deftos, L. J. (1974). Parathyroid hormone. Calcitonin, vitamin D, bone and bone mineral metabolism. In *Duncan's diseases of metabolism* (ed. P. K. B. Bondy and L. E. Rosenberg), pp. 1225–1430. W. B. Saunders, London.

Price, P. A., Poser, J. W. and Raman, N. (1976). Primary structure of the gamma-carboxyglumatic acid-containing protein from bovine bone. *Proceedings of the National Academy of Sciences, USA* **73**, 3374–5.

Raisz, L. G., Dietrich, J. W., Simmons, H. A., Segberth, H. W., Hubbard, W. N. and Oates, J. A. (1977*a*). Effects of prostaglandin endoperoxidases and metabolites on bone resorption in vitro. *Nature, London* **267**, 532–5.

—— Luben, R. A., Mundy, G. R., Dietrich, J. W., Horton, J. E., and Trammel, C. L. (1975). Effect of osteoclast activating factor from human leucocytes on bone metabolism. *Journal of Clinical Investigation* **56**, 408–13.

—— Simmons, H. A., Gworek, S. C. and Eilon, G. (1977*b*). Studies on congenital osteopetroses in microphthalmic mice using organ cultures: impairment of bone resorption in response to physiologic stimulators. *Journal of Experimental Medicine* **145**, 857–65.

—— —— Sandberg, A. L., and Canalis, E. (1980). Direct stimulation of bone resorption by epidermal growth factor. *Endocrinology* **107**, 270–3.

Rasmussen, H and Bordier, P. L. (1974). *The physiological and cellular basis of metabolic bone disease*. Williams & Wilkins, Baltimore.

Reeve, J. (1977). Disorders of plasma calcium. *Hospital Update* **3**, 19–30.

Rodan, G. A. and Martin, T. (1981). The role of osteoblasts in hormonal control of resorption: a hypothesis. *Calcified Tissue International* **33**, 349–51.

Rubin, P. (1964). *Dynamic classification of bone dysplasias*. Year Book Medical Publishers Inc., Philadelphia, London.

Russell, R. G. G. and Fleisch, H. (1976). Pyrophosphate and disphosphonates. In *The biochemistry and physiology of bone*. Vol. 4. (ed. G. Bourne), pp. 61–105. Academic Press, New York/London.

Samuelsson, B., Granston, E., Green, K., Hamberg, M., and Hammarstrom, S. (1975). Prostaglandins. *Annual Review of Biochemistry* **44**, 669–95.

Schapiro, F. (1982). Ultrastructural abnormalities of osteoclasts in malignant recessive osteopetrosis. *Archives of Pathology and Laboratory Medicine* **106**, 425.

—— Glimcher, M. J., Holtrop, M. E., Tashjian, A. H., Brickley-Parsons, D. and Kenzora, J. E. (1980). Human osteopetrosis: a histological, ultrastructural and biochemical study. *Journal of Bone and Joint Surgery* **62A**, 384–400.

Schofield, B. H., Levin, L. S., and Doty, S. B. (1974). Ultrastructure and lysosomal histochemistry of ia rat osteoclast. *Calcified Tissue Research* **14**, 153–60.

—— —— —— (1978). Activation of osteoclasts in the incisor absent (ia) rat. In *The mechanism of localised bone loss* (ed. J. E. Horton, T. M. Trapley, and W. F. Davis), pp. 442–3. Information Retrieval Inc, Washington and London.

Schour, I., Bhaskar, S. N., Greep, R. O., and Weinman, J. P. (1949). Odontoma-like formation in a mutant strain of rats. *American Journal of Anatomy* **85**, 73–112.

—— Massler, M., and Greep, R. O. (1944). Hereditary dental morphogenesis imperfection. A genetic study of the teeth of albino rats. *Journal of Dental Research* **23**, 194.

Scott, B. L. (1967). The occurrence of specific cytoplasmic granules in the osteoclast. *Journal of Ultrastructure Research* **19**, 417–31.

Sela, J., Bab, I., and Deol, M. S. (1982). Patterns of matrix vesicle calcification in osteomalacia of gyro mice. *Metabolic Bone Disease and Related Research* **4**, 129–34.

Smith, R. (1979). *Biochemical disorders of the skeleton*. Butterworth, London/Boston.

Sokoloff, L. (1967). Articular and musculoskeletal lesions of rats and mice. In *Pathology of laboratory rats and mice* (ed. E. Cotchin and F. F. C. Roe). F. A. Davis Co, Philadelphia.

—— Crittenden, L. B., Yamamoto, R. S., and Jay, G. E. (1962). The genetics of degenerative joint disease in mice. *Arthritis and Rheumatism* **5**, 531–46.

Solcia, E. Rondini, G, and Capella, C. (1968). Clinical and pathological observations on a case of newborn osteopetrosis. *Helevetica Paediatrica Acta* **23**, 650.

Strumpf, M., Kowalski, M. A., and Mundy, G. R. (1978). Effects of gluco-corticoids on osteoclast activating factor. *Journal of Laboratory and Clinical Medicine* **92**, 772–8.

Suda, T., Testa, N. G., Allen, T. D., and Jarrett, O. (1983). Effect of hydrocortisone on osteoclasts generated in cat bone marrow cultures. *Calcified Tissue International* **35**, 82–6.

Sutton, J. S. and Weiss, L. (1966). Transformation of monocytes in tissue culture into macrophages, epitheloid cells and multinucleated giant cells. *Journal of Cell Biology* **28**, 30–3.

Sweet, H. O. and Green, M. C. (1981). Progressive ankylosis, a new skeletal mutation in the mouse. *Journal of Heredity* **72**, 87–93.

Tasjian, A. (1975). Prostaglandins, hypercalcaemia and cancer. *New England Journal of Medicine* **293**, 1317–18.

—— Holman, E. L., Antoniades, H. N., and Levine, L. (1982). Platelet derived growth factor stimulated bone resorption via a prostaglandin mediated mech-anism. *Endocrinology* **111**, 118–24.

—— Howland, B. G., Melvin, K. E. W., and Hill, C. S. (1970). Immunoassay of human calcitonin. Clinical measurements, relation to serum calcium and studies in patients with medullary carcinoma. *New England Journal of Medicine* **283**, 890–5.

—— and Levine, L. (1978). Epidermal growth factor stimulates prostaglandin production and bone resorption in cultured mouse calvaria. *Biochemical and Biophysical Research Communications* **85**, 966–75.

—— Tice, J. E., and Sides, K. (1977). Biological activities of prostaglandin analogues and metabolites in bone organ culture. *Nature* **206**, 645–7.

Tavassoli, M and Crosby, W. H. (1968). Transplantation of marrow to extra-medullary sites. *Science* **161**, 54–6.

Tetevessian, A. (1973). Effect of parathyroid extract on blood calcium and osteoclast count in mice. *Calcified Tissue Research* **11**, 251–7.

Thesingh, C. W. (1983a). Distribution of osteoclast precursor cells throughout the embryonic fetal and newborn mouse. *Calcified Tissue International* **35**, Suppl. A6.

—— (1983b). Fusion failure of osteoclast precursor cells in fetal long bones of the osteopetrotic mutant microphthalmic (mi) mouse. *Calcified Tissue International* **35**, Suppl. A6.

Thompson, R. C., Gaull, G. E., Horrowitz, S. J., and Schenk, R. K. (1969). Hereditary hypophosphatasia. *American Journal of Medicine* **47**, 209–19.

Thybery, J. (1974). Electron microscopic studies on the initial phases of calcification in guinea pig epiphyseal cartilage. *Journal of Ultrastructural Research* **46**, 206–18.

Tomlinson, S. and O'Riordon, J. L. H. (1978). The parathyroids. *British Journal of Hospital Medicine* **19**, 40–53.

Toyama, K., Moutier, R., and Lamedin, H. (1974). Resorption osseuse après parabiose chez les rats op (ostéopetrose). *Comptes Rendus de l'Académie des Sciences, Paris* **D278**, 115–19.

Triffit, J. T., Gebauer, U., Ashton, B. A., Owen, M. E., and Reynolds, J. J. (1976). Origin of plasma Alpha H S -glycoprotein and its accumulation in bone. *Nature.* **262**, 226–7.

Vaes, G and Nichole, S. (1963). Bone metabolites in a mutant strain of rats which lack bone resorption. *American Journal of Physiology* **205**, 461–6.

Vaughan, J. (1975). *The physiology of bone*, 2nd edn. Clarendon Press, Oxford.

Verdy, M., Beaulieu, R., Demers, L., Sturtridge, W. C., Thomas, P., and Kumar, M. A. (1971). Plasma calcitonin activity in a patient with thyroid medullary carcinoma and her children with osteopetrosis. *Journal of Clinical Endocrinology and Metabolism* **32**, 216–22.

Walker, D. G. (1966a). Elevated bone collagenolytic activity and hyperplasia of parafollicular cells of the thyroid gland in parathormone treated grey lethal mice. *Zeitschrift für Zellforschung und Mikroskopische Anatomie* **72**, 100–24.

—— (1966b). Counteration to parathyroid therapy in osteopetrotic mice as revealed by the plasma calcium levels and ability to incorporate ^{3}H proline into bone. *Endocrinology* **79**, 836–42.

—— (1971). The induction of osteopetrotic changes in hypophysectomised thyroparatyroidectomised and intact rats of various ages. *Endocrinology* **89**, 1389–406.

—— (1972). Congenital osteopetrosis in mice cured by parabiotic union with normal sibs. *Endocrinology* **91**, 916–20.

—— (1973a). Experimental osteopetrosis. *Clinical Orthopaedics and Related Research* **97**, 158–74.

—— (1973b). Osteopetrosis in mice cured by temporary parabiosis. *Science* **180**, 875.

—— (1975a). Bone resorption restored in osteopetrotic mice by transplants of normal bone marrow and spleen cells. *Science* **190**. 784–5.

—— (1975b). Control of bone resorption by haematopoietic tissue. *Journal of experimental Medicine* **142**, 651–63.

—— (1975c). Spleen cells transmit osteopetrosis in mice. *Science* **190**, 785–7.

Warshawski, H., Goltzman, D., Rouleau, M. F., and Bergeron, J. J. M. (1980). Direct in vivo demonstration by autoradiography of specific binding sites for calcitonin in skeletal and renal tissue of rats. *Journal of Cell Biology* **85**, 682–94.

Watchorn, E. (1938). Some biochemical data on the grey lethal mouse. *Journal of Genetics* **36**, 171–6.

Wezeman, F., Kuettner, K. E., and Horton, J. E. (1979). Morphology of

178   *The genetics of the skeleton*

osteoclasts in resorbing fetal rat bone explants: effects of PTH and AIF in vitro. *Anatomical Record* **194**, 311–23.

Wiktor-Jedrzejezack, W., Ahmed, A., Szczylik, C., and Skelly, R. R. (1982). Haematological characterisation of congenital osteopetrosis in the op/op mouse. Possible mechanism for abnormal macrophage differentiation. *Journal of experimental Medicine* **156**, 1516–27.

Williams, T. F. and Winters, R. W. (1972). Familial (hereditary) vitamin D resistant rickets with hypophosphatemia. In *The metabolic basis of inherited disease* (ed. J. B. Stanbury, J. B. Wyngaarden, and D. S. Fredrickson) McGraw pp. 1465–85. McGraw Hill, New York.

Wong, G. L. and Cohn, D. (1974). Separation of parathyroid hormone and cacitonin-sensitive cells from non-responsive bone cells. *Nature* **252**, 713–15.

—— Luben, R. A. and Cohn, D. V. (1977). 1,25 dihydroxycholecalciferol and parathormone: effects on isolated osteoclast like cells. *Science* **197** 663–5.

Wong, K. M., Zika, J., and Klein, L. (1983). Direct measurements of basal bone resorption in microphthalmic mice in vivo. *Calcified Tissue International* **35**, 562–5.

Zawisch, C. (1947). Marble bone disease, a study of osteogenesis. *Archives of Pathology* **43**, 55–75.

5. The axial skeleton

The postcranial axial skeleton of mammals is largely mesodermal in origin. The classic description of the development of the axial skeleton for the mouse is that of Dawes (1930) but more recently the classic view of segmentation has been challenged (Verbont 1976).

Grüneberg (1963) identified four groups of mutations of the axial skeleton affecting the notochord, unsegmented mesoderm, segmentation, and sclerotome differentiation, respectively. Verbont's views (Verbont 1976) necessitate reinterpretation only in the areas of unsegmented mesoderm and sclerotome differentiation. In this chapter I therefore propose merely to summarize and update the data given by Grüneberg and to re-examine critically mutations affecting unsegmented mesoderm and the process of segmentation.

THE PRIMITIVE STREAK AND THE NOTOCHORD

The notochord is formed from the mammalian primitive streak as the notochordal process, which extends forwards in the midline. Although located at first between ectoderm and endoderm, the notochordal process becomes incorporated into the roof of the gut, then freed again from the anterior end backwards as the notochord proper. Three stages are recognized in this development in man: first the notochordal process fuses with the underlying endoderm, second the fused ectodermal–endodermal region regresses, opening the notochordal lumen to the yolk sac below, and third the remaining notochordal tissue rounds up, losing contact with the endoderm below. In other species the details of this process differ; in rodents, for example, the notochord has no lumen. In the tail region of the mouse the definitive notochord arises as a separate structure from the primitive streak, and is thus never incorporated into the tail gut.

This rather complex series of events depends for its successful completion on three separations. First the separation of the notochordal process from the overlying ectoderm, second the separation of the notochord from the paraxial mesoderm on either side, and third the separation, in the trunk region, of the notochord from the gut roof. A number of mutant genes affect these separations and consequently the fate of the notochord.

MUTANTS AFFECTING THE PRIMITIVE STREAK AND NOTOCHORD

Danforth's short tail (Sd, Chr 2)

This semidominant mutant (Dunn *et al.* 1940) is lethal in most homozygotes and many heterozygotes due to kidney anomalies arising from both defective ureters and kidney mesenchyme (Gluecksohn-Waelsch and Rota 1963). Surviving heterozygotes have good viability and fertility but reduced numbers of caudal vertebrae and some kinking of the tail. Tails may be absent in extreme individuals and the third or fourth sacral vertebra missing. The bodies of all vertebrae are reduced (Grüneberg 1953) and one or both kidneys reduced or absent. Homozygotes are more severely affected and, in addition, have an imperforate anus and absent rectum and sometimes uretha and bladder. The abnormalities can be traced back to a defective notochord (Grüneberg 1958a) which, from nine days onwards is increasingly reduced and abnormal in the posterior trunk and tail region. Soon after its formation in this area it breaks down in the heterozygote to leave only occasional scattered cell nests; in the homozygote dissociation is virtually complete (Figs. 5.1 and 5.2.). In the trunk vertebral bodies develop without the 'scaffolding' of the notochord and are consequently reduced in size. In the tail extensive pycnosis occurs in somites and neural tube. The nucleus pulposus of the intervertebral discs is absent or reduced and the development of the annulus fibrosus is rudimentary. The abnormalities of the gut, and of the urinogenital system can also be traced back to the abnormality of the notochord.

Although Grüneberg (1958a, 1963) suggested a primary role for the abnormal notochord in Sd mice, Gluecksohn-Schoenheimer (1945) suggested that notochordal degeneration is secondary to degenerative changes in the mesoderm of the tail. Paavlova *et al.* (1980) used

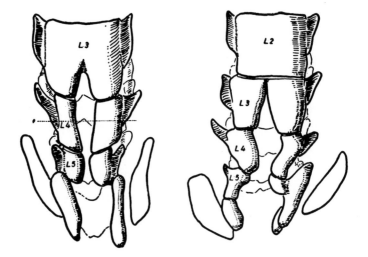

Fig. 5.1.   Ventral view of the lower vertebral column of a newborn Sd/Sd mouse. (From Grüneberg 1963.)

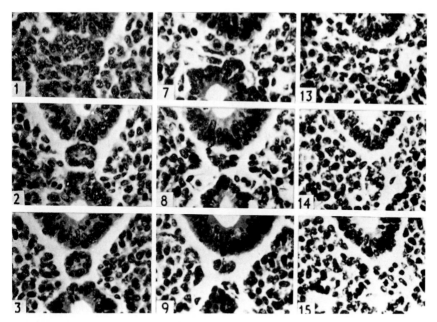

Fig. 5.2. Transverse sections through the tails of +/+ (left), Sd/+ (centre), and Sd/Sd (right) mice showing defective notochord. (From Grüneberg 1958*a*.)

histochemical means to stain the notochord of Sd mice and normal littermates. They found abnormalities in the histochemistry of the notochord as early as nine days of development. Normal notochord cells had PAS-positive diastase-resistant granules in their cytoplasm which were much decreased in abnormals. Abnormal notochord cells also failed to acquire the ability to promote chondrogenesis (Cooper 1965). Mesenchyme surrounding these abnormal notochords at 9 and 10 days was normal, but on day 11 the mesodermal cells failed to align themselves as they would in the presence of a normal notochord. Paavola *et al.* concluded that Grüneberg rather than Gluecksohn-Schoenheimer had the correct hypothesis and the abnormal notochord is primary.

Center *et al.* (1982) found that the notochordal abnormality also extended to the perinotochordal sheath, this structure being more abnormal in Sd/Sd than in Sd/+ heterozygotes. Jurand (1974) pointed out that the formative role of the basement membrane which forms the perinotochordal sheath in the development of the normal notochord.

Pintail (Pt, Chr 4)

Pintail (Hollander and Strong 1951; Berry 1960, 1961) is semidominant, heterozygotes having tails of variable lengths, usuallly with kinks near the end and a threadlike tip. Homozygotes have shorter tails, are smaller, and have a higher preweaning mortality, although survivors are healthy and fertile. The notochord is seen to be abnormal at 10 days of embryonic development and proved to have a reduced rate of mitosis. This leads to a smaller than normal notochord and hence to reduced intervertebral discs (Fig. 5.3).

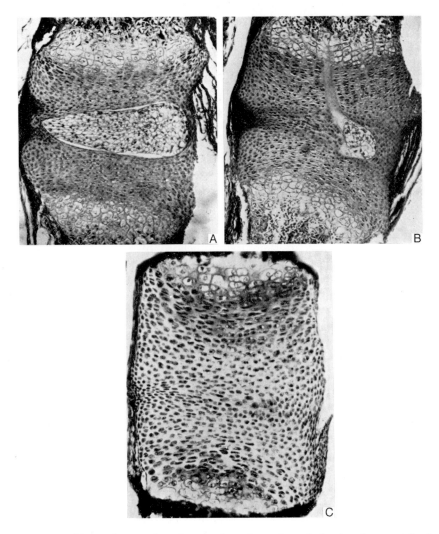

Fig. 5.3. Photomicrographs of sagittal sections through lumbar intervertebral bodies of (A) newborn +/+; (B) Pt/+; and (C) Pt/Pt mice. (From Berry 1961.)

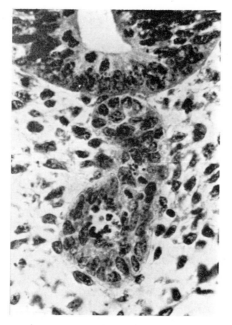

Fig. 5.4. Thickening of the posterior part of the notochord, cell degeneration, and mitosis in a 10.75 day truncate embryo. (From Theiler 1959.)

Truncate (tc, Chr 6)

Truncate (Theiler 1957, 1959*a*) is a recessive with incomplete penetrance. Homozygotes have short or absent tails and may have missing lumbar or sacral vertebrae. Characteristically the vertebral column has interrupted regions separated by virtually normal vertebrae, or the bony tail ends suddenly. The embryonic lesion is first seen at 9.5–10 days when, in some areas, the sclerotomic cells which normally migrate from somites to midline to form vertebrae degenerate; in these degenerate areas the notochord is seen to be interrupted (Fig. 5.4). In tailless individuals the notochord terminates suddenly in a knob-like swelling. Theiler was not sure whether the abnormality of the notochord led to the pycnosis of the sclerotomal cells or vice versa. Grüneberg (1963) favoured the former interpretation.

Vestigial tail (vt, Chr 11)

Vestigial was originally described as a gene affecting the unsegmented mesoderm of the tail (Grüneberg 1955*a*, 1957, 1963). But, like Sd which was also thought to have a primary effect on this tissue (Gluecksohn-

Schoenheimer 1945), the defect really seems to reside in the primitive streak (Grüneberg and Wickramaratne 1974). Vestigial tail (Heston 1951) reduces the tail to a short stump or a thin filament, with ankyloses between sacral and proximal caudal vertebrae and bilateral centres of ossification in lumbar vertebrae (the latter a common consequence of reduced size). In 9.5. day vt/vt embryos neural tissue is hived off the ventral side of the neural tube: this rounds up, acquires a secondary lumen, and branches to form multiple neural tubes in the tail. The ventral ectodermal ridge of the tail (Grüneberg 1956), a transitory structure of columnar epithelium sometimes likened to the apical ectodermal ridge of the limbs, is reduced. These two abnormalities are contemporaneous. Grüneberg and Wickramaratne (1974) were able to look at litters of embryos known to be all of the genotypes vt/vt and +/+, respectively, before the morphological abnormalities described above became apparent. They found that, although the length of the posterior half of the body at 9 days was normal in vt/vt, the hind gut or tail gut did not extend as far caudally as in normals, and the area of the tail bud was reduced. vt/vt has a reduced primitive streak which distally 'fails to fill the surrounding epidermis' (Fig. 5.5), a fact previously interpreted as oedema (Grüneberg 1957). The notochord, where it appears, is less firmly knit than usual and nodules of notochordal origin are present between notochord and neural tube. This is considered sufficient cause for the abnormal neural tube, VER, and subsequent anomalies.

Bent-tail (Bt, Chr X)

Bent-tail (Garber 1952*a,b*) is sex-linked, with reduced viability in hemizygous males and homozygous females (Grüneberg 1955*c*). The effects described by Grüneberg (1955*c*, 1963) are largely confined to the tail, which is shortened and kinked in a dorsoventral plane. Bn/– males have 4–5 fewer caudal vertebrae than normal. The tail skeleton consists of a mixture of normal and abnormal vertebrae; abnormal ones are characteristically shaped like a bow tie, obviously a consequence of ossification from two centres, or are shortened and have non-parallel ends, so that they are slightly curved in a dorsoventral plane. Morphologically 'normal' vertebrae are reduced in size. The embryology of the condition has not been described. However, Bn affects some minor skeletal variants (Grüneberg 1955*a,c*), perhaps the most significant of which is the interfrontal bone. Johnson (1976) was able to show that seven out of eight genes known to increase the frequency of occurrence of the interfrontal bone, or its size, also affected the neural tube. Lyon (in Johnson 1976) noted that Bn/– males and Bn/Bn females often had an open sacral neural tube and cranioschisis *in utero*. It appears, as Grüneberg (1963) suggested, that the bent-tail gene interferes with the supply of raw material from

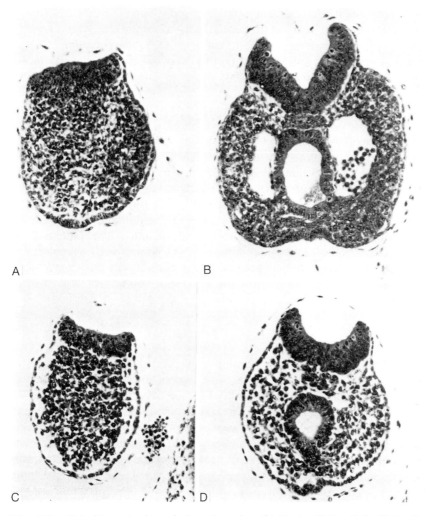

Fig. 5.5. Primitive streak and cloacal region (A,B +; C,D vt/vt). Note the reduced area of the vestigial embryo, reduced primitive streak in (C), and loose paraxial mesoderm in (D). (From Grüneberg and Wickramaratne 1974.)

which the segments are ultimately formed. The involvement of the neural tube suggests, although in the absence of embryological studies it cannot be proven, that the defect may lie in the primitive streak.

Porcine tail (pr, Chr ?)

MacNutt (1967, 1969) described this recessive gene with variable expression. The tail is affected by a range of abnormalities, and may have only a slight permanent bend or form a tightly coiled stump. The defect has been traced to 'unequal growth of the primitive streak' at the level of the seventh caudal vertebra, causing the tail to curve dorsally; the pigtailing effect is caused later by the caudal musculature. The tail usually consists of six normal caudal vertebrae and 8–15 vestiges.

BRACHYURY AND THE T LOCUS (CHR 17)

The T locus is very complex and, although first described as a series of genes affecting the tail, now has quite other connotations, as it resembles the major H-2 histocompatibility locus in man; there are also effects on fertility. The wider effects of the locus have been extensively reviewed (Bennett 1975; Klein and Hammerberg 1977; Sherman and Wudl 1977).

Mutations at this locus comprise a series of semidominants (T, T^x) and a series of recesives (t^n). In general T^x/T^x, t^n/t^n are lethal. T^x/t^n tailless, $T^x/+$ short-tailed, and $t^n/+$ normal, although exceptions occur. Segregation ratios of $t^n/+$ and T^x/t^n males are vey abnormal with an excess of t^n offspring since t^n bearing sperm fertilize a disproportionately large number of eggs.

Brachyury (T)

This allele, first described by Dobrovolskia-Zavadskaia (1927) produces heterozygotes with shortened tails, which characteristically end abruptly as if amputated, or tailless individuals. T/T homozygotes lack the posterior part of the body and die at 10.5 days. Abnormalities of thoracic vertebrae are seen from T6 (the sixth thoracic vertebra) onwards, vertebrae having double centres of ossification or being split longitudinally through the centrum, which in some cases is completely absent. Vertebral arches may be open dorsally and commonly synostoses in the tail region give rise to tail kinks. The column ends abruptly, often in the middle of a vertebra at any level below the lumbar region. T affects both total and regional vertebral numbers (Pennycuick 1980).

The development of the abnormality (Chesley 1935; Grüneberg 1958*b*; Theiler 1961) involves the notochord which, although of normal length, is retained by either the roof of the hind gut or the cloaca (Fig. 5.6). Often separation of the notochord and overlying neural tube is delayed and, when it finally occurs, neural material remains attached to the dorsal aspect of the notochord. In older embryos the tail notochord is secondarily incorporated in either neural tube or tail gut. Composite structures formed of neural tube plus notochord or notochord plus tail gut are thus seen in various regions. The result, at around 11 days, is a constriction cutting off the distal part of the tail, followed by pycnosis. In the lethal homozygote a notochord is demonstrated only with difficulty, closely associated with

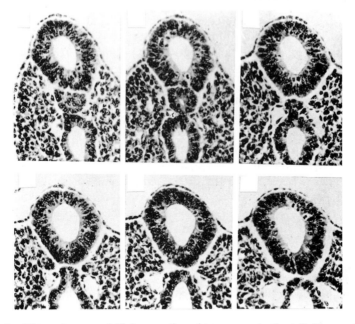

Fig. 5.6. T/+ embryo aged 10 days sectioned transverse to the tail. These selected serial sections show how the notochord may be periodically incorporated into the roof of the gut. (From Grüneberg 1958*b*.)

either neural tube or gut roof. The deficiency may be confined to the notochord, but it seems likely that the abnormalities of the homozygote reflect abnormality of the whole primitive streak. Bennett (1958) found that in culture T/T somites do not form cartilage even in association with a normal neural tube and that 'organizer relationships' in this area are disturbed. However Jacobs-Cohen *et al.* (1983) showed that T/T somites are able to undergo chondrogenesis, although the resulting somites are small and malformed. The apparent contradiction is a methodological one. Bennett (1958) used horse serum, since shown to be inhibitory to chondrogenesis, and trypsin which may have caused a loss of sclerotomal material from somites.

Wittman *et al.* (1972) looked at the ultrastructure and histochemistry of the regressing T/+ tail tip with the idea of identifying any concentrations of acid phosphatase (a lysosomal enzyme) in the degenerating region. They found numerous lysosomal bodies in the regressing, filamentous end of the tail tip, which they suggested were taking part in terminal, or near terminal stages of involution or digestion.

Electron microscopy of 8–9 day T/T embryos (Spiegelman 1976) shows that neuroepithelial cells form abnormal associations with other cell types showing specialized contact zones with notochord and somite cells; in these areas the basement membrane of neuroepithelial cells is defective. Cultures of disaggregated T/T embryonic cells form smaller aggregates than normal, but more of them (Yanagisawa and Fujimoto 1977). Since alleles at the T locus are known to affect cell surface components, these authors suggested that T/T might have different adhesiveness from +/+, as is indeed the case; we are reminded immediately of some of the mutants described in earlier chapters which also effect cell adhesion.

Yanagisawa and Kitamura (1975) looked at mitotic activity in the T neural tube at nine days of gestation. They found that mitotic index was very variable in T/T embryos, especially in the ventral part of the tube, perhaps the location where the effect of the gene is greatest, although differences between T/+ and +/+ were also found in the ependymal layer. In wild type mice mitotic densities were highest dorsally and lowest laterally: in T/+ this pattern was disordered.

Yangisawa *et al.* (1981) looked at early T embryos. The first abnormality that they were able to identify was that one in four embryos from T/+ × T/+ matings, but none in +/+ × +/+ showed a delay in the transition from cuboidal to squamous endoderm. At 7 and 8 days one-quarter of the embryos in these litters had a low mesoderm:ectoderm ratio and were classified as T/T individuals. In 6, 7, and 8-day-old T/T embryos, cells proliferated at normal rates, suggesting that the morphological abnormalities seen in 8-day-old T/T individuals had not arisen by anomalies of cell proliferation, as perhaps suggested by the earlier findings of Yanagisawa and Kitamura (1975). At 8 days, when normal and abnormal cell number was 5×10^4 cells per embryo a change in regional distribution of mesoderm and ectoderm was seen along the anteroposterior axis of the T/T embryo. This comprised an increase in the number of cells in the primitive streak (confirming the observations of Chesley (1935) and Spiegelman (1976) that the primitive streak is bulky in these embryos), i.e. the posterior 15 per cent of the embryo. This was allied to a reduction in the number of mesodermal cells in the region immediately ahead. At 8.75 days many pycnotic cells were seen in the posterior part of T/T individuals: cell cycle time was unaffected.

Yanagisawa *et al.* suggested that these findings indicate normal mitosis but a defective allocation of cells to primitive streak or mesoderm—cells destined for the epiblast are 'stuck' in the primitive streak and proper migration of invaginated cells is impeded.

The combination T/t (Gluecksohn-Schoenheimer 1938; Grüneberg 1958*b*) produces a situation essentially comparable to T/+ with the notochord incorporated into the neural tube.

Curtailed (T^C)

This semidominant allele of T (Searle 1966) is a little more extreme than T itself. $T^C/+$ always has a reduced tail (on some backgrounds the tail of T/+ can be almost normal) and lacks the odontoid peg of the atlas, whilst T/+ never does. The homozygote has a greater posterior reduction of the body (so extreme as to include the forelimb buds), open neural folds or kinked neural tube in the trunk, and distended pericardium.

Harwell T (T^h)

$T^h/+$ mice (Lyon 1959) are indistinguishable from T/+, but the T^h/T^h homozygote dies earlier, at 8 days, when it is represented only by a small cone of tissue in an embryo sac of half the normal size.

Hairpin-tail (T^{Hp})

Hairpin-tail is an allele whose greatest interest lies in its non-skeletal effects, since it is one of the very rare conditions in vertebrates where the phenotype of the heterozygote is dependent upon whether the gene was transmitted from the father or the mother. This is not a simple maternal effect: $T^{Hp}/+$ mice of type 1 and 2 can grow side by side in the same uterus, yet one will die and the other live. Johnson (1974, 1975) postulated post-reductional gene action to account for this. Hairpin-tail also differs from other T^x alleles in that the combination T/T^{Hp} is short-tailed and viable rather than lethal. $T^{Hp}/+$ type 1 mice ($T^{Hp}/+^1$, which inherited the gene from the father) are viable with variable shortening and kinking of the tail. $T^{Hp}/+$ type 2 ($T^{Hp}/+^2$, which inherited the gene from the mother) are oedematous, with short or kinked tail and often postaxial polydactyly. This version of $T^{Hp}/+$ is lethal at birth or earlier. T^{Hp}/T^{Hp} is a pre-implantation lethal. $T^{Hp}/+^1$ embryos show abnormaliites from day 10 of development when they have a persistently open neural tube, a wavy neural tube, or a malformed tail bud. On day 11 the tip of the tail is characteristically duplicated in the dorsoventral plane and the neural tube either interrupted in the sacral region or widened, with lozenge-shaped inclusions in its roof. On day 12 the tail is thinned distal to a constriction and by day 14 has become a mere filament, which often drops off, leaving a blunt tail end. In sections the trunk notochord is poorly differentiated at 10 days, consisting of too few cells which lie too close to the neural tube (Fig. 5.7). In the tail transverse sections show the neural tube duplicated, swollen, or absent (Fig. 5.8). The thoracic neural tube often has a characteristic ventral outgrowth of its outer ependymal layer.

Some $T^{Hp}/+^2$ embryos can be distinguished externally at 10 days by their irregular neural tube, blunt tail, and swollen pericardium. By day 13 they have postaxial polydactyly of the forefeet, or of all feet and are very oedematous, with a large umbilical hernia. From day 16 there is extensive subcutaneous haemorrhage. The $T^{Hp}/+^2$ notochord is also deranged at 10 days and often doubled or trebled in the tail.

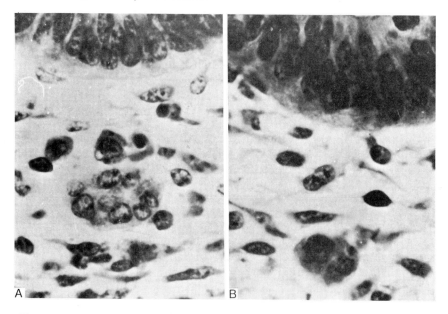

Fig. 5.7. Transverse sections through the tail of (A) +/+ and (B) $T^{Hp}/+^1$ embryos showing abnormal notochord. (From Johnson 1974.)

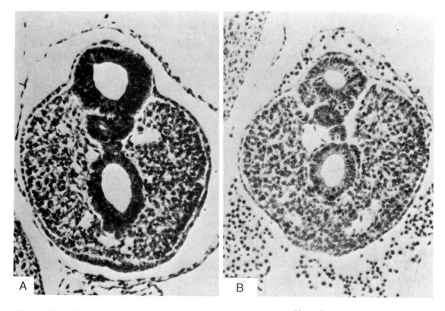

Fig. 5.8. Transverse sections through the tail of two $T^{Hp}/+^2$ embryos (A,B) aged 12 days showing multiple neural tubes. (From Johnson 1974.)

These abnormalities are consistent with those described in T and T^C although somewhat more extreme; the presence of two types of hetero-zygote is almost unique: could it be that the t locus acts in the egg as well as the sperm? $T^{Hp}/+^2$ embryos can be rescued (Bennett 1978) if aggregated with a normal embryo to form a chimera, but the chimeras, if female, do not pass on T^{Hp} to their viable offspring. Hairpin-tail is thought to be a small deletion including both T and the adjacent quaking (qk) locus (Bennett 1975).

T-Oak Ridge (T^{Or})

Six distinct mutants have been described under this name (Bennett *et al.* 1975). Serological studies suggest that these also comprise a short deletion, not extending to qk. They all behave genetically like T and cannot be distinguished from T^h on the basis of phenotype. Two that were studied histologically T^{1Or} and T^{3Or} are indistinguishable.

T-Orleans (T^{Orl})

This semidominant allele arose spontaneously (Moutier 1973). It is a deletion involving T and qk and is indistinguishable from T^{Hp} except that it lacks a maternal effect (Bennett 1975). Homozygous embryos form only trophoblast and inner parietal endoderm and die before 10 days of gestation (Erickson *et al.* 1978).

T ALLELES AND CHROMOSOME REARRANGEMENT

Since some of the T alleles mentioned above embrace deletions it is perhaps pertinent to ask if all T effects are due to similar chromosomal aberrations. Jaffe (1952), Geyer-Duszynska (1964), and Wrathall (1974) all looked at T itself, and the latter extended his study to Fu, Sd, Lp, and Cd (see below). At the level of the light microscope no deletions, inversions, or duplications were seen. This does not, of course, invalidate the findings with respect to T^{Hp}, T^{Or}, or T^{Orl}.

The t alleles

These form a series of complementation groups (Erickson *et al.* 1978) with differing effects on transmission via sperm, segregation, etc. which need not concern us here. Some complementation groups form a tailless or short-tailed T/t heterozygote, whilst in others this genotype has a normal tail. Lethal homozygotes die at around implantation and often fail to produce differentiated extra-embryonic ectoderm (for examples see Bennett 1975; McLaren 1976). At least one allele (t^6) seems to be a generalized cell lethal when homozygous (Wudl *et al.* 1977). In another complementation group (t^9) abnormals have a massive primitive streak at 8 days of gestation, and defective mesoderm (Spiegelman and Bennett 1974; Figs. 5.9 and 5.10). The cells of the primitive streak are rounded rather than stellate and do not form normal intercellular contacts. They subsequently fail to produce normal mesoderm. Embryos transplanted to testes produce epitheliomas devoid of mesodermal derivatives (Bennett 1975; McLaren 1976). In a third group (t^{12}) homozygotes die in the early-to-late morula stage (Hillman and Hillman 1975) with excessive cytoplasmic fluid, mitochondrial abnormalities, binucleated cells, and cellular lipid droplets. They fail to undergo normal compaction and so can be recognized as early as the eight-cell stage (Granholm and Johnson 1978). t^{12} is also regarded as a probable cell lethal since cells of the genotype t^{12}/ t^{12} do not survive past their normal point of death in culture (Wudl *et al.* 1977) or in chimeras (Mintz 1964).

The t^{w1} (t-wild 1) complementation group dies over a long period between 9 days and birth with a degeneration of the ventral part of the neural tube and brain (Bennett 1975). Homozygotes of the t^{w2} group are lethal over an even longer time-scale, from implantation to adulthood; ectoderm and primitive streak are affected (Bennett 1975; Bennett and Dunn 1969). t^{w5}/t^{w5} is a cell lethal in culture (Wudl *et al.* 1977); *in vivo*

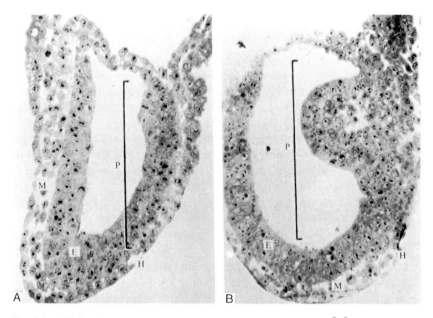

Fig. 5.9.   Light micrographs of 1 μm sections of (A) + and (B) t^9/t^9 embryos. Note the enlarged primitive streak (P) in the abnormal. (From Spiegelman and Bennett 1974.)

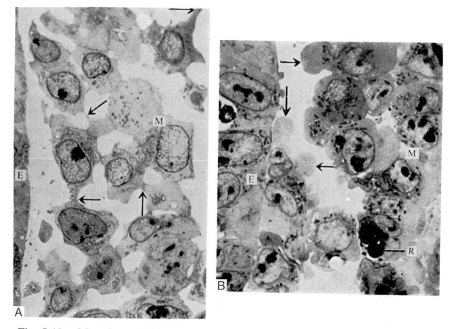

Fig. 5.10. Mesoderm of 8-day embryos (A) normal, (B) t^9/t^9. Note that normal mesoderm cells (M) are stellate; t^9/t^9 have extensive areas of cell surface apposition and broad cytoplasmic processes (arrowed). (From Spiegelman and Bennett 1974.)

death follows the degeneration of the embryonic ectoderm at day 6.5, although extra-embryonic areas live on for several days. t^{w73}, which is so far unique, fails to form a close association between the trophectoderm and ectoplacental cone and the uterine deciduum at implantation.

Most of the t alleles, however, belong to the t^v, viable-t type. These form tailless compounds with t, and homozygotes are viable, fertile, and have normal tails. However t^{AE5} homozygotes are short-tailed (Vojtiskova *et al.* 1976). Hoshino *et al.* (1979) suggested that Tal (tail anomaly lethal) in the rat may be analagous to the T locus in the mouse.

In summary we may say that the T region is very complex and embraces what probably constitute several loci which may have several different functions or several interrelated functions; the way in which t alleles affect the surface properties of sperm, for instance, may be another aspect of the changes in adhesiveness seen in the T/T embryo.

It is clear that the largest deletions T^{Hp} and T^{Or} have the most drastic effects on the developing embryo and a table of time of effect could be constructed (probably without profit) for the other alleles. It is also clear that the locus acts in some cases as a cell lethal and in others probably as a partial cell lethal which disrupts the relationship of the primitive streak and its derivatives.

Tail short (Ts, Chr 11)

Tail short (Morgan 1950) is a gene of very variable expressivity. On the same inbred strain background Ts heterozygotes are heterogeneous, and changing the genetic background modifies the effects of the gene dramatically; the heterozygote may become totally lethal, show an absent tail and reduced viability, or exhibit extensive normal overlapping.

Deol (1961) described the morphology and development of these mice. The tail length in his stock varied from a stump to seven-eighths normal. The vertebral column has fusions and dyssymphyses, most commonly in the neck and lower thorax (Fig. 5.11). The fourth sacral vertebra is caudalized, and fusions are common in caudal vertebrae, which are irregular. Effects of Ts on the head and limbs are described elsewhere (Chapters 6 and 8).

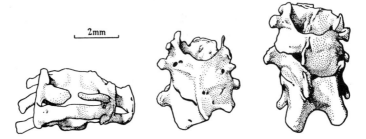

Fig. 5.11.   Abnormalities of (A) cervical and (B,C) thoracic vertebrae in a Ts/+ mouse. (From Deol 1961.)

Ts/Ts is lethal, apparently at about the time of implantation. The Ts/+ embryo was first identified at 8 days on the basis of an abnormal yolk sac containing a smaller than normal number of blood islands. At 9 days litters segregating for Ts are very variable in size, and Ts/+ can be reliably identified as they are anaemic; Deol noted a reduced amount of blood in the vessels of 9-day-old embryos. At 10–11 days the neural tube and notochord are seen to change shape, thickness, and position several times along the length of the tail (Fig. 5.12). At 12–13 days the notochord is much thicker than normal and the neural tube much thinner in the tail; the abnormality of the neural tube also extends to the trunk. The neural canal may be obliterated in places. The neural tube often loses its central position and moves from side to side of the median axis. The notochord also wanders, to such an extent that it may touch the epidermis of the tail. Changes in position of these structures occurred rather rapidly, in the course of a few transverse sections.

Deol suggested that, although anaemia was the first abnormality seen, skeletal effects were not a consequence of it, but a distinct entity. Paterson

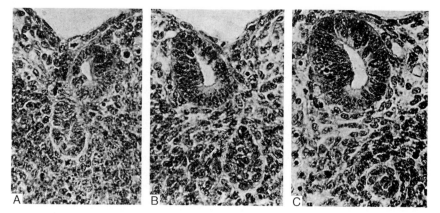

Fig. 5.12. Transverse sections through the tail of a 12-day Ts/+ embryo. (From Deol 1961.)

(1980) used linear descendants of Deol's stock to look at earlier developmental effects. He found that Ts/Ts on this background became abnormal as morulae, and that they could be identified at 3.5 days *post coitum* by low mean cell number (13.7 vs. 30.3), absence of contraction, and a weak cytoplasmic staining with haematoxylin. Nucleolar morphology was also abnormal. At 4.5 days Ts/Ts were still pale-staining, with a reduced cell number, but four out of 12 studied had contracted to form small blastocysts. Inner cell mass and trophoblast were not clearly differentiated however. Paterson agreed with Deol that Ts/Ts dies at around the time of implantation, although abnormalities can be seen earlier, with retardaton of cell division dating from the third cleavage and cell division proceeding at an ever-decreasing rate until 4.5 days, when a few homozygotes start to disintegrate. Pale-staining probably indicates a failure to increase RNA/protein synthesis, a feature of normal embryos at this stage of development. This may also account for the strangely 'ragged' nucleoli.

Matta (1981) found that half the 7-day embryos from litters producing 50 per cent Ts/+, 50 per cent +/+ individuals were retarded and that calculations of the mean tissue volume, mean cell volume, and mean cell number in the three germ layers showed that all cell layers were affected, but the mesoderm most severely. Embryos isolated at 3.5 days had reached the blastocyst stage (30/30 from controls, 48/98 from Ts/+ × Ts/+ matings). Those which had not (26 morula stage, 24 early blastocyst stage) were identified as putative Ts embryos. A day later 25 of these had stopped developing and were still at the morula stage; these were identified as Ts/Ts. Of the others 47 were retarded, but appeared to be developing normally (Ts/+) and 26 developed at the same rate as controls. This is a remarkably good fit to the expected 1:2:1 ratio.

Crinkly tail (cy, Chr 4)

Crinkly tail (Johnson and Wallace 1979) is a recessive with imperfect penetrance. The region between the ninth and fourteenth caudal vertebrae is affected, with one, or rarely two, short abnormal lengths, expressed as one or more misshapen vertebrae. Total vertebral number is unaffected. The abnormality is seen in 11-day embryos when the tail tip, instead of curling ventrad, turns dorsad. There is no sign of spina bifida nor of somite irregularity. Later (13–14 days) the tail is seen to be bent laterally. In sections minor irregularities can often be seen near the tail tip of 11-day-old individuals (Fig. 5.13) when the tissue of the primitive streak is misallocated briefly, so that two tail guts are formed over a short length of tail. This gene adds little to the understanding of tail abnormalities; either the short double tail gut mechanically interferes with the normal near-helix of the embryonic tail, or it indicates a local malaise in the primitive streak.

Looptail (Lp, Chr 1)

Looptail (Strong and Hollander 1949) was considered by Grüneberg (1963)

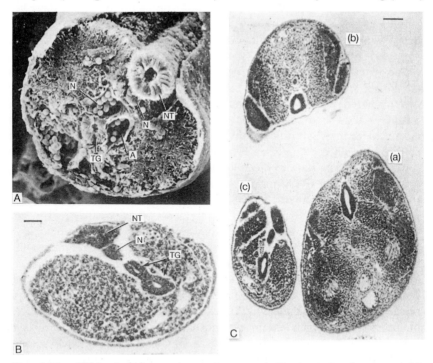

Fig. 5.13. SEM of (A) a cy/cy embryo aged 11 days showing no visible abnormality and (B,C) of 11-, 13-day-old cy/cy embryos showing doubling of the tail gut (TG). A: aorta; N: notochord; NT: neural tube. (From Johnson and Wallace 1979.)

to be a disorder of segmentation. However, there is evidence that this is secondary to a previous abnormality of the primitive streak. When heterozygous this semidominant gene produces a loop or twist in the mid-region of the tail. Accompanying this are various neural abnormalities. Many heterozygotes have enlarged lateral ventricles of the brain (Van Abeelen and Raven 1968) with abnormal surface cells (Wilson and Michael 1975) and slight behavioural abnormalities (Van Abeelen 1968) including head shaking (Strong and Hollander 1949).

Homozygotes have open neural folds in the region of the brain and often complete craniorachischisis is seen. The homozygote is very foreshortened, and the back concave rather then convex. Stein and colleagues (Stein and Rudin 1953; Stein *et al.* 1960) showed an increased mitotic rate in the hindbrain and mesencephalon, but this may be secondary to the non-closure of the neural tube. Wilson and Center (1974) showed that this increase was due to an increase in the length of the cell cycle in Lp/Lp embryos (Fig. 5.14).

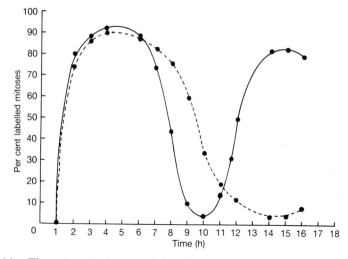

Fig. 5.14. The cell cycle in normal (continuous line) and Lp/Lp (broken line) midbrain at 11 days. (From Wilson and Center 1974.)

In Lp/Lp mice aged 9–9.5 days the shortening and regression of the primitive streak is retarded (Smith and Stein 1962). Normally the streak should shorten by 40 per cent between the time of neural tube closure and the 13–14 somite stage, with further shortening by the time the midgut closes. The failure of this shortening to happen in Lp/Lp embryos seems to result in a lack of the normal elongation of neural tissues and somites, leading to non-closure of the former and abnormality and irregularity of the latter which leads to later skeletal disturbances. The abnormality of the primitive streak could also account for the larger than normal bulk of the looptail notochord and smallness of the somites; some paraxial meso-dermal cells could end up in the notochord. The abnormalities of axial skeleton and ribs of homozygotes, whose thoracic vertebrae have centra abnormal in size, shape, and position and whose neural arches often fuse with those of adjacent vertebrae, are presumably secondary, either to the abnormal primitive streak, somites, or neural tube.

Wilson and Finton (1980) demonstrated that, as in Splotch (Wilson and Finton 1979, and see below), gap junctional vesicles are abnormal in looptail embryos. They are more numerous in the dysraphic hindbrain and lumbosacral spinal cord than in normal controls, but similar in number to normals in the cervicothoracic cord.

Copp and Wilson (1981) looked at the mesoderm adjacent to the dysraphic hindbrain and found that the extracellular matrix and mesen-chyme were separated from the neural basement membrane and showed unusual staining reactions suggesting that a larger than normal amount of chondroitin sulphate was present in the extracellular matrix. The basement membrane of the neural tissue also showed an excess of chondroitin sulphate and other sulphated glycosaminoglycans. The authors suggested that this could indicate insufficient support for the neural folds by the surrounding mesenchyme, and cited this as a possible cause of the failure of the neural tube to close.

Urinogenital (ur, Chr ?)

Dunn and Gluecksohn-Schoenheimer (1947) and Gluecksohn-Waelsch and Kamell (1955) described a mutation leading to shortening of the skull and cleft palate (see Chapter 6) which also involved the remainder of the axial skeleton and the urinogenital system. Fitch (1957) studied the development of ur/ur mice and noted short, wide centra to the vertebrae with a tendency to ossification from twin centres, a fourteenth pair of ribs and hence 14 thoracic vertebrae, and a corresponding reduction in the lumbar spine and irregular attachment of ribs to the sternum.

At 10.5 days the tail somites are short, broad, and unevenly crowded together; ur is also smaller than normal at this stage. Grüneberg (1963) considered that the defect may be traceable to the primitive streak. Unfortunately the gene is now extinct, so no experimental verification of this supposition can be made.

PATHOLOGY OF THE NOTOCHORD

Jurand (1974) studied the development of the normal mouse notochord. The points which are of especial interest here concern the role of the notochord in the further development of the embryo. Jurand pointed out the role of the basal lamina of the notochord, formed at the time of its separation from the archenteron, is to allow the former to become a separate anatomical entity, probably protecting the notochordal cells, helping them to form a coherent organ, and preventing the notochord from adhering to neighbouring structures. These roles are later taken over by the pericordial sheath. It cannot be coincidental that in Sd the notochord and perichordal sheath are abnormal, and that in Brachyury (T) in cell adhesion is abnormal and that the characteristic feature of the surviving individuals in lethal genotypes is abnormal adhesion between the noto-chord and its neighbours, leading to further abnormalities. In Pt and tc the notochord does not seem to be abnormally sticky, but simply degenerates, or is of less than normal size. More modern work thus vindicates the grouping made by Grüneberg (1963) and his suspicions about the adhesiveness of the notochord.

THE UNSEGMENTED PARAXIAL MESODERM

The primitive streak produces undifferentiated mesenchyme stellate cells fairly closely packed which give rise to both notochord and paraxial mesenchyme. Grüneberg (1963) discussed two genes, vestigial tail and bent-tail, as defects of this tissue. However vestigial tail is now known to be an abnormality of the primitive streak (Grüneberg and Wickramaratne 1974) and bent-tail could be either a defect of the paraxial mesoderm, or more likely another primitive streak abnormality. We cannot therefore confidently identify a mutant in which unsegmented paraxial mesoderm is primarily affected.

DISORDERS OF SEGMENTATION AND SCLEROTOME DIFFERENTIATION

We must ask ourselves two questions about the developing somites. First, how do the cells which are to become the somites come to lie as a homogeneous tissue—the sclerotome— around the notochord and neural tube, and secondly how does the regular pattern of loose and dense cell bands necessary for vertebral chondrogenesis, the vertebral blastemata, arise within this tissue? The answers to both these questions can be sought in mutants in which these processes are palpably upset, presumably as a result of mis-interaction between notochord, neural tube, and sclerotome cells.

Origin of the sclerotome

The sclerotome is defined as containing those cells in the somite that will ultimately contribute to the vertebral cartilages. The origin of the mouse sclerotome (Flint 1977) is similar to that of the chick (Langman and Nelson

1968). In the primary mesenchyme, after ingression through the primitive streak, cells are orientated at random. The somite appears as a pseudostratified epithelial wall—the dermatome—with cells and nuclei orientated towards the sclerocoele, a central cavity: considerable organization thus occurs, either generated within the somite, or imposed from without. Langman and Nelson showed that dermatome nuclei only divide when they are adjacent to the sclerocoele, migrating to the periphery before they enter S phase. Some division products escape into the sclerocoele, forming the sclerotome, where little division subsequently takes place. The myotome forms later, also as a product of dividing dermatome cells.

The sclerocoele expands as the bulk of the sclerotome increases: the pseudostratified epithelial wall facing notochord and neural tube is stretched to form a laminar epithelium which eventually breaks down, releasing sclerotome cells into the space between neural tube and dermatome. The originally segmental sclerotomes then unite to form a homogeneous tissue. The mechanism of the breakdown of the dermatome wall adjacent to the neural tube presents an unsolved problem. The environment here enhances the formation of chondroitin sulphate by sclerotome cells, and this may also render the dermatome wall unstable (Lash 1968a,b).

Flint and his co-workers studied mouse somites with the transmission electron microscope, cells migrating from cultured somites by time lapse cine-microscopy, and the behaviour of sclerotome cells by scanning electron microscopy. These studies were made with reference to the amputated mutant in the mouse.

Amputated (am, Chr 8)

Amputated is a recessive, lethal at birth and having major effects on the body axis, which is shortened, and the vertebrae, which are distorted and fused (Flint and Ede 1978).

In the normal somite at 9.5 days Flint and Ede found that sclerotome cells were evenly distributed throughout the sclerocoele, of roughly equal size, and randomly orientated. Numerous long cytoplasmic filopodia stretched from cell to cell. In amputated the cells were unevenly distributed (Fig. 5.15), aggregated in small clumps separated by large intercellular gaps not usually crossed by filopodia. These differences, seen by light microscopy were even more apparent using the scanning electron microscope (Fig. 5.16). Amputated cells have numerous filopodia which form a complex network adhering to the surface of cells or to one another, or can fall back and adhere to themselves, forming intricate knots. Extensive areas of cell contact were seen by transmission electron microscopy. Gap junctions between cells (Trelstad *et al.* 1967) were more

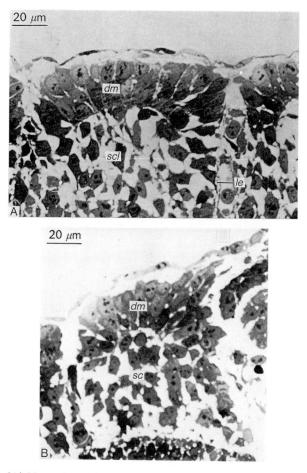

Fig. 5.15. (A) Normal and (B) amputated 9.5-day somites. dm, dermatome; le, laminar epithelium; sc, sclerotome. (From Flint and Ede 1978.)

extensive in am/am. Desmosomes like *macula adherans diminuta* were more common in am than in normal littermates and involved greater lengths of cell membrane: the desmosome gap was also 12–15 nm (cf. 20–5 nm in normals) and the junction rather more complex.

Differences were also found in the presomitic mesoderm adjacent to the primitive streak, with the intercellular space in am consisting of few large spaces, whereas in normal mice there are many small spaces; amputated cells again had extensive areas of cell contact.

The amputated body axis is shortened from the eighth day onwards: the normal number of somites is cut off, each containing a less than normal number of cells. This relative lack of somite cells is not made good in later

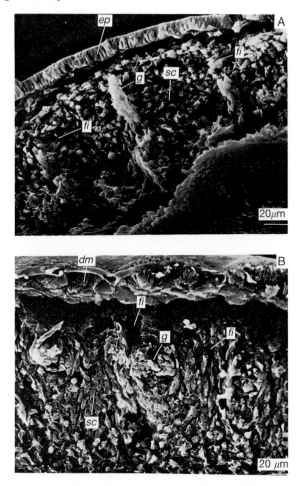

Fig. 5.16. SEM of sclerotome in (A) normal and (B) amputated embryos aged 10 days. dm: dermatome; ep: epithelium; fi: fissure; g: ganglion; sc: sclerotome. (From Flint and Ede 1978.)

stages. Flint and Ede considered the fault to be contained in amputated mesoderm and pointed out the similarity to both the T locus (see above) and talpid[3] (see Chapters 2 and 3).

Flint *et al.* (1978) returned to the problem of specification of somite numbers, which the amputated mouse apparently does at the expense of somite cell number. They found that the most recently formed amputated somite at a given time contained a mean of 134 ± 23 cells against 277 ± 7 in normals. Since somites grow after formation largely by increase in cell number, rather than change in cell density, somite size is clearly regulated at the time of somite formation, by specification of cell number.

Flint (1977) and Flint and Ede (1978) showed that cell mobility in amputated is decreased (the corollary of increased cell adhesion) and supposed that this affects primitive streak regression, and hence somite development. Flint *et al.* (1978) proposed that the length of the unsegmented paraxial presomitic mesoderm specifies somite cell number. They suggested that two bands of paraxial mesenchyme are created on either side of the notochord by the passage of the regressing node of the primitive streak through the primary mesenchyme. Then 'as each new group of cells is recruited into the presomitic mesenchyme band' the cells start to synthesize a morphogen. A computerized model of such a system is presented, showing that somites can be generated. The fault probably lies with the present author, but he is unclear as to what specifies a 'group' of recruited cells—surely the paraxial mesenchyme is a continuum? Nevertheless, the model is an intriguing one.

The talpid³ chick

Ede and Kelly (1964) discussed the axial abnormalities of the talpid³ mutant in the chick. This is of interest, as we know that cell adhesion is also abnormal in this mutant (see Chapter 2). They found a variety of fairly late-occurring abnormalities of the vertebrae secondary to a normal notochord and slightly abnormal neural tube, the latter probably due to mechanical pressure exerted by the surrounding mesenchyme. Somite formation is normal but the further development of the somite is deranged, the myotome and sclerotome cells mingling indiscriminately rather than segregating. Cartilaginous vertebrae are very malformed, and adjacent vertebrae are fused (Fig. 5.17): Ede and Kelly attributed this to the inability of pre-cartilage cells to segregate properly, a condition described elsewhere in this mutant.

Rib-fusions (Rf, Chr ?)

Rib fusions is a semidominant gene (Mackensen and Stevens 1960). The heterozygote is mildly affected and has a reduced viability. The tail may be kinky, a condition only noted when the mouse is 7–10 days old, and 118 out of 139 sampled had some fusion of the ribs. There is occasional fusion between adjacent vertebrae. Homozygotes are severly affected and usually die *in utero*; the remainder are stillborn with a shortened axial skeleton terminating in the sacral region. Little attempt is seen at the formation of individual vertebrae. The bodies of Rf homozygotes are short and wide. They can be identified at 9 days by the absence of somites, or by the presence of only a slight suggestion of their formation, and a wavy neural tube (Theiler and Stevens 1960). The anterior neuropore also fails to close. Aggregations of mesenchymal cells may be seen, but they never achieve

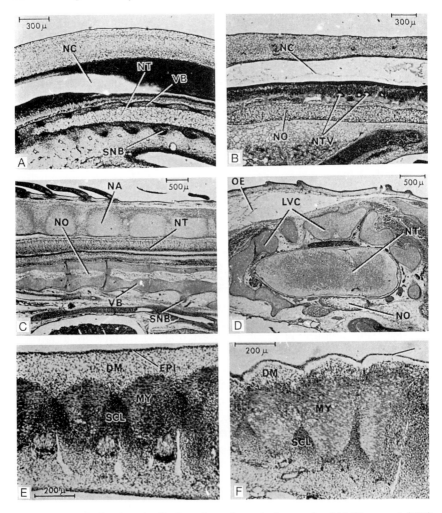

Fig. 5.17. Median longitudinal sections through the trunk of (AC) normal (BD) talpid[3] embryos aged 9 days (D slightly oblique) and vertical longitudinal sections through (E) normal and (F) talpid[3] embryos aged 4.5 days. DM: dermal mesachyme; EPI: epidermis; LVC: lateral vertebral cartilages; MY: myotome; NA: neural arch; NC: neural canal; NO: notochord; NT: neural tube; SCL: sclerotome; SNB: sub notochordal bar; VB: vertebral body. (From Ede and Kelly 1964.)

the continuous epithelial arrangement typical of early somite formation. The sclerotomic boundaries are irregular and the spinal nerves unevenly spaced, with spinal ganglia often united.

In fact the final phenotype of Rf/Rf is very similar to that seen in amputated (Fig. 5.18); small size of somites and virtually no somites at all seem to produce a very similar final phenotype.

Pudgy (pu, Chr 7)

Pudgy (Grüneberg 1961) which is possibly a recurrence of another condition (stub, Dunn and Gluecksohn-Schoenheimer 1942) is a recessive gene which affects only the axial skeleton. The tail is reduced and the vertebral column of the trunk represented by a jumble of irregular vertebrae and hemi-vertebrae (Fig. 5.18). Many vertebral arches are open dorsally, and kyphoses, lordoses, and scolioses are common. An occasional vertebra may be almost normal in shape. Ribs and sternum are correspondingly affected. The somites of 9–11 day pudgy embryos are

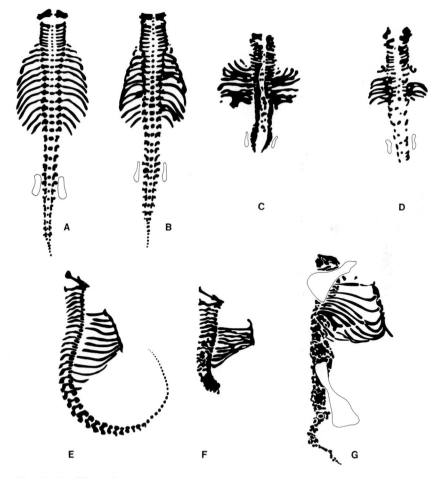

Fig. 5.18. The axial skeletons of amputated, rib fusions, malformed vertebrae, and pudgy. Posterior views: (A) normal; (B) Rf/+; (C) Rf/Rf; and (D) Mv/Mv. Lateral views: (E) normal; (F) an/am; and (G) pu/pu. Various ages.

irregular in shape and size and the width of adjacent segments is not constant; intersegmental fissures are often not parallel. Unsegmented paraxial mesoderm is a little irregular, and in the tail this block of mesoderm is not broken up into somites. A rudimentary segmentation occurs later, at the time of sclerotome formation, but this is usually irregular.

Malformed vertebrae (My, Chr ?)

Theiler *et al.* (1975) described another axial mutation in the mouse. Mv is semidominant, producing a mere tail kink and an occasional fused rib in heterozygotes. The homozygote has a short stub of a tail and a very disturbed axial skeleton and thoracic basket. Homozygotes usually die at around birth, and the rare survivors do not breed. The axial skeleton resembles Rib fusions in having fused ribs and asymmetric fusion of neural arches in cervical, thoracic, and lumbosacral regions. Ossification centres of vertebral bodies are irregular in size and alignment. Ribs are extensively fused and reduced to six to 10 pairs. Tails do not extend much beyond the pelvic girdle. These abnormalities can be traced back to 11 days of gestation when intersegmental vessels are irregular, and the ventral edges of thoracic somites are abnormal. In the cervical region myotomes are irregular, an in the lumbosacral region extensive fusions of somites are seen. At 9 days one potential Mv/Mv was seen to have only one very small cervical somite.

Crooked tail (Cd, Chr 6)

The semidominant crooked gene (Morgan 1954) affects both axial skeleton and dentition (see Chapter 7). Heterozygotes may be unaffected, or have abnormal caudal, sacral, and lumbar vertebrae. Homozygotes show no normal overlapping and are more severely affected. Fifty per cent of homozygotes die early in development. Eighteen per cent of the remainder are exencephalic and die at or around birth. Survivors are small, with microphthalmia, crooked tails, and a nervous movement of the head, as well as tooth and coat defects. Vertebral defects are similar to those of heterozygotes and include wedge-shaped half vertebrae and ossification from twin centres giving scoliosis or, in the tail, kinks. Matter (1957) studied the development of the crooked vertebral column and found somites of unequal size, segmental fissures not parallel, and irregular fusion between adjacent somites (Fig. 5.19 and 5.20). The total number of somites produced is normal, as in amputated (see above) Pycnosis was present in the somites, but the other axial structures (neural tube, notochord) seemed normal, although we must remember the exencephaly present in approximately one-third of crooked embryos.

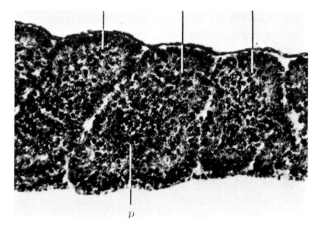

Fig. 5.19. Longitudinal section through the tail of a Cd/+ embryo aged 12.5 days. Note the irregular segmentation and pycnosis (p). (From Matter 1957.)

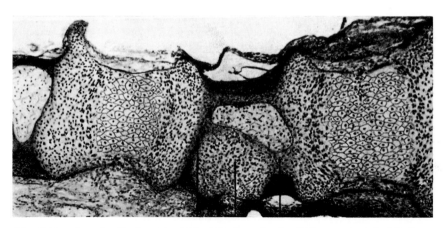

Fig. 5.20. Longitudinal section through the tail of a Cd/+ mouse aged 3 days showing vertebral irregularities. (From Matter 1957.)

Rachiterata (rh, Chr 2)

Varnum and Stevens (1974) and Theiler *et al.* (1974) described a recessive condition producing short tails with distal kinks. Affected mice have a missing or abnormal axis, and fused ribs, thoracic and lumbar vertebrae. Homozygotes have six cervical vertebrae (instead of the normal seven), the last bearing occasional cervical ribs. The malformations of the axial skeleton can be traced back to disturbances in the somites first seen on the eleventh day. Somites at 11 days are irregular, and shortened in the anteroposterior axis or fused in the thoracolumbar region. At 12 days the sclerotome 'halves' are fused and spinal nerves interconnected. The atlas and axis appear normal at this age, but the *anlagen* of only six cervical vertebrae are seen. The abnormality of the axis was traced to the intercalation of a supernumerary neural arch between atlas and axis which Theiler considered not to be secondary to the somite irregularity. It seems likely, however, that the condition is best explained as a result of suppression of a cervical vertebral body rudiment, the arch persisting and causing later modification of the arch of the axis.

MUTATIONS AFFECTING THE NEURAL TUBE

In another group of abnormalities defects of the vertebrae can be seen to be the probable sequelae of pre-existing defects of the neural tube. Of course we shall probably wrongly place some primitive streak defects into this category, but the following mutants have no described defect other than in the developing neural tissue.

Curly tail (ct, Chr ?)

Curly tail represents that bane of the geneticist, the recessive gene with irregular manifestation. Grüneberg (1954*a*) was not quite sure that a single major gene rather than a constellation of minor ones exists, because on most backgrounds overlaps (normal-looking mice breeding as ct/ct) are commoner than manifesting abnormals. Embury *et al* (1979) supported the idea of a single gene. Recessivity was shown by outcrossing: no abnormal mice were seen in the F_1. Six generations of breeding from normal overlaps within a homozygous curly tail stock did not reduce the incidence of phenotypically curly-tailed mice.

The phenotypic expression of ct (when it can be seen) is similar to looptail (see above) with tail twists, reduction of the tail to a coiled 'bun', and a massive spina bifida, or smaller spina bifida occulta. Occasional exencephaly is also seen. In earlier stages there is a lumbosacral spina bifida which may extend to the tail, and the tail neural tube continues to grow out as a flat plate: the growth of ventral tail structures is not affected, and the tail curls dorsally into a bun or twist. Sometimes a persistently open posterior neuropore can be seen at earlier stages. This was further investigated by Copp *et al.* (1982) who were able to ascertain the status of the posterior neuropore in a given embryo by injecting a small quantity of toluidine blue into the closed thoracolumbar portion of the neural tube. Caudally dye was lost from the tube if the neuropore was open. Visual examination was not an accurate means of classification (Fig. 5.21). They found that posterior neuropore closure started between 30 and 34 somites and in control embryos (A/Strong strain) 90 per cent had closed at the 34-

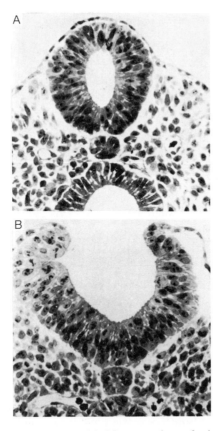

Fig. 5.21.   Transverse plastic embedded 2 μm sections of ct/ct embryos. Although both individuals (A,B) appeared to have closed neural tubes on visual inspection, dye injection and serial sectioning demonstrated the open neural tube in (B). (From Copp *et al.* 1982.)

somite stage and 100 per cent at the 39-somite stage. In the ct stock 57 per cent of embryos produced from phenotypically abnormal parents still had open neuropores at 39 somites compared to 36 per cent of embryos from parents genetically ct/ct but phenotypically normal. Sixty per cent of embryos from ct × ct matings later developed tail flexions.

The persistently open neuropore was not a consequence of general retardation, as Copp *et al.* were able to show. The tail defect in ct is thus likely to be a specific consequence of late neuropore closure, or of an earlier defect which brings about this failure of the neuropore to close.

In abnormal ct tails the neural tube and its canal occupied more space than in straight tails (Fig. 5.22), although the total neurectodermal tissue was not increased. Clearly the bulk of the neural tube is increased at

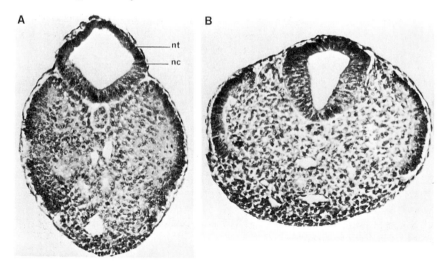

Fig. 5.22. Comparable sections through (A) ct/ct embryo with flexion defect, (B) ct/ct embryo with a straight tail. Note the bulk of the neural tube and its canal in (A). (From Copp *et al.* 1982.)

flexures. Copp *et al.* considered that the tail defect is consequent upon this local overexpansion of neural tube; delayed posterior neuropore closure together with continued normal development of structures in the tailbud impose mechanical stresses on the developing trunk and tail, leaving a flexion deformity. In some cases neural fold closure in the lumbosacral region is rendered impossible by the imbalance in rates of development between neural folds and non-neural structures and spina bifida results.

Curly tail spina bifida resembles that seen in man in a number of ways. A marked excess of females is found amongst exencephalic ct mice, together with polyhydramnios, high levels of amniotic fluid alphafetoprotein (Adinolfi *et al.* 1976), and distinctive rapidly adhering cells in the amniotic fluid (Embury *et al.* 1979). Vitamin A has a larger teratogenic effect on ct than on normal mice (Seller *et al.* 1979) on day 8, but the situation is reversed on day 9. Seller (1983). noted that the incidence of neural tube defects is decreased by exposure to certain teratogens which also decrease DNA synthesis and hence the rate of cell division. These findings are discussed further in Chapter 9.

Shaker short tail (sst, Chr ?) and shaker short (st, Chr ?)

This recessive mutation (Wahlsten *et al.* 1983) produces an abnormality of the tail, which either ends bluntly with a thin terminal filament or is shortened. A fluid-filled bleb is also seen in some cases in the parietal region of the skull. Survivors to the age of 10 days (and survival is rather poor, especially in mice with blebs) develop an abnormal righting reflex and are found to have abnormalities of the cerebellum.

This phenotype resembles that of shaker short (st, Dunn 1934), now extinct. Dunn described these mice at birth as having either a very short tail, or a three-quarter length tail ending in a slender filament and a lesion near the parieto-occipital suture. Older mice showed 'erratic circus movements' and ataxia of the head. Bonnevie (1936) described the origin of the head bleb, and defects of the inner ear leading to deafness.

From the point of view of the axial skeleton we must await a fuller description of the development of sst. However, the presence of abnormalities of the brain and the absence of ventral pigmentation (sst/sst has a white belly spot) indicates a probable neural tube defect: we obviously cannot guess as to the involvement or otherwise of the notochord and/or primitive streak.

Splotch (Sp, Chr 1)

This locus is rather mutable (Dickie 1964). The semidominant splotch mutant (Russel 1947) is named from white belly and tail spotting in the heterozygote. Homozygotes die at 13 days *in utero* with malformations resembling those seen in curly tail, rachischisis in the lumbosacral region and overgrowth of the neural tissue of the hindbrain, reduction or absence of spinal ganglia and their derivatives, and abnormal tail morphology. Auerbach (1954) showed that the neural crest of Sp/Sp individuals, together with adjacent regions, failed to develop pigment when transplanted to chick coelom or the anterior chamber of the mouse eye, suggesting a neural crest defect. In fact this defect probably extends to the whole neural tube. Wilson (1974) showed that the increased numbers of mitotic figures seen in the Sp/Sp neural tube were a consequence of a longer time spent in division as part of a lengthened cell cycle—a very similar result to that seen in looptail (Wilson and Center 1974). Posterior and anterior neuropore closure is delayed in Sp/Sp (Dempsey and Trasler 1983). Wilson and Finton (1979) observed gap junctional vessels in the lumbosacral region of Sp/Sp embryos aged 9 days—a region of the neural tube which is likely to remain open subsequently. The significance of these structures, normally regarded as a means of cellular communication, is not clear. However, Wilson and Finton cited several studies implicating them in loss of cellular contact and these may be pertinent to non-closure of the neural tube.

Dempsey and Trasler (1983) looked at the early morphology of Sp/Sp, Sp/+, and +/+ littermates of 15–24 somites and found good evidence of delay in closure of the neural tube in splotch embryos. Homozygotes had a delay in closure of both anterior and posterior neuropores and 50 per cent of heterozygotes showed delayed closure of the anterior neuropore.

Morris and O'Shea (1983) looked at the neuroepithelial development of splotch mice using TEM and SEM (Fig. 5.23). They found that in the region of non-closure of the neural tube there was degeneration and malalignment of neuroepithelial cells. Neuroepithelial cell processes made lateral progress in these areas rather than maintaining their normal contacts with the luminal surface. Intercellular space was considerably increased and there were many ectopic cell processes in basal regions of affected areas. It is suggested that these changes in cell–cell adhesion could be the visible result of the previously described increase in cell cycle time and increase in number of gap junctional vesicles in splotch.

A less extreme allele, delayed splotch (Sp^d, Dickie 1964) has caudal rachischisis only when homozygous, and survives to birth.

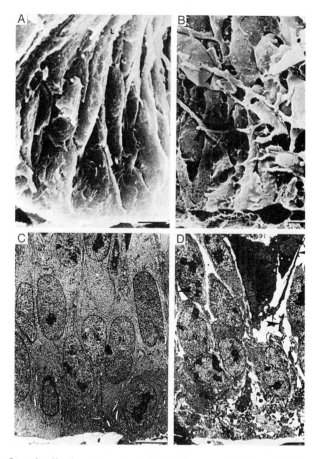

Fig. 5.23. Longitudinal sections through the neuroepithelium of the hindbrain of (A) normal and (B) Sp/Sp embryos. (C,D) TEM through similar areas. (From Morris and O'Shea 1983.)

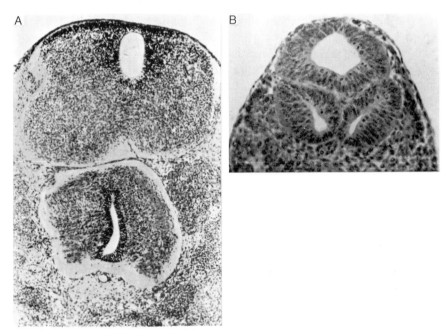

Fig. 5.24. Fused embryos (A) aged 14.5 days and (B) 13 days, respectively, showing duplication of the neural tube. (From Theiler 1959*b*; Theiler and Gluecksohn-Waelsch 1956.)

Fused, kinky, and knobbly (Fu, FU^Kb, Chr 17)

Fused, its allele kinky, and knobbly, which may also be allelic, are situated very close to the T region, so close in fact that t^h20, a deletion, covers the Kb locus (Lyon and Bechtol 1977). Fused (Reed 1937), the original allele, is semidominant and variable in expression. Homozygotes and heterozygotes may have kinked tails or be normal. Offspring of Fu/+ or Fu/Fu mothers are less liable to express the character than those of +/+ sibs. Homozygotes and heterozygotes occasionally show circling and are deaf (Dunn and Caspari 1945). The tails of fused individuals are shortened and kinked, and bifurcated. One is reminded of T^Hp, but the bifurcation here is in the lateral rather than the vertical plane although Theiler (1959*b*) illustrated a fused embryo with vertically split neural tube and one of those illustrated by Theiler and Gluecksohn-Waelsch is equivocal (Fig. 5.24). Thoracic and lumbar fusion of vertebrae is also found, with occasional lumbosacral or caudal spina bifida. Embryos (Theiler and Gluecksohn-Waelsch 1956) can be seen to be abnormal from the nine-day stage with irregular folding of the neural tube from the level of the fifth or sixth somite; in the tail the neural tube tends to form ventral diverticulae which run parallel to the neural tube proper and distally the tube may branch into up to four parallel structures which run on to the tail tip and may be associated with a forked tail. The notochord is said not to be involved. The somites may be irregular next to irregular regions of the neural tube, but many neural tube abnormalities are not accompanied by irregularities of the somites. The abnormality of the neural tube is therefore likely to be primary.

Kinky (Fu^{ki}) is very similar but the defects present have been traced to an earlier stage. Dunn and Caspari (1945) found several possible cross-overs between fused and kinky but Dunn and Gluecksohn-Waelsch (1954) found none, and the two are now considered alleles. Kinky heterozygotes are similar to those of fused (Caspari and David 1940) but bifurcations of the tail are rare. Many show abnormalities of the inner ear (Deol 1966). Kinky homozygotes are inviable (Gluecksohn-Schoenheimer 1949) and show twinning and tissue hyperplasia on day seven: the condition is lethal at day eight to ten. The branching of the neural tube in fused may represent a low-grade twinning abnormality.

Knobbly (Lyon 1977) has a kinky tail and is lethal at the small mole stage of embryogenesis when homozygous.

Duplicatus posterior

Duplicatus posterior was first described by Danforth (1925). More recent studies (Danforth and Center 1967) suggested that at least two genes are involved. Duplicatus posterior specimens may have two rudimentary extra hind legs, and a duplication of the phallus, anus, and urethal orifice (Fig. 5.25). The condition is included here because of the findings of Center (1969). The earliest identified anomaly was a 10 day old embryo with a bifid tail. This proved to have a duplicated tail gut, but normal neural tube and notochord. At 11 days embryos with duplicated rear limb buds also had duplicated neural tubes: duplication of the notochord was not seen. At 12.5 and 14 days the story was similar, with duplication of the neural tube but not the notochord. The tails of older embryos (15.5 days) showed overgrowth and distortion of a retained tail gut (the tail gut has normally regressed by this stage).

Center suggested that subdivision of the neural tube may occur anterior to the posterior neuropore and this may lead to duplication of posterior structures; a low grade of this anomaly, perhaps happening a little later, could lead only to transitional duplication of the tail neural tube and gut. She reminds us of the duplicated neural tube in fused and kinky and the fact that no duplications of axial structures other than the neural tube have been seen in this mutant, although both fused and duplicatus posterior disrupt the development of the urinogenital system.

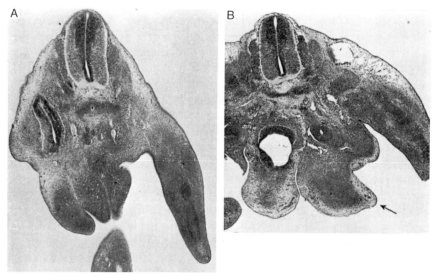

Fig. 5.25. (A) Duplicatus posterior embryo aged 14 days and showing an accessory neural tube. (B) In a 12.5-day embryo an accessory hindlimb (arrow) and accessory genital papilla were present. (From Center 1969.)

Screwtail (sc, Chr ?)

This gene, now extinct, was described by MacDowell *et al.* (1942) and Bryson (1945). Homozygotes have widespread abnormalities of the skeleton which include short, kinked tails and defective vertebral centra, which may result in rather sharp kinks in the thoracic or lumbar spine and a short, broad, unsegmented sternum. Abnormalities of screwtail teeth and jaws are described in Chapter 7. Bryson described the poor growth of the ribs which fails to bring the bilateral precursors of the sternum into opposition and the failure of subsequent sternal segmentation.

FURTHER DIFFERENTIATION OF THE SCLEROTOME

The primitive metameric segmentation of the somites is not strictly reflected in the vertebrae, which develop between the primitive somites. The lack of exact correlation between somite and vertebra has lead to many theories of vertebral development, extensively reviewed by Verbont (1976). Most of these devolve about the concept of a resegmentation or *Neugliederung* of material so that vertebral boundaries are determined by a shift of half a segment in comparison with somite boundaries.

Although a full review of this is unecessary (Verbont having done this so well) a basic understanding of this theory is necessary to understand the development of the vertebrae.

Remak (1855) first postulated the Neugliederung hypothesis. He identified somites—*Urwirbeln*—from which both musculature and axial skeleton would develop, and the notochord, which he also thought to be involved in vertebral development, although he did not know what part it played. He correctly noted that the axial skeleton was first seen lateral to the neural tube, progressing later to the axial region. He also identified the '*Ruckentafel*' or muscle plate (myotome) and inner '*Wirbelkernmasse*' (sclerotome). The *Wirbelkernmasse* then developed into two zones, the caudal part, which condensed into an opaque zone to become vertebral arch and rib and the cranial part which became transparent and 'developed' into spinal nerve, ganglia, and nerve fibres.

In the midline Remak described the origins of the vertebral bodies as segmental thickenings of the blastemata around the notochord to form primitive vertebral bodies separated from each other by narrow fissures. The metamerism of these bodies coincided with that of the somites. The development of bodies and arches is thus such that the spinal nerve belonging to each segment is cranial to the vertebral arch, exactly the opposite of the final situation when the arch meets the vertebral body cranially and the nerve is caudal. Remak 'solved' this paradox when he observed the appearance of dark lines in the centre of the primitive vertebral bodies, and the disappearance of the former fissures between them. The new dark lines were the *anlagen* of vertebral discs, marking the new division between vertebrae.

Kolliker (1861, 1884) saw the primitive vertebral bodies enclosing the notochord and spinal cord as an unsegmented mass, interrupted only by spinal ganglia, and from which the vertebral bodies later developed. The axial blastemata was not therefore resegmented, but merely segmented along the definitive vertebral boundaries.

Froriep (1883, 1886) denied the necessity for resegmentation, since the arch primordia run obliquely to insert anteriorly into the vertebral body. These findings, however, are based on the cervical region only and revolve around the '*Spange*', the *anlagen* of the cervical ribs; his theory is thus not general.

In 1888 Von Ebner gave a simplified view of vertebral development; he saw a double segmentation occurring at an early stage in ring snake embryos. This was achieved by the appearance of a fissure (of Von Ebner) at the midpoint of the somite. He assumed (incorrectly) that this fissure would allow a resegmentation like that proposed by Remak. Von Ebner thought that tissue cranial to this fissure was incorporated into the skeleton of the vertebra ahead, whilst the darker tissue of the caudal half became the anterior part of the vertebra behind, crucially different from the ideas of Remak. Von Ebner went on to use the presence of his fissure to reject the idea of an unsegmented parablast around the notochord and to suggest (again incorrectly) that the intervertebral fissure was concerned with the development of intervertebral joints.

Von Ebner pointed out that the use of different fixations produced a different appearance in the fissure, but was inclined to see the *absence* of the fissure under certain conditions of fixation as an artefact. Corning (1891) thought similarly; it was not until much later that the *presence* of the fissure was considered artefactual.

The influence of Von Ebner's ideas was immense, perhaps because they were simple enough to be readily understood. The search for, and the finding of the intervertebral fissure in many other species followed swiftly. The distinction between lateral and axial events was lost sight of. Despite the work of such champions of the axial blastema as Williams (1908), a modification of Von Ebner's views, as exemplified by Bardeen (1900, 1905*a*,*b*, 1908), became received wisdom; the somite separates into dermatome, myotome, and sclerotome. The sclerotome separates into a caudal condensed (dark) zone and a cranial, looser (light) zone, separated by Von Ebner's fissure. Recombination of these 'sclerotomites' leads to an alternation of muscular and skeletal tissue. The relationship between arch and vertebral body, rib and disc is obscure.

This concept was retained, with various modifications, such as those of Sensenig (1949), who suggested that Von Ebner's fissure was not quite at the division between light and dark tissue. Williams (1959) and Wake (1970) took resegmentation as an established fact. Blechschmidt (1957) objected to the idea of recombination between tissue blocks but was inconsistent in considering the origin of Von Ebner's fissure as due to movement between blocks of tissue. He did, however, reinstate the very real difference between axial and lateral structures. Baur (1969) also denied resegmentation. Based on material from man, other mammals, reptiles, and birds he suggested the following scheme.

Tissue originating from the somites fuses into a uniform, continuous blastema in which no segmentation is visible, not even into sclerotomes, a concept which he rejects. The vertebra arises in the unsegmented tissue and is in its definitive position from its first formation. *Neugliederung* does not occur. This renders superfluous the problem of the relationship of vertebral body, intervertebral disc, vertebral arch, and ribs to the sclerotome. Baur, like Flint (1977) argued that the fissure of Von Ebner is an artefact.

Verbont (1976) also rejected resegmentation, on the basis of a large series of sheep embryos. His rejection revolved largely about the different developmental processes found laterally and in the midline. The earliest phases of development in Verbont's material show an unsegmented phase in somitic mesoderm. Later segmentation is found in lateral areas only and alternation is provided by tissue surrounding the peripheral spinal nervous system (perineural mass). The 'sclerotome' thus represents only part of the skeletal tissue of each somite, the remainder originating axially and non-segmentally. The vertebral column arises as an unsegmented axial mass, segmented later by a process divorced from lateral segmentation in both time and space. Neither the fissure of Von Ebner nor the median intersegmental fissure was ever seen by Verbont.

The amputated (am) mouse and sclerotome differentiation

Flint (1977) argued similarly that, since the dorsal root ganglia are well developed within the sclerotome prior to the appearance of banding and since these structures and their associated ventral nerves are massive, the passage of these structures through the developing sclerotome is enough to account for the appearance of dense and loose bands in the sclerotome. Like Verbont he considered the fissure of Von Ebner to be artefactual.

It should be noted, however, that Keynes and Stern (1984) showed that the directing of the growth of nerve axons through the anterior part of each somite can be shown to be a property of the somitic mesoderm and not of the neural tube. If a portion of the neural tube is rotated through 180°, nerves still grow through the anterior half of each somite; if the *mesoderm* is rotated through 180° the path of nerves is changed so that they run through the originally rostral tissue. This suggests that there is a difference between anterior and posterior somite tissue which can be read by developing nerves.

Normal dorsal root ganglia remain separate and distinct structures throughout their development. In amputated they are practically fused caudally from the tenth day of development onwards. There is good correspondence between these fusions and distributions between light and dark sclerotome bands and later fusions between vertebrae. Ventral nerves associated with dorsal root ganglia exactly separate both the regular sclerotomes of the normal mouse and the irregular sclerotomes of amputated. Flint was also able to demonstrate dense tissue between ventral nerve and myotome which is not accounted for on Sensenig's (1949) model; if dense tissue arises as a consequence of mechanical stress this tissue is in exactly the right spot.

Flint argued that the induction of cartilage is heralded by the amplification of the synthetic pathway leading to chondroitin sulphate synthesis and that this is initiated by the neural tube, especially the ventral part. Ganglionic differentiation is similarly promoted by the neural tube. The last cells to leave the neural crest and enter the somite become the dorsal root ganglion. Norr (1973) showed that differentiation of these cells is induced by the somite which in turn is conditioned by the ventral neural tube. Four mouse genes have been suggested to involve errors in resegmentation; if we abandon this concept we must look again at their development.

Tail-kinks (tk, Chr 9)

Tail-kinks homozygotes (Grüneberg 1955*b*) are recognized by their shortened kinky tails. The whole vertebral column is involved, with tail cervical and thoracic regions most affected. Although vertebral number is normal the pelvic girdle is a little displaced, so that the presacral vertebral number is reduced. Cervical and upper thoracic vertebrae are in two or three parts with bodies separated from neural arches, which often fail to meet in the midline; neural arches of consecutive vertebrae may be fused dorsally (Fig. 5.26). From T4 on, the articular processes of the vertebrae are absent or represented as separate entities but on the whole the column is more normal. Ribs, especially the superior ones, are also affected. The tail consists of a mixture of normal and abnormal vertebrae in irregular sequence.

Fig. 5.26. Dorsal, lateral, and ventral views of the cervical region of a tk/tk mouse. (From Grüneberg 1955*b*.)

Embryologically at 10 days the formation of dense banding in the sclerotomes is delayed. Grüneberg interpreted this as a failure of the *Neugliederung*: if we follow Verbont and abandon resegmentation then we can suggest merely that sclerotome development is delayed. At 11–12 days the mesenchymal condensations of the vertebrae are reduced, especially the perichordal discs which give rise to the primitive intervertebral discs and contribute to the vertebrae. At 13 days there is secondary ingress of cerebrospinal fluid into the pia-arachnoid space. At 14 days the tail can first be seen to be shortened.

Undulated (un, Chr 2)

un homozygotes (Wright 1947) have an axial skeleton abnormal along the whole of its length. There is often a large kyphosis at the level of T10. Vertebral processes are reduced throughout the axial skeleton and vertebrae tend to ossify from twin centres (Grüneberg 1950). These changes and the reduction in size of tail vertebrae seem to be the result of small vertebral precursors. At 11 days mesenchymal condensations are reduced in size and the sclerotomic fissure is indistinct. Grüneberg (1954b) suggested that material ahead of this indistinct fissure is misallocated, and does not contribute to the posterior part of the developing vertebra but is retained in the intervertebral disc. This would explain the characteristic reduction of the caudal parts of the vertebrae and, as this material also contributes to transverse processes and ribs, vertebral arches, and spinous processes, vertebral anomalies elsewhere in the spine. However, the explanation relies on the theory of resegmentation; if we abandon this, then the detailed explanation of the syndrome must be rethought.

Flexed tail (f, Chr 13)

Flexed tail (Hunt *et al.* 1933) is regarded primarily as a gene causing a transitory embryonic anaemia. Some individuals (but not all) have tail defects dating from day 14 (Kamenoff 1935) when cartilage cells on the periphery of the developing intervertebral discs fail to become fibrous; cartilage thus persists and ossifies uniting some vertebrae. The link between the anaemia and the failure of the cartilage cells to differentiate into fibrocartilage (if any) is unknown.

Diminutive (dm, Chr 2)

The association of tail-kinks and anaemia might be dismissed as coincidental in flexed mice, but for the appearance of a similar association in diminutive (Stevens and Mackenson 1958). Diminutive homozygotes have short kinked tails, usually with an additional presacral vertbera and extra ribs at the cervical or lumbar end of the thorax. The vertebrae are malformed and ribs may be fused. They also have a macrocytic anaemia (Bannerman *et al.* 1973) due to a bone marrow which contains only half the normal number of colony-forming units. We have already noted that Ts embryos are anaemic due to the formation of small blood islands (see above).

OTHER CONDITIONS

Besides the genes discussed above many others affect the mouse tail, and have not been considered in detail either because they have not been fully described, or because involvement of the axial skeleton seems trivial. To complete any catalogue of tail mutants in the mouse the reader should consult the entries for Hk, jy, kw, mea, ol, Q, st, tc, and us in Green (1981).

AXIAL ABNORMALITIES—AN OVERVIEW

The loci discussed here are not the only ones which affect the axial skeleton but they do allow us to draw some conclusions as to development. Many known genes affect the tail because a crooked mouse stands out in a litter; for the same reason many coat colour genes are known. When the involvement extends to the higher parts of the vertebral column there is potential clinical interest; even if the gene is only manifested in the tail, the processes involved are of interest in pinpointing the possible anomalies of the vertebral column.

Perhaps the most interesting pointer to a basic mechanism is in amputated, talpid, and T where we see again the influence of abnormal cell adhesion on development. We have already seen the effects of this on other parts of the skeleton, and it perhaps should not surprise us to find it here also.

If we arrange the mouse genes into a table (Table 5.1) we can see that they group quite readily according to the structures involved. T, vt, Lp, cy, pr, for instance, all affect the primitive streak, and, secondarily, the involvement is seen in its derivatives, the neural tube, notochord, and tail gut. We may imagine here perhaps a mechanical cause for skeletal upset, since neither sclerotomes nor somites are described as abnormal, and these are known to depend upon a normal primitive streak for normal development.

In a second group of disorders, including am, tc, Rf, pu, rh, Cd, the somites are the first structures seen to be abnormal: this invariably leads to upsets in the sclerotomes and consequent vertebral abnormalities. In this group the neural tube is sporadically involved.

In a third group, ct. Sp, Fu, Fu^{ki}, Fu^{Kb}, and duplicatus posterior, the neural tube is consistently mentioned as abnormal; in Sd and Pt the notochord seems to be at fault. In neither group has the anomaly been traced back further to the primitive streak.

In the fourth and last group, a small one including tk and un, sclerotome differentiation seems to be abnormal.

We may thus see a pattern, although not necessarily a profound one. If the sclerotome alone is deficient, then sclerotome formation may be the primary upset. We must distinguish this from genes where abnormal sclerotomes are the result of pre-existing somite abnormalities. Similarly the notochord or neural tube may be involved in the legacy of abnormal vertebrae in its own right, or as a result of a pre-existing primitive streak defect.

The existence of normal vertebrae interspersed with very abnormal ones (as in tail kinks for example) is remarkable. One explanation is that of Moore and Mintz (1972). They looked at chimeras formed between two inbred strains of mice and found that in some cases a single vertebra could

TABLE 5.1. *Structures affected by various tail genes in the mouse*

Mutant	Primitive streak	Neural tube	Notochord	Tail gut	Somites	Sclerotomes
vt	+	+	+	+		
T	+	+	+	+		
Lp	+	+	+			
cy	+			+		
pr	+					
Sd			+			
Pt			+			
tc					+	+
am					+	+
pu					+	+
Mv					+	+
rh					+	+
Rf			+		+	+
Cd			+		+	
ct		+				
Sp		+				
Fu, Fu^{k1}, Fu^{Kb} Q						
Ts		+		+		
Dup post.		+		(+)		
Bt		?				
tk						+
un						+

be identified as belonging to one or other of the parental strains. In other cases the unit seemed to be a left or right half vertebra, since some were asymmetrical. In yet other cases vertebrae were intermediate between parental strains. Moore and Mintz interpreted these findings as suggesting that a quarter vertebra is formed from a single progenitor cell or clone. In fact they offer no formal proof that the unit is smaller than a half vertebra. No evidence is offered to suggest that anterior and posterior halves of a single vertebra are of different strain origin; the idea of four clones per vertebra clearly owes much to the classical theory of vertebral resegmentation. However the idea that a half vertebra may be formed from a clone of cells explains why half vertebrae may be abnormal, and why occasional normal vertebrae are found interspersed as units in an otherwise abnormal vertebral column.

The situation in man

The chief clinical interest in axial defects in man concerns spina bifida, usually associated with an abnormal neural tube. The formation of the neural tube by the meeting of the neural folds ceases in all species at the posterior neuropore. The tube at this stage is still relatively short, and further extension is by one of two processes. In rodents, opossum, and pig, extension is by intrinsic growth, although in the latter, with its somewhat reduced tail this process occurs only after the closure of the posterior neuropore (Hughes and Freeman 1974). In man and the chicken (Lemire 1973; Criley 1969) the neural tube is prolonged caudally by the addition of new cells; between these irregular cavities arise, to give the appearance of a forked neural tube, or one with several lumina. Tails in various species may thus vary in (a) the method of prolongation of the lumen and (b) the position of the posterior neuropore. More importantly human embryos may have a transient phase with an irregular neural tube. This was described by Ikeda (1930) and Tarlow (1938); Lendon and Emery (1970) undertook a quantitative study. They found that in a sample of 11 necropsies of children aged mainly 0–1 year, from which cases of CNS malformation had been excluded, 45 per cent had a major forking of the cord at some level. A common major anomaly was the dorsoventral or lateral duplication of the central canal. Ikeda (1930) had found an incidence of 34 per cent forking. Emery and Lendon (1973) looked at a further series of 100 cases of meningomyelocele and found that the commonest abnormality cranial to the open plaque was a total or partial duplication of the cord; such cords had either double canals (5 per cent) or partial or total duplication (31 per cent). Caudal to the plaque the corresponding figures were 25 and 27 per cent. Lendon and Emery suggested a proportion of children express a tendency to cord duplication and that in extreme cases this extends more cranially than usual and perhaps leads to spina bifida. The relatives of children with meningomyelocele often have spina bifida occulta (Lorber and Levick 1967). Naik *et al.* (1978) looked at the vertebrae and ribs of children with meningomyelocele and found that absence of twelfth ribs, fusion of vertebral arches and bodies, and hemivertebrae were common at levels other than that of the lesion.

Hughes and Freeman (1974) pointed out that the spontaneous incidence of spina bifida in man may be as high as 3.6 per thousand (Elwood 1972) and that the chick is susceptible to defects of the spinal cord or sacrum caused by teratogens (Romanoff 1972). They suggested than many of these may arise by the persistence of several neural cavities in the tail region and suspected that at least some human conditions of late onset may be similarly formed.

REFERENCES

Adinolfi, M., Beck, S., Polani, P. E., and Seller, M. J. (1976). Levels of alpha-fetoprotein in amniotic fluids of mice (curly tail) with neural tube defects. *Journal of Medical Genetics* **13**, 511–13.

Auerbach, R. (1954). Analysis of the developmental effects of a lethal mutation in the house mouse. *Journal of experimental Zoology* **127**, 305–29.

Bannerman, R. M., Edwards, J. A., and Pinkerton, P. H. (1973). Hereditary disorders of the red cell in animals. *Progress in Haematology* **8**, 131–79.

Bardeen, C. R. (1900). Costo-vertebral variation in man. *Anatomische Anzeiger* **18**, 377–82.

—— (1905*a*). The development of thoracic vertebrae in man. *American Journal of Anatomy* **4**, 163–74.

—— (1905*b*). Studies on the development of the human skeleton. *American Journal of Anatomy* **4**, 265–302.

—— (1908). Early development of the cervical vertebrae and the base of the occipital bone in man. *American Journal of Anatomy* **8**, 181–6.

Baur, R. (1969). Zum Problem der Neugliederung der Wirbelsäule. *Acta anatomica* **72**, 321–56.

Bennett, D. (1958). In vitro study of cartilage induction in T/T mice. *Nature* **181**, 1286.

—— (1975). The T-locus of the mouse. *Cell* **6**, 441–54.

—— (1978). Rescue of a lethal T/t locus genotype by chimaerism with normal embryos. *Nature* **272**, 539.

—— and Dunn, L. C. (1969). Genetical and embryological comparisons of semilethal t alleles from wild mouse populations. *Genetics* **61**, 411–22.

—— —— Spiegelman, M., Artzt, K., Cookingham, J., and Schermerhorn, E. (1975). Observations on a set of radiation induced dominant T-like mutations in the mouse. *Genetical Research* **26**, 95–108.

Berry, R. J. (1960). Genetical studies on the skeleton of the mouse. XXVI. Pintail. *Genetical Research* **1**, 439–51.

—— (1961). Genetically controlled degeneration of the nucleus pulposus in the mouse. *Journal of Bone and Joint Surgery* **43B**, 387–93.

Blechschmidt, E. (1957). Die Entwicklungsbeweglungen der Somiten and ihre Bedeutung für die Gliederung der Wirbelsäule. *Zeitschrift für Anatomie und Entwickslungsgeschichte* **120**, 150–72.

Bonnevie, K. (1936). Vererbhare Gehirinanomalie bei kurzschwänzigen Tranzmäusen. *Acta Pathologica et microbiologica scandanavica, Suppl*, **26**, 20–7.

Bryson, V. (1945). Development of the sternum in screw tail mice. *Anatomical Record* **91**, 119–41.

Caspari, E. and David, P.R. (1940). The inheritance of a tail abnormality in the house mouse. *Journal of Heredity* **31**, 427–31.

Center, E. M. (1969). Morphology and embryology of duplicatus posterior mice. *Teratology* **2**, 377–88.

—— Spiegelman, S. S., and Wilson, D. B. (1982). Perinotochordal sheath of heterozygous and homozygous Danforth's short tail mice. *Journal of Heredity* **73**, 299–300.

Chesley, P. (1935). Development of the short tailed mutant in the house mouse. *Journal of experimental Zoology* **70**, 429–59.

Cooper, G. W. (1965). Induction of somite chondrogenesis by cartilage and notochord: a correlation between inductive activity and specific stages of cytodifferentiation. *Developmental Biology* **12**, 185–212.

Copp, A. J., Seller, M. J., and Polani, P. E. (1982). Neural tube development in mutant (curly tail) and normal mouse embryos: the timing of posterior neuropore closure. *Journal of Embryology and experimental Morphology* **69**, 151–67.

Copp, S. N. and Wilson, D. B. (1981). Cranial glycosaminoglycans in early embryos of the looptail (Lp) mutant mouse. *Journal of Craniofacial Genetics and Developmental Biology* **1**, 253–60.

Corning, H. K. (1891). Uber die sogenannte Neuliederung der Wirbelsäule und über das Schicksal de Urwirdbelhöhle bei Reptielen. *Morphologisches Jahrbuch* **17**, 611–22.

Criley, B. (1969). Analysis of the embryonic sources and mechanisms of development of posterior levels of chick neural tubes. *Journal of Morphology* **128**, 465–502.

Danforth, C. H. (1925). Hereditary doubling suggesting anomalous chromatin distribution in the mouse. *Proceedings of the Society for experimental Biology and Medicine* **23**, 145–7.

—— and Center, E. M. (1967). Genetical and embryological basis of the duplicatus posterior manifestation in the mouse. *Genetics* **56**, 554.

Dawes, B. (1930). The development of the vertebral column in mammals, as illustrated by its development in *Mus musculus. Proceedings of the Royal Society* **B218**. 115–70.

Dempsey, E. E. and Trasler, D. G. (1983). Early morphological abnormalities in Splotch mouse embryos and predisposition to gene and retinoic acid induced neural tube defects. *Teratology* **28**, 461–72.

Deol, M. S. (1961). Genetical studies on the skeleton of the mouse. XXVIII. Tail short. *Proceedings of the Royal Society* **155**, 78–95.

—— (1966). The probable mode of gene action in circling mutants of the mouse. *Genetical Research* **7**, 363–71.

Dickie, M. M. (1964). New splotch alleles in the mouse. *Journal of Heredity* **55**, 97–101.

Dobrovolskaia-Zavadskaia, N. (1927). Sur la mortification spontanée de la chez la souris nouveau-née et sur l'existence d'un charactere héréditaire 'non viable'. *Comptes rendus des Séances de la Société de Biologie* **97**, 114–16.

Dunn, L. C. (1934). A new gene altering behaviour and skeleton in the house mouse. *Proceedings of the National Academy of Sciences USA* **20**, 230–2.

—— and Caspari, E. (1954). A case of neighbouring loci with similar effects. *Genetics* **30**, 543–68.

—— and Gluecksohn-Schoenheimer, S. (1942). Stub, a new mutation in the mouse. *Journal of Heredity* **33**, 235–9.

—— —— (1947). A new complex of hereditary abnormalities in the house mouse. *Journal of experimental Zoology* **104**, 25–51.

—— —— and Bryson, V. (1940). A new mutation in the mouse affecting spinal column an urogenital system. *Journal of Heredity* **31**, 343–8.

—— and Gluecksohn-Waelsch, S. (1954). A genetical study of the mutation 'fused' in the house mouse, with evidence concerning its allelism with a similar mutation 'kink'. *Journal of Genetics* **53**, 383–91.

Ede, D. A. and Kelly, W. A. (1964). Developmental abnormalities in the trunk and limbs of the talpid[3] mutant of the fowl. *Journal of Embryology and experimental Morphology* **12**, 339–56.

Elwood, J. H. (1972). Major cns nervous system malformations notified in Northern Ireland 1964–1968. *Developmental Medicine and Child Neurology* **14**, 731–9.

Embury, S., Seller, M. J., Adinolfi, M., and Poloni, P. E. (1979). Neural tube defects in curly tail mice. I, Incidence, expression, and similarity to the human condition. *Proceedings of the Royal Society* **B206**, 85–94.

Emery, J. L. and Lendon, R. G. (1973). The local cord lesion in neurospinal dysraphism (meningomyelocele). *Journal of Pathology* **110**, 83–96.

Erickson, R. P., Lewis, S. E., and Slusser, K. S. (1978). Deletion mapping of the t complex of chromosome 17 of the mouse. *Nature* **274**, 163–4.

Fitch, N. (1957). An embryological analysis of two mutants in the mouse both producing cleft palates. *Journal of experimental Zoology* **136**, 329–57.

Flint, O. P. (1977). Cell interactions in the developing axial skeleton in normal and mutant mouse embryos. In *Vertebrate limb and somite morphogenesis* (ed. D. A. Ede, J. R. Hinchliffe, and M. Balls), pp. 463–84. Cambridge University Press.

—— and Ede, D. A. (1978). Cell interactions in the developing somite: in vivo comparisons between amputated (am/am) and normal mouse embryos. *Journal of Cell Science* **31**, 275–91.

—— —— Wilby, O. K., and Proctor, J. (1978). Control of somite number in normal and amputated mutant mouse embryos: an experimental and a theoretical analysis. *Journal of Embryology and experimental Morphology* **45**, 189–202.

Froriep, A. (1883). Zur Entwicklungsgeschichte der Wirbelsäule, insbesondere des Atlas und Epistopheus und der Occipitalregion. I. Beobachtungen am Hühner-embryonen. *Archiv für Anatomie und Physiologie* 177–234.

—— (1886). Zur Entwicklungsgeschichte der Wirbelsäule, insbesondere des Atlas und der Occipitalregion. II. Beobachtungen am Säugerierembryonen. *Archiv für Anatomie und Physiologie*, pp. 69–150.

Garber, E. D. (1952a). Bent-tail a dominant, sex linked mutation in the mouse. *Proceedings of the National Academy of Sciences, USA* **38**, 876–9.

—— (1952b). A dominant, sex linked mutation in the house mouse. *Science* **116**, 89.

Geyer-Duszynska, I. (1964). Cytological investigation on the T locus in *Mus musculus* L. *Chromosoma* **15**, 478–502.

Gluecksohn-Schoenheimer, S. (1938). The development of two tail-less mutants in the house mouse. *Genetics* **23**, 573–84.

—— (1945). The embryonic development of mutants of the Sd strain in mice. *Genetics* **30**, 29–38.

—— (1949). The effects of a lethal mutation responsible for duplications and twinning in mouse embryos. *Journal of experimental Zoology* **110**, 47–76.

Gluechsohn-Waelsch, S. and Kamell, S. A. (1955). Physiological investigations of a mutation in mice with pleiotropic effects. *Physiological Zoology* **28**, 68–73.

—— and Rota, R. T. (1963). Development in organ tissue culture of kidney rudiments from mutant mouse embryos. *Developmental Biology* **7**, 432–44.

Granholm, N. H. and Johnson, P. M. (1978). Identification of eight cell t^{w32} homozygous lethal mutants by aberrant compaction. *Journal of experimental Zoology* **203**, 81–8.

Green, M. C. (1981). *Genetic variants and strains of the laboratory mouse*. Fischer Verlag, Stuttgart, New York.

Grüneberg, H. (1950). Genetical studies on the skeleton of the mouse. II. Undulated and its 'modifiers'. *Journal of Genetics* **50**, 142–73.

—— (1953). Genetical studies on the skeleton of the mouse. VI. Danforth's short tail. *Journal of Genetics* **51**, 317–26.

—— (1954a). Genetical studies on the skeleton of the mouse. VIII. Curly tail. *Journal of Genetics* **52**, 52–67.

—— (1954b). Genetical studies on the skeleton of the mouse. XII. The development of undulated. *Journal of Genetics* **52**, 441–55.

—— (1955*a*). Genetical studies on the skeleton of the mouse. XV. Relations between major and minor variants. *Journal of Genetics* **53**, 515–35.

—— (1955*b*). Genetical studies on the skeleton of the mouse. XVI. Tail-kinks. *Journal of Genetics* **53**, 536–50.

—— (1955*c*). Genetical studies on the skeleton of the mouse. XVII. Bent-tail. *Journal of Genetics* **53**, 551–62.

—— (1956). A ventral ectodermal ridge of the tail in mouse embryos. *Nature* **177**, 787–8.

—— (1957). Genetical studies on the skeleton of the mouse. XIX. Vestigial tail. *Journal of Genetics* **55**, 181–94.

—— (1958*a*). Genetical studies on the skeleton of the mouse. XXII. The development of Danforth's short tail. *Journal of Embryology and experimental Morphology* **6**, 124–48.

—— (1958*b*). Genetical studies on the skeleton of the mouse. XXIII. The development of brachyury and anury. *Journal of Embryology and experimental Morphology* **6**, 424–43.

—— (1961). Genetical studies on the skeleton of the mouse. XXIX. Pudgy. *Genetical Research* **2**, 384–93.

—— (1963). *The pathology of development.* Blackwell, Oxford.

—— and Wickramaratne, G. A. de S. (1974). A re-examination of two skeletal mutants, vestigial tail (vt) and congenital hydrocephalus (ch). *Journal of Embryology and experimental Morphology* **31**, 207–22.

Heston, W. E. (1951). The vestigial tail mouse. A new recessive mutation. *Journal of Heredity* **42**, 71–4.

Hillman, N. and Hillman, R. (1975). Ultrastructural studies of t^{w32}/t^{w32} mouse embryos. *Journal of Embryology and experimental Morphology* **33**, 685–95.

Hollander, W. F. and Strong, L. C. (1951). Pintail, a dominant mutation linked to brown in the house mouse. *Journal of Heredity* **42**, 179–82.

Hoshino, K., Oda, S., and Kamegama, Y. (1979). Tail anomaly lethal Tal: a new mutant gene in the rat. *Teratology* **19**, 27–34.

Hughes, A. F. and Freeman, R. B. (1974). Comparative remarks on the development of the tail cord among higher vertebrates. *Journal of Embryology and experimental Morphology* **32**, 355–63.

Hunt, H. R., Mixter, R., and Permar, D. (1933). Flexed-tail in the mouse *Mus musculus*. *Genetics* **18**, 335–66.

Ikeda, Y. (1930). Beitrage zur normalen und abnormalen Entwicklungsgeschichte des caudalen Abschnittes des Ruckenmarlies bei menschlichen. *Zeitschrift für Anatomie und Entwicklungsgeschichte* **92**, 380–430.

Jacobs-Cohen, R. J., Spiegelman, M., and Bennett, D. (1983). T/T somite mesoderm is able to differentiate into cartilage in vitro. *Cell Differentiation* **12**, 219–23.

Jaffe, J. (1952). Cytological observations concerning inversion and translocation in the house mouse. *American Naturalist* **86**, 101–4.

Johnson, D. R. (1974). Hairpin-tail: a case of postreductional gene action in the mouse egg? *Genetics* **76**, 795–805.

—— (1975). Further observations on the hairpin-tail (T^{hp}) mutation in the mouse. *Genetical Research* **24**, 207–13.

—— (1976). The interfrontal bone and mutant genes in the mouse. *Journal of Anatomy* **121**, 507–13.

—— and Wallace, M. E. (1979). Crinkly-tail, a mild skeletal mutant in the mouse. *Journal of Embryology and experimental Morphology* **53**, 327–33.

Jurand, A. (1974). Some aspects of the development of the notochord in mouse embryos. *Journal of Embryology and experimental Morphology* **32**, 1–33.

Kamenoff, R. J. (1935). Effects of the flexed tail gene on the development of the house mouse. *Journal of Morphology* **58**, 117–55.

Keynes, R. J. and Stern, C. D. (1984). Segmentation in the vertebrate nervous system. *Nature* **310**, 786–9.

Klein, J. and Hammerberg, C. (1977). The control of differentiation by the T complex. *Immunological Reviews* **33**, 70–104.

Kolliker, A. (1861). *Entwicklungsgeschichte des Menschen und der höheren Thiere.* Engleman, Leipzig.

—— (1884). *Entwicklungsgeschichte des Thiere*, 2nd edn. Engleman, Leipzig.

Langman, J. and Nelson, G. R. (1968). A radiographic study of the development of the somite in the chick embryo. *Journal of Embryology and experimental Morphology* **19**, 217–26.

Lash, J. W. (1968a). Chondrogenesis: genotypic and phenotypic expression. *Journal of Cellular and Comparative Physiology* **72**, (Suppl. 1), 35–46.

—— (1968b). Somite mesenchyme and its response to cartilage induction. In *Epithelial–Mesenchymal interactions* (ed. R. Fleischmajer and R.E. Billingham), pp. 165–72. Williams & Wilkins, Baltimore.

Lemire, R. J. (1973). Variations in development of the caudal neural tube in human embryos (Horizons XIV–XXI). *Teratology* **2**, 361–70.

Lendon, R. G. and Emery, J. L. (1970). Forking of the central canal in the equinal cord of children. *Journal of Anatomy* **106**, 499–505.

Lorber, J. and Levick, R. K. (1967). Spina bifida cystica: incidence of spina bifida occulta in parents and controls. *Archives of Diseases in Childhood* **42**, 171–3.

Lyon, M. F. (1959). A new dominant T allele in the house mouse. *Journal of Heredity* **50**, 140–2.

—— (1977). *Mouse News Letter* **56**, 37.

—— and Bechtol, K. B. (1977). Derivation of mutant t-haplotypes of the mouse by presumed duplication or deletion. *Genetical Research* **30**, 63–76.

MacDowell, E. C., Potter, J. S., Loanes, T., and Ward, E. N. (1942). The manifold effect of the screw tail mouse mutation. *Journal of Heredity* **33**, 439–49.

Mackenson, J. A. and Stevens, L. C. (1960). Rib-fusions, a new mutation in the mouse. *Journal of Heredity* **51**, 264–8.

MacNutt, W. (1967). Porcine tail, a new mutation in the house mouse. *Anatomical record* **157**, 286.

—— (1969). Developmental anomalies associated with the porcine (pr) gene in the mouse. *Anatomical Record* **163**, 340.

Matta, C. A. (1981). Genetic background and the effects of the gene Tail-short in the mouse. PhD thesis, University of London.

Matter, H. (1957). Die formale Genese einer vererbten Wirbelsäulenmissbildung am Beispiel der Mutante Crooked-tail der Maus. *Revue Suisse de Zoologie* **64**, 1–38.

McLaren, A. (1976). Genetics of the early mouse embryo. *Annual Review of Genetics* **10**, 361–88.

Mintz, B. (1964). Formation of genetically mosaic mouse embryos and early development of 'lethal (t^{12}/t^{12})-normal' mosaics. *Journal of experimental Zoology* **157**, 273–92.

Morgan, W. C. (1950). A new tail short mutation in the mouse. *Journal of Heredity* **41**, 208–15.

—— (1954). A new crooked tail mutation involving distinctive pleiotropism. *Journal of Genetics* **52**, 354–73.

Moore, W. J. and Mintz, B. (1972). Clonal model of vertebral column and skull development derived from genetically mosaic skeletons in allophenic mice. *Developmental Biology* **27** 55–70.

Morris, G. L. and O'Shea, K. S. (1983). Anomalies of neuropithelial cell associations in the Splotch mutant embryo. *Developmental Brain Research* **9**, 408–10.

Moutier, R. (1973). *Mouse News Letter* **49**, 42.

Naik, D. K., Lendon, R. G. and Barson, A. J. (1978). A radiographical study of vertebral and rib malformations in children with myelomeningocele. *Clinical Radiology* **29**, 427–30.

Norr, S. C. (1973). In vitro analysis of sympathetic neuron differentiation from chick neural crest cels. *Developmental Biology* **34**, 16–38.

Paavola, L. G., Wilson, D. B., and Center, E. M. (1980). Histochemistry of the developing notocord, pericordal sheath and vertebrae in Danforth's short tail (Sd) and normal C57B1/6 mice. *Journal of Embryology and experimental Morphology* **55**, 227–45.

Paterson, H. F. (1980). In vivo and in vitro studies of the early embryonic lethal tail short (Ts) in the mouse. *Journal of experimental Zoology* **21**, 247–56.

Pennycuick, P. R. (1980). Total and regional vertebral numbers and lumbo-sacral morphology in mice. Strain differences and the effects of the brachyury gene. *Journal of Heredity* **71**, 93–9.

Reed, S. C. (1937). The inheritance and expression of fused, a new mutation in the house mouse. *Genetics* **22**, 1–13.

Remak, R. (1855). *Untersuchungen über die Entwicklung der Wirbelthiere* Reimer, Berlin.

Romanoff, A. (1972). *Pathogenesis of the avian embryo.* Wiley-Interscience, New York.

Russel, W. L. (1947). Splotch, a new mutation in the house mouse *Mus musculus.* *Genetics* **32**, 102.

Searle, A. G. (1966). Curtailed, a new dominant T-allele in the house mouse. *Genetical Research* **7**, 86–95.

Seller, M. J. (1983). The cause of neural tube defects: some experiments and a hypothesis. *Journal of Medical Genetics* **20**, 164–8,

—— Embury, S., Adinolfi, M., and Polani, P. E. (1979). Neural tube defects in curly tail mice. II. Effects of maternal administration of Vitamin A. *Proceedings of the Royal Society of London* **B206**, 95–107.

Sensenig, E. C. (1949). The early development of the human vertebral column. *Contributions to Embryology* **33**, 23–41.

Sherman, M. I. and Wudl, L. R. (1977). T complex mutations and their effects. In *Concepts in mammalian embryogenesis* (ed. M. I. Sherman), pp. 136–234. MIT Press, Cambridge Massachusetts.

Smith, L. J. and Stein, K. F. (1962). Axial elongation in the mouse and its retardation in homozygous looptail mice. *Journal of Embryology and experimental Morphology* **10**, 73–87.

Spiegelman, M. (1976). Electron microscopy of cell associations in T locus Mutants. In *Embryogenesis in mammals.* Ciba Foundation Symposia **40**, 199–266.

—— and Bennett, D. (1974). Fine structural study of cell migration in the early mesoderm of normal and mutant mouse embryos (T locus t^9/t^9). *Journal of Embryology and experimental Morphology* **32**, 723–38.

Stein, K. F., Lievre, F., and Smaller, C. G. (1960). Abnormal brain differentiation in the homozygous looptail embryo. *Anatomical Record* **136**, 324–5.

—— and Rudin, I. A. (1953). Development of mice homozygous for the gene for looptail. *Journal of Heredity* **44**, 59–69.

Stevens, L. C. and Mackensen, J. A. (1958). The inheritance of a mutation in the mouse affecting blood formation, the axial skeleton and body size. *Journal of Heredity* **49**, 153–60.

Strong, L. C. and Hollander, W. F. (1949). Hereditary loop tail in the house mouse, accompanied by imperforate vagina and with craniorachischisis when homozygous. *Journal of Heredity* **40**, 329–34.

Tarlow, I. M. (1938). Structure of the filum terminale. *Archives of Neurology and Psychiatry, Chicago* **40**, 1–17.

Theiler, K. (1957). Boneless tail, ein recessives autosomales Gene der Hausmaus. *Archiv der Julius Klaus-Stiftung für Vererbungsforschung, Sozianthropologie und Rassenhygiene* **32**, 474–81.

—— (1959a). Anatomy and development of the 'truncate' (boneless) mutation in the mouse. *American Journal of Anatomy* **104**. 319–43.

—— (1959b). Schwanzmutanten bei mausen. *Zeitschrift für Anatomie und Entwicklungsgeschichte* **121**, 155–64.

—— (1961). Genetisch bedingte Choraschadigungen bei der Maus. *Archiv der Julius-Klaus Stiftung für Vererbungsforschung, Sozianthropologie und Rassenhygiene* **36**, 118–125.

—— and Gluechsohn-Waelsch, S. (1956). The morphological effects and the development of the fused mutation in the mouse. *Anatomical Record* **125**, 83–104.

—— and Stevens, L. C. (1960). The development of rib-fusions, a mutation in the house mouse. *American Journal of Anatomy* **106**, 171–83.

—— Varnum, D. S., Southard, J. L., and Stevens, L. C. (1975). Malformed vertebrae: a new mutant with the 'Wirbel–Rippen' syndrome in the mouse. *Anatomy and Embryology* **147**, 161–6.

—— —— and Stevens, L. C. (1974). Development of rachiterata, a mutation in the house mouse with 6 sacral vertebrae. *Zeitschrift für Anatomie und Entwicklungsgeschichte* **145**, 75–80.

Trelstad, R. L., Hay, E. Z., and Revel, J. P. (1967). Cell contact during early morphogenesis in the chick embryo. *Developmental Biology* **16**, 78–106.

Van Abeelen, J. H. F. (1968). Behavioural ontogeny of looptail mice. *Animal Behaviour* **16**, 1–4.

—— and Raven, S. M. J. (1968). Enlarged ventricles in the cerebrum of loop tailed mice. *Experientia* **24**, 191–2.

Varnum, D. S. and Stevens, L. C. (1974). Rachiterata: a new skeletal mutation on chromosome 2 of the mouse. *Journal of Heredity* **65**, 91–3.

Verbout, A. J. (1976). A critical review of the 'Neugliederung' concept in relation to the development of the vertebral column. *Acta Biotheoretica* **25**, 219–58.

Vojtiskova, M., Viklicky, V., Voracova, B., Lewis, S. E., and Gluecksohn-Waelsch, S. (1976). The effects of a t-allele (t^{AE5}) in the mouse on the lymphoid system and reproduction. *Journal of Embryology and experimental Morphology* **36**, 443–51.

Von Ebner, V. (1888). Urwirbel und Neugliederung der Wirbelsäule. *Sitzungsberichte der Akademie der Wissenschaften Wein* **III/97**, 194–206.

Wahlsten, D., Lyons, J. P., and Zagaja, W. (1983). Shaker short tail, a spontaneous neurological mutant in the mouse. *Journal of Heredity* **74**, 421–5.

Wake, D. B. (1970). Aspects of vertebral evolution in the modern Amphibia. *Formo et Functio* **3**, 33–60.

Williams, E. E. (1959). Gadow's arcualia and the development of tetrapod vertebrae. *Quarterly Review of Biology* **34**, 1–32.

Williams, L. W. (1908). The later development of the notochord in mammals. *American Journal of Anatomy* **8**, 251–84.

Wilson, D. B. (1974). Proliferation in the neural tube of splotch (Sp) mutant mouse. *Journal of Comparative Neurology* **154**, 249–56.

—— and Center, E. M. (1974). The neural cell cycle in the looptail (Lp) mutant mouse. *Journal of Embryology and experimental Morphology* **32**, 697–705.

—— and Finton, L. A. (1979). Gap junctional vesicles in the neural tube of the Splotch (Sp) mutant mouse. *Teratology* **19**, 337–40.

—— —— (1980). Early development of the brain and spinal cord in dysraphic mice: a transmission electron microscopic study. *Genetical Research* **7**, 86–95.

Wilson, D. B. and Michael, S. D. (1975). Surface defects in ventricular cells of brains of mouse embryos homozygous for the looptail-gene: scanning electron microscopic study. *Teratology* **11**, 87–98.

Wittman, K. S., Krupa, P. L., Pesetsky, I., and Hamburgh, M. (1972). Electron microscopy and histochemistry of tail regression in the Brachyury mouse. *Developmental Biology* **27**, 419–24.

Wrathall, C. R. (1974). Cytological analysis of seven dominant genes affecting the neural tube in mice. *Journal of Heredity* **65**, 58–9.

Wright, M. E. (1947). Undulated: a new genetic factor in *Mus musculus* affecting the spine and tail. *Heredity* **1**, 137–41.

Wudl, L. R., Sherman, M. I., and Hillman, N. (1977). Nature of lethality of t mutations in embryos. *Nature* **270**, 137–40.

Yanagisawa, K. O. and Fujimoto, H. (1977). Differences in rotation mediated aggregation between wild type and homozygous Brachyury (T) cells. *Journal of Embryology and experimental Morphology* **40**, 277–83.

—— and Urushihara, H. (1981). Effects of brachyury (T) mutation on morphogenetic movement in the mouse embryo. *Developmental Biology* **87**, 242–8.

—— and Kitamura, K. (1975). Effects of the brachyury (T) mutation on mitotic activity in the neural tube. *Developmental Biology* **47**, 433–8.

6. The skull, face, and palate

DEVELOPMENT OF THE HEAD REGION

Mutations affecting the head region may be divided for our purposes into four groups, those affecting the face, the palate, the nose, and the skull. In order to understand the effect of these mutants a brief summary of the development of these regions is necessary.

FACE, PALATE, AND NOSE

The face

The face develops around a shallow ectodermal depression, the stomodeum, closed by the buccopharyngeal membrane. The latter later ruptures to give continuity between the foregut and the amniotic cavity. The stomodeum is surrounded by a number of mesodermal swellings (Fig. 6.1). Cranially is the unpaired frontal prominence; laterally are two pairs of swellings derived from the ventral parts of the first pharyngeal arches—the maxillary and mandibular processes. The maxillary swellings lie above the stomodeum and abut on to the frontal process. The mandibular pair lie below the stomodeum and unite in the midline.

In the frontal prominence above the stomodeum bilateral nasal placodes can be distinguished. These each become flanked by a medial and a lateral nasal swelling which unite above and below the placode, so forming a shallow depression, the nasal pit. The medial nasal swellings will become the middle parts of the nose and upper lip, the premaxilla and the primary palate, and the lateral nasal swellings will form the alae of the nose.

The nasal pits are temporarily continuous ventrally with the stomodeum, and are then closed off as medial and lateral nasal processes unite beneath them. The line of fusion between the lateral nasal process and the maxillary process is continuous with that between the maxillary and frontal processes and thus runs from upper lip to the inner corner of the developing eye as the nasolacrimal groove, later to become the nasolacrimal duct.

Primary palate

The medial nasal swellings eventually meet in the midline, as a result of differential growth, and fuse to form the intermaxillary segment (Fig. 6.2). This contributes to three major structures. A labial part forms the philtrum

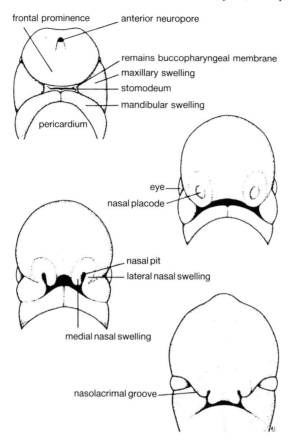

Fig. 6.1. The development of the face.

of the upper lip, an upper jaw component carries the four upper incisor teeth, and a palatal part forms the primary palate. The latter is a triangular horizontal shelf with its apex directed posteriorly which separates the nasal pits above from the stomodeum below. Along the midline of its upper surface it is continuous with the centre of the frontal process; this continuation later becomes relatively thinner and contributes to the nasal septum.

Secondary palate

This originates as paired shelf-like outgrowths which appear on the deep surface of each maxillary process (Fig. 6.2). When first formed these

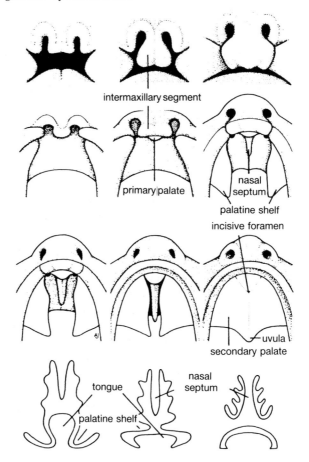

Fig. 6.2.   The development of the primary and secondary palate.

palatine processes are directed medially and downwards on either side of the developing tongue. Later they reorientate so as to become horizontal, from the posterior extremity forwards. Fusion occurs between the free edges of the palatine processes in the midline, between each process and the free edge of the primary palate anteriorly, and between the newly formed secondary palate and the infereior margin of the nasal septum. The point of union of primary and secondary palate is marked by the incisive foramen.

Before the formation of the secondary palate the oronasal membrane (separating nasal pits from pharynx) ruptures to form the primitive choanae. With the formation of the secondary palate this communication is more posterior, via the definitive choanae.

The skull

The skull is best regarded as comprising two units, the chondrocranium whose elements first develop in cartilage and the dermal component (skull vault) whose bones ossify directly in mesenchyme. The chondrocranium of mammals comprises the cranial base and capsules surrounding the inner ears and nasal organs. The cranial vault and upper facial skeleton, apart from some bones around the nose, are made up of dermal bone.

THE MUTANTS

Mutations acting in the head region may affect the skeleton via a series of different routes. First a group of mutants is identifiable which grossly affect the maxillary arch (and possibly other pharyngeal arches as well). This leads either to the formation of split-faced individuals (when the maxillary arches fail to meet in the midline as in Ts, Ph, and the ta^3 chick) or mice with grossly abnormal facial regions due either to the reduction or excessive size of the maxillary arch (far, pc, Xt). Second, and perhaps an extension of this group, are dancer and twirler, with regular cleft lip and palate. Third, a group of inbred strains also show a low incidence of sporadic cleft lip and palate [CL(P)] in the absence of any single major locus. The frequency of this CL(P) may be enhanced by teratogens. Face shape is also altered in a series of more subtle ways.

Fourth, we find a heterogeneous group of mutants and inbred lines with cleft palate alone, in the apparent absence of any change in shape of the pharyngeal arches or the face.

The fifth and final group is one in which the proportions of the skull are affected, and which may or may not lead secondarily to a cleft palate. For example the skull proportions of all chondrodysplastic mutants are changed but only in some cases is a cleft present. This modification of skull proportions is perhaps best considered as a low grade of abnormality which may or may not exceed the threshold beyond which clefting occurs.

MAJOR ABNORMALITIES OF THE FACIAL PROCESSES

Extra-toes (Xt, Chr 13)

Extra-toes (Johnson 1967) is a semi-dominant gene with multiple effects, only some of which concern us here. The homozygote, which dies at around the time of birth, can first be recognized at nine days of gestation by overgrowth of the first pharyngeal arch. By 10 days the arch has divided into maxillary and mandibular portions, the former being much enlarged and the latter of normal size. The external nasal processes are reduced in size. By 13 days the maxillary process partially or totally covers the eyes, and the ectoderm of its anterior face is thicker than normal. A little later the enlarged maxillary region carries six rather than five rows of follicles for mystacial vibrissae, with more follicles per row than usual (Fig. 6.3). Some Xt/Xt individuals are exencephalic. The increased size of the maxillary arch leads to abnormalities in the skull. The lower jaw is apparently normal, and the palate closed, although the nasal chamber is misshapen. In an allele of extra-toes, brachyphalangy (Xt^{bph}, Johnson 1969), the nasal processes were noted to be widely separated in some homozygotes with unilateral or bilateral cleft lip. Some heterozygotes had a central face bleb. The proportion of these individuals was increased by treatment with trypan blue (Johnson 1970). A similar phenotype which shares many abnormalities with Xt was described by Scott (1937, 1938) in the guinea pig.

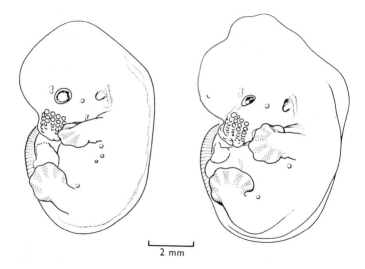

Fig. 6.3. Camera lucida drawings of normal (left) and Xt/Xt (right) embryos aged 13 days. (From Johnson 1967.)

It seems clear that overgrowth of the maxillary portion of the first pharyngeal arch is the primary culprit in this group of abnormalities, and leads on to certain of the other defects by purely mechanical means; the large maxillary processes overgrow the eye and impinge on the space normally occupied by the nasal cavity. As to this reason for this outgrowth there is clearly an upset in the ectodermal–mesodermal interaction in this mutant, the clearest clue being the overdeveloped limb buds (see Chapter 8). It may well be that this upset extends to the interaction between ectoderm and ectomesenchyme in the developing head. Another point of view is that abnormalities of brain and CNS in this mutant which often lead to non-closure of the neural tube could extend to the neural crest which supplies ectomesenchyme for the pharyngeal arches. Johnson (1967, 1969) noted that both extra-toes and brachyphalangy produce belly spots—an indication of possible involvement of the trunk neural crest.

Patch (Ph, Chr 5)

Patch is a semidominant gene producing an extensive belly spot when heterozygous and lethal in its homozygous form (Grüneberg and Truslove 1960). The only known skeletal effect in the heterozygote is on the interfrontal bone (see below). The homozygote can first be recognized at nine days by a wavy neural tube and irregularities of the somites. Most homozygotes at this stage also have inflated hearts, or normal hearts within an inflated pericardium. Many 9-day-old Ph homozygotes have accumulations of clear fluid flanking the notochord and enlarged dorsal aortae and umbilical arteries. At 10–11 days this waterlogging is even more evident and some embryos simply have a wavy neural tube, while others are grossly retarded with wavy neural tubes, irregular somites and often an enormously inflated heart/pericardium. Grüneberg and Truslove described one embryo at this age, which was sectioned, as having a bleb which had just started to form in the concavity between the nasal pits. 'Once formed, the bleb becomes a mechanical hinderance to the movement of the two halves of the nose towards the midline. The cleft face phenotype found later [Fig. 6.4] thus results from the prevention of movement towards the midline.'

At 12–13 days superficial blebs are found elsewhere on the embryo as well and the central face bleb runs under the epithelium of the primitive palate to finish somewhere in front of the pituitary. There is a cleft palate. In older embryos the cleft face individuals are uniform in appearance (Fig. 6.5), with the two halves of the nose widely separated from each other by a cleft occupied by a bleb. Large paired blebs also flank the neural tube. Grüneberg and Truslove concluded that the essential abnormality off patch is an excess of 'water content (hydrops)' developing at 8–9 days.

Erickson and Weston (1983) noted that patch homozygotes can first be identified by the failure of the neural tube to close and the appearance of fluid-filled blebs lateral to the neural tube. They report that Ph neural crest cells migrate, into these fluid-filled spaces, several days earlier than in normal embryos.

Truslove (1977) described a second allele patch-extended (Ph^e) with a more severe effect on pigmentation in the heterozygote. One split-faced Ph^e/Ph^e survived until birth, and one split-face was found in embryos aged 14 days derived from the mating Ph/+ × Ph^e/+, a putative Ph/Ph^e individual.

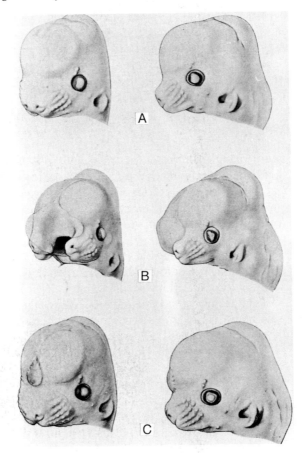

Fig. 6.4. Heads of 13-day embryos from the Patch stock. (A) Normal; (B) Ph/Ph with cleft face; (C) ? Ph/+ with face bleb. (From Grüneberg and Truslove 1960.)

Tail short (Ts, Chr 11)

The split-face phenotype has also been reported in Ts/+ individuals (Matta 1981). In his study of the mutant Morgan (1950) noted that when Ts/+ individuals were outcrossed to a number of inbred strains the F_1 varied greatly. Matta outcrossed Ts/+, long kept on its original BALB/c background, to strong A, C57BL, C3H, and to normal mice from the patch stock (Grüneberg and Truslove 1960) in such a way that the background was sequentially varied by small amounts. Most of the effects of Ts/+ were described by Deol (1961; see Chapter 5), but interestingly Matta's outcrosses also produced a split-faced phenotype similar to patch (Fig. 6.6) Matta found no correlation between split-face and exencephaly, the latter being most frequent on 50 per cent C3H background where the incidence

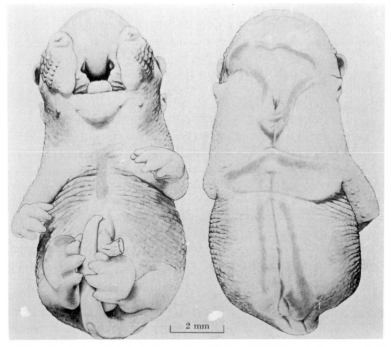

Fig. 6.5. Ventral and dorsal views of a 16-day-old Ph/Ph cleft face individual. (From Grüneberg and Truslove 1960.)

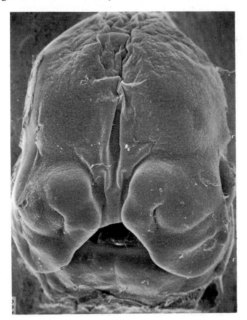

Fig. 6.6. Scanning electron micrograph of a Ts/+ splitface embryo aged 15 days.

of split-face was lowest. On 50 per cent A background split-face was very common and exencephaly very rare. In fact these conditions are almost mutually exclusive as she saw only two individuals which had both defects.

Matta also found that split face was not correlated with hare lip; if split face were an extended manifestation of hare lip one would expect them to be positively correlated. However many individuals with 75 per cent C57BL background were found to die at birth with a median hare lip and cleft palate—possibly a halfway stage between split-face and CL(P).

It is also interesting to note that oedema was only mentioned by Matta in connection with outcrosses to the Patch genetic background. Here Ts/+ animals are usually split-faced (64/80) or exencephalic (14/80) and 'tend to be oedematous'. It seems that either the split-face in Patch is fundamentally different from that in Ts, or that the interpretation of Grüneberg and Truslove needs to be revised. Matta certainly mentioned no central face bleb mechanically separating the maxillary processes of the first arch. Oedema is a common finding in late Xt/Xt embryos (Johnson 1967) but not a causal factor; could the Ph genetic background be particularly susceptible to oedema?

Both Ph/+ and Xt^{bph}/+ mice have a small central face bleb at 13 days (Johnson 1969). Grüneberg and Truslove (1960) rejected the idea that this was associated with the patch gene but Johnson (1969) was able to reclassify their data to show that face bleb incidence varied in reciprocal matings (being higher with a +/+ mother in both stocks; fluid volume in inbred strains of mice is also patroclinal, Johnson 1971) which removed their objection. In both mutants and in Ts the interfrontal is affected which suggests a change in skull proportions. The sequence affected skull proportions attracts face bleb may hold for Ph/Ph as well as Ph/+ and Xt^{bph}/+, and perhaps (Matta gives no clue) for Ts/+ on an appropriate background.

Skull proportions and interfrontal bones

The heterozygous forms of Xt, Ph, and Ts all have slightly abnormal skulls and an increased incidence of interfrontal bones. This seems unlikely to be coincidental. Deol (1961) noted two types of Ts/+ in his stock. In the 'ordinary' Ts/+ the skull was little affected except that the frontal bones were wider than normal, a condition associated with the presence of an interfrontal bone in 92 per cent of Ts/+ but only 12 per cent of normal littermates. A short-snouted form of Ts/+ had a shortening amounting to almost 20 per cent, without associated narrowing (Fig. 6.7). The frontals in this short-snouted form were even broader, the nasal processes of the maxillae reduced, and the nasals short and malformed, with oblong apertures of various sizes. The frontals were rather loosely joined in the mid-line and often incomplete medially. Nasals and frontals were often fused.

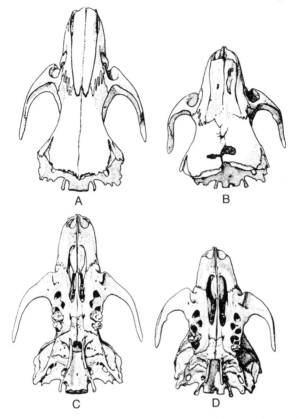

Fig. 6.7. Dorsal and ventral views of (A,C) normal and (B,D) short-snouted Ts/+ skulls. (From Deol 1961.)

In Ph/+, Grüneberg and Truslove (1960) also found a large interfrontal, attributed to a wide skull. Johnson (1967) found a similar effect in Xt/+.

The presence of an interfrontal bone is a known variant in inbred strains of mice (Truslove 1952). Johnson (1976) investigated the presence of an interfrontal associated with mutant genes. In the absence of an interfrontal an index of the skull proportions of a given individual (width at maxillary-frontal suture/length between basisphenoid and nasal spine) varied widely. If an interfrontal was present, then its size was positively correlated with skull width/length; the size of the interfrontal depended on the relative width of the skull. Of the eight loci known to affect the presence of the interfrontal (bh, Bn, fi, 1st, Ph, Ts, ur, Xt), five, Bn, fi, Ph, Ts, and Xt, (all those available as papain preparations) affected skull proportions. It is also perhaps more than coincidental that seven of these eight genes (the aetiology of ur being unknown) also affect the development of the neural tube.

First arch malformation (far, Chr ?)

McLeod *et al.* (1980) described this recessive mutation which affects the secondary palate along with other cranial bones, largely those originating from the first branchial arch. The defects can be traced back to day 12 of gestation. Palatal shelves are deficient and later become bizarre polypoid structures. The trigeminal nerve in the maxillary process seems abnormal, and maxillary vibrissae pads and lower eyelid are clearly deficient. By day 16 affected individuals can be recognized in cartilage/bone preparations (Fig. 6.8). The zygomatic arch is thickened and flattened against the side of the head and ossification centres for the squamosal and alisphenoid are missing. The mandible lacks most of its coronoid process. At birth the premaxilla is misshapen and connections between premaxilla and maxilla, and maxilla and frontal bones are indistinct or absent. The zygomatic bone is around five times normal thickness and the squamosal almost entirely absent. There are other smaller defects in the ear ossicles and the styloid apparatus. The secondary palate is cleft, but primary palate and lip are intact.

McLeod *et al.* point out that most of the defective bones have their origin in the first arch, and more particularly in the maxillary process. The abnormality of the mandible is confined to its area of articulation, which

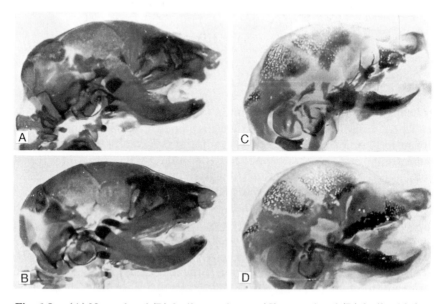

Fig. 6.8.   (A) Normal and (B) far/far newborns; (C) normal and (D) far/far 16-day embryos stained for skeletal tissues and cleared. (From McLeod *et al.* 1980.)

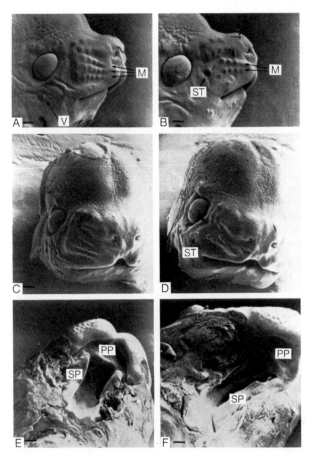

Fig. 6.9. Scanning electron micrograph of 13-day-old (A,C) normal and (B,D) far/far embryos and 14-day-old (E) normal and (F) far/far palates. PP, primary palate; SP, secondary palate; ST: skin tag. (From Juriloff and Harris 1983.)

may be secondarily affected. The involvement of the styloid cartilage suggests that the second arch may also be involved.

Juriloff and Harris (1983) looked further at the development of far. They found that far/far differed from normal mice in overall face shape (Fig. 6.9) being flattened and deficient in the maxillary region, especially in the area derived from the nasolacrimal groove. The three rows of mystacial vibrissae which originate from the maxillary process were disrupted and deficient. Other vibrissae appeared normal. Most newborn far (72 per cent) had a 'skin tag' on one or both sides of the face.

The palatal development was disrupted with irregular palatal shelves reduced in size and elevated above the tongue (Fig. 6.10). Ossifications in

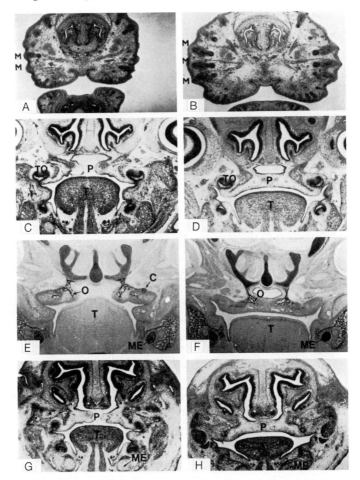

Fig. 6.10. Frontal sections of (A,C,E,G) normal and (B,D,F,G) far/far embryos aged 15–16 days. (From Juriloff and Harris 1983.)

the palatal shelves are abnormal in shape and an abnormal rod of cartilage is seen in some embryos lateral to the palate. Meckel's cartilage and lower jaw appear normal.

At 14 days all far palatal shelves are lobed irregularly in their anterior part and deficient posteriorly.

Juriloff and Harris concluded that far acts in that part of the first arch destined to become the maxillary facial process. The gene must act before 12 days (when individuals can be classified as far), but presumably after 9 when maxillary and mandibular processes arise. The possible primary defect in far is most plausibly explained if we specify either a defect in maxillary epithelium, a local deficiency in neural crest migration, or a quantitative defect in the mesenchymal matrix.

Amputated (am, Chr 8)

Flint and Ede (1978) described the development of the face in amputated mice, which, as we have already seen (Chapter 2) have a basic defect involving cell adhesion.

The nasofrontal region of amputated mice is first seen to be abnormal at 10.5 days by scanning electron microscopy, when the olfactory pit is less deeply invaginated than in normal littermates (Fig. 6.11). Flint and Ede

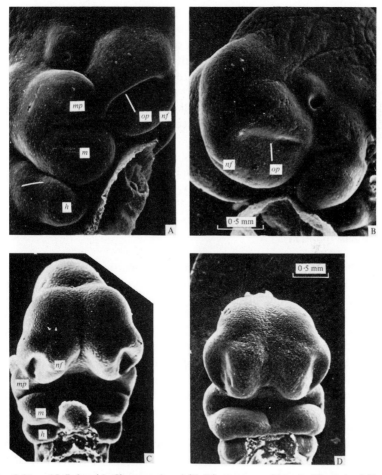

Fig. 6.11. 10.5-day (A,C) normal and (B,D) amputated heads. Note the failure of the nasal pit to invaginate in amputated; *h*: hydro arch; *m*: mandibular arch; *mp*: maxillary process; *nf*: nasal fold; *op*: olfactory pit. (From Flint and Ede 1978.)

therefore looked in detail at the nasofrontal region to determine the aetiology of the abnormality at cellular level. They argued that the cells of the mesenchyme of the face are largely made up of neural crest derivatives. and that, although a direct test was not possible since amputated embryos could not be identified on the basis of their earliest visible abnormality (shortening of the body axis) until after the neural crest has migrated, there was no indirect evidence of deficient neural crest. Dorsal root ganglia are of normal size and the dermal skeleton of the head is not deficient. Mandibular, maxillary, and nasal portions of the skeleton are shortened, but present. Cell density of nasofacial mesenchyme at 9.5 days is also normal: this suggests that amputated facial development is not initially disadvantaged by a shortage of mesenchymal cells.

Cell proliferation rate, expressed as mitotic index, was also normal over the period 9.5–10.5 days in amputated nasofrontal region, as well as in the mandibular arch and a control region not involved in facial growth. The facial mesenchyme in amputated did, however, show changes at the ultrastructural level, with increased areas of cell contact. This was more striking at 10.5 than at 9.5 days, when the cell density in facial mesenchyme had fallen in both normal and abnormal embryos. Flint and Ede concluded that, as elsewhere, this increase in cell contact is the basic defect in amputated facial development.

Flint (1980) extended these findings, with essentially similar results, to the palatal shelves of amputated, which invariably has a cleft palate. Again he found no difference in mitotic index, but an increase in the area of cell contact between mesenchyme cells.

Phocomelia (pc, Chr ?)

Sisken and Glueksohn-Waelsch (1959) described the development of the recessive phocomelia in the mouse. We have already referred to this gene as a possible systemic abnormality of the mesenchymatous skeleton (Chapter 2). Sisken and Glueksohn-Waelsch noted that at 13.5 days some phocomelic individuals could be recognized by a smaller than normal head with a notable reduction in the size of the mandible. In material aged 11.5–13 days abnormals could not be distinguished.

In sections Fitch (1957) noted that all the subsequent abnormalities of the pc head can be traced to the stage of mesenchymal condensations. There is a reduction in the size of the condensations which are to become the mandible and maxilla and an accessory pair of condensations are seen, which later become rods of cartilage ventral to the paries nasi, and which are ultimately thought to interfere with palatal closure. By day 15 the reduced size of maxilla, premaxilla, and mandible is clearly seen and the head is smaller and narrower than normal: the chondrocranium is not proportionally reduced, although it is abnormal in many areas. The newborn pc/pc was described by Gluecksohn-Waelsch *et al.* (1956). The head is narrow and pointed with a wide palatal cleft. In alizarin clearance preparations the nasal bones are seen to be absent, the premaxilla reduced, the maxilla partially absent, and the mandible reduced in length and width. The squamosal was absent or reduced in 50 per cent of animals studied. The frontals were always parted, the palatines absent, and the presphenoid reduced or absent.

It is clear that phocomelia is a gene with complex effects and should not be looked at other than as a whole. Disproportionate dwarfing and polydactyly or syndactyly are present as well as defects of the head region. The late appearance of the abnormalities of bones derived from the first pharyngeal arch suggests that arch tissue as such is not abnormal; but it may well be that some misallocation of tissues could occur at the membranous skeleton stage. The blastemal abnormalities in the skull, reduction in the blastema of the maxilla and in Meckel's cartilage, and the allocation of tissue to the 'cartilage bars' which prevent normal palatal closure (and which may correspond to the extensions of the paries nasi seen in normal mice, Fitch 1957), must be considered in conjunction with the small irregular extra cartilages seen ventral to the nasal septum and the characteristic polydactyly/syndactyly. The abnormality here is clearly systemic and the involvement of the pharyngeal arch tissues secondary.

Shorthead (sho, Chr ?)

Like phocomelic this recessive mutation (Fitch 1961*a,b*) produces a wide palatal cleft and shortened limbs. Shorthead homozygotes are small in size at birth, and cyanosed, with a foreshortened rounded head. The limbs are stocky. The lower jaw is shortened and there is a wide median palatal cleft. Nasal and frontal bones are reduced and the frontal fontanelle has a characteristic saddle shape. An interfrontal bone is often present and all dorsal sutures of the skull are widened. Premaxillae, maxillae, and sphenoid are small and of unusual shape. The vomers are widely separated. Shorthead embryos can be identified at 12 days after fertilization by their foreshortened head and slightly shortened tail. The groove separating the two sides of the face is shallow and wide at 13 days, and the lower jaw flatter than normal. The facial abnormalities are, according to Fitch, based upon the reduced growth of the facial processes, although a group of embryos characterized by small maxillary and mandibular processes at 11 days could not be distinguished from retarded embryos found in control litters.

In the 15-day-old sho/sho individual the tongue is seen to be shorter than the palatal shelves which extend beyond the tongue and partially enclose it anteriorly. Fitch considered that the abnormally small tongue of these individuals is wedged in so firmly that no further palatal closure can occur.

If the tongue is experimentally moved (a difficult procedure in sho/sho), palatal movement follows. Fitch pointed out that tongue growth is reduced and that the tongue develops partly from the mandibular processes: in many other cleft palate syndromes the tongue, by contrast, protrudes.

The talpid (ta) chick

We have referred to talpid in earlier sections but the abnormalities of the head are worth considering individually. In talpid[3] (Ede and Kelly 1964) the eyes are drawn together in the midline and may partially fuse. There is no upper beak, or at best a small peg above the mouth and the nasal process forms a plaque above or between the eyes (Fig. 6.12). The lower beak is often represented by a single midline protrusion. Ede and Kelly pointed out that the abnormalities of the head region of talpid are consequent upon the failure of the prechordal mesoderm to separate into lateral strips (which become the visceral arches) and a central prechordal plate. Work on amphibia (Mangold 1931; Adelmann 1934) has shown that removal of the central prechordal mesoderm or its derangement by lithium treatment led to cyclopia. Ede and Kelly suggested that a similar state of affairs obtains in talpid[3]. The failure of separation also accords well with the excessive cell adhesion which seems to be near the primary abnormality of the talpid syndrome.

VISCERAL ARCH ABNORMALITIES AND HEAD DEFECTS

In the group of mutants described above we see a series of varied abnormalities of the pharyngeal arches. In talpid, the most extreme, the

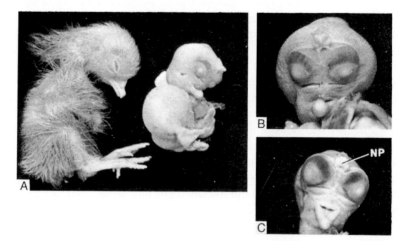

Fig. 6.12.   (A) Normal (left) and talpid embryos aged 14 days. (B,C) Frontal views of talpid embryos aged 14 and 11 days, respectively. (From Ede and Kelly 1964.)

mesoderm which is destined to become the cores of the pharyngeal arches remains in the midline rather than separating into two lateral masses; in Xt the separation occurs but the masses are too large. Whether this is due to too much material being allocated at an early stage, whether the mesoderm is defective, or whether a normal amount of material is later acted upon by abnormal ectoderm is a moot point. The extreme polydactyly and the occurrence of supernumerary vibrissae suggest the latter. The question of mesodermal/ectodermal interaction has only been fully worked out in the limb bud, and not there in the mouse.

In Ph and Ts the reverse situation is apparently found, with the first pharyngeal arch being too small, or failing to move in a normal way. Interestingly the problem of the size of the arch in Ph and Ts has not been tackled; it does not look excessively small. There is evidence (see Chapter 2) that cell adhesion is abnormal in Ts as in talpid. In this case cell adhesion and morphological movement may be linked in Ts also, and this could be reasonably be extended to include Patch in the absence of evidence to the contrary.

In phocomelia, far, and shorthead it is clear, that the material of the first arch is deficient, leading to absence or partial absence of pharyngeal arch derivatives.

CLEFT LIP AND CLEFT PALATE

Dancer (Dc, Chr 19)

Dancer (Deol and Lane 1966; Trasler 1969) appeared in the C57BL stock. All homozygotes die with cleft lip and cleft palate, but penetrance in heterozygotes is incomplete. The cleft lip is thought to be due to reduction of the volume of mesenchyme in the nasal process (Trasler and Leong 1976) Heterozygotes have a white head spot and show circling and head-tossing behaviour, but are not deaf. The utricular macula is absent and there are defects in the bony and membranous labyrinths in the vestibular region, probably consequent upon a reduction of the vestibular ganglion and its nerves.

Twirler (Tw, Chr 18)

Twirler arose in a multiple recessive stock (Lyon 1958) and has a similar constellation of behavioural abnormalities to those seen in Dancer when heterozygous, based on irregularites of the semicircular canals and reduction or absence of otoliths. Homozygotes have cleft lip and palate. Juriloff (1978) suggested that both these mutants affect the same developmental pathway.

It is not clear whether dancer and twirler form a separate subgroup, or should be considered an extension of those mutants with abnormal pharyngeal arches.

Cleft lip and palate in inbred strains

Spontaneous cleft lip and palate is low in most mouse strains (Kalter 1978; Loevy 1968); the exceptions are all derived from the A strain (Table 6.1). Different sublines of A have different overall incidences of the condition. Congenic lines of A (Snell and Stimpfling 1966; Staats 1972) will differ from each other only at the H-2 histocompatibility locus, and might therefore be expected to produce the same frequency of cleft lip and palate. The importance of the H-2 locus in this context is discussed below. The L line (Trasler *et al.* 1978) is not definitely known to have A ancestry but this is possible, and often assumed.

TABLE 6.1. *Cleft lip and palate (CL/P) in inbred strains*

Percentage CL(P)	Strain	Reference
10	A/J,A/HeJ,A/WySn,A/St	Loevy (1968); Kalter (1975); Staats (1972); Miller (1974)
c. 10	Congenic A	Snell and Stimpfling (1966); Staats (1972)
25	CL/Fr	Staats (1972)
2	L line	Trasler *et al.* (1978)

After Juriloff (1978).

Two question arise from these findings: what is the genetic basis between 'A-like' strains (with a high CL(P) incidence) and 'non-A-like' strains (with a low incidence of CL(P) and what is the genetic basis of the difference between the various 'A-like' strains with high incidences?

Outcrossing the 'A-like' strains

F_1 embryos between A and C57BL, C3H, BALB/c do not have cleft lip and palate (Loevy 1968; Davidson *et al.* 1969). This rules out the presence of both a dominant gene and a pure maternal effect. Backcrossing of F_1 to A (first backcross) reintroduces the tendency at a lower rate (Loevy 1968; Davidson *et al.* 1969; Francis 1973). The difference between the incidence in the parental strain and the first backcross reflects the number of genes involved, and the observed differences lead us to suspect that no more than four loci are involved.

Read (1936*a*,*b*) asked if any of these loci were essential for cleft lip and palate or whether they were all permissive. He outcrossed a cleft-lip-liable stock to a non-liable stock. Males of the second generation were backcrossed to liable females, hypothetically homozygotes for the 'single gene' (hp/hp). If this single gene were operating, then half the males should be hp/hp and half +/hp. Of 18 tested males, 10 produced CL(P) and eight did not. This is consistent with the single gene of low penetrance hypothesis, but multiple loci could also give the same result. The single gene hypothesis was not favoured (Read 1936*b*; Grüneberg 1952) because the percentage of clefts seen was lower than expected and data for other generations of the cross were inconsistent.

Juriloff (1977) performed what was essentially a repeat of Read's experiments, using A and C57BL and found a lower rate of segregation in backcrosses than expected for a single locus; the fit for two or three loci was much better. Davidson (1963) and Trasler (1960) suggested that the inheritance of CL(P) in crosses between A and C57BL involved three or more major genes as well as maternal factors.

Maternal factors

In backcrosses the frequency of CL(P) often shows a maternal effect component. In crosses between A, or CL/F and C57BL the frequency of abnormal offspring is higher in genetically equivalent mothers (Read 1936*a*; Francis 1973) whilst in crosses to C3H and A/St (Loevy 1968) the frequency is higher in hybrid mothers. Bornstein *et al.* (1970) showed that this effect was not cytoplasmically transmitted.

The difference in incidence of CL(P) between A/J and CL/Fr is entirely due to a maternal effect (Juriloff 1977). The frequency of CL(P) produced by CL/Fr mothers was consistent no matter what the genotype of the fetus. The frequency produced by A/J mothers was also unaffected by fetal genotype. F_1 mothers and first backcross mothers were like A/J. First backcrosses to CL/Fr were intermediate between the parental strains. These results are consistent with a single gene effect, dominant in A. There was no evidence of sex linkage. The situation here is obviously different from that obtaining in crosses between A, C57BL, C3H, and BALB/c described above.

Facial proportions and cleft lip

The abnormalities of Xt, far, pc, and Ts facial processes are gross and lead to considerable deformity. Is it possible that lesser upsets or variations in facial morphology could lead to less marked abnormalities in facial structure? We have seen how the genes and factors controlling CL(P) behave in crosses between strains susceptible to the condition and those not, but what do these genes actually do? Could they affect facial proportions? Trasler (1968) looked at the shape of the developing face in A and C57BL mice and found that they differed in well-defined particulars (Fig. 6.13). C57BL embryos had nasal pits that were relatively further

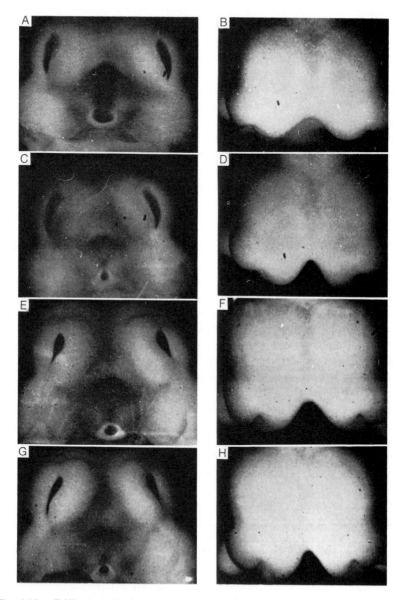

Fig. 6.13. Differences in face shape between (A,B,E,F) C57BL and (C,D,G,H) A/J mouse embryos. (From Trasler 1968.)

apart and their medial nasal processes diverged at a greater angle. Fusion of medial and lateral nasal processes occurred at the same time in both strains, and somite counts were similar. A/J mice destined to have a cleft lip were first recognizable at the 10-somite stage because the epithelium of lateral and medial nasal processes did not fuse behind the nasal pit. Since this fusion was always present at a corresponding stage in C57BL and its absence would logically lead to a cleft lip this characteristic was supposed to be a forerunner of the abnormal condition. Sometimes asymmetrical conditions, forerunners of unilateral clefts, were seen. Juriloff and Trasler (1976) measured facial characteristics in strains selected for varied susceptibility to 6-aminonicotinamide induced cleft lip (see Chapter 9). The most susceptible of these had the smallest distance between nasal pits, although head size was not affected. Trasler and Mackado (1979) selected new lines of A/J with high and low specificity to cleft lip and found that the L line, with 9 per cent incidence of cleft lip had a narrow internasal pit distance which was like the original A/J strain (12 per cent CL) whereas the M line, with 0 per cent cleft lip had a wider internasal pit distance like the resistant C57BL. It was found that adults of these various strains also showed changes in skull proportions, premaxilla length being especially important; discriminant analysis of skull measurements allowed these authors to find a particular facial shape complex associated with pre-disposition to cleft lip.

This is obviously important in man. Fraser and Pashagan (1970) suggested that the normal parents of cleft lip children might have characteristic facial dimensions and the work of Coccaro *et al.* (1972), Erickson (1974), and Kurisu *et al.* (1974) supported this idea as do the varying incidences of cleft lips in different racial groups.

Millikovsky *et al.* (1982) also looked at facial development in C57BL and CL/Fr strains (the latter derived partly from A/J, partly from MSL, and selected for high incidence of abnormality) using the scanning electron microscope. They also found consistent differences in face shape between the susceptible strain and C57BL, which correlate well with those of Trasler (Fig. 6.14). Using the higher resolution of the SEM they looked at the region posterior to the nasal pit where Trasler had reported the failure of fusion in susceptible embryos (Fig. 6.15). They found that in normal mice a 'primary fusion area' consisting of 20–30 cells first lost their peripheral microvilli, then showed temporary ridges followed by filopodia and 'flattened cell ridges' which filled the fusion area from the bottom of the cleft upwards. This sequence lasts about 12 hours and covers the 6–10-somite stage of development. This epithelial activity was absent or reduced in CL/Fr mice; A/J mice, although they had unusual facial geometry generally showed primary fusion. CL/Fr mice underwent a later secondary fusion brought about by processes from individual epithelial cells bridging the space between medial and lateral nasal processes and which allowed

254   *The genetics of the skeleton*

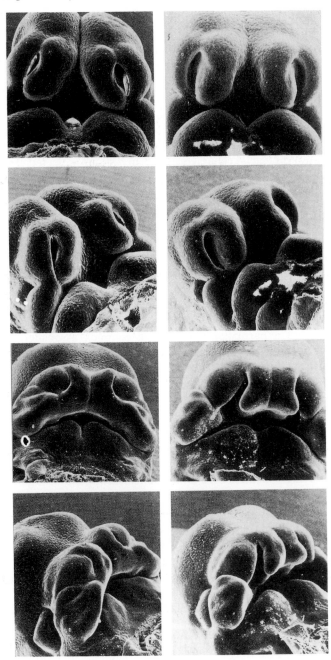

Fig. 6.14. Scanning electron micrograph of C57BL/6J (left column) and CL/Fr (right column). Upper 4 at the 8, lower 4 at the 18 tail somite stage. (From Millikovsky *et al.* 1982.)

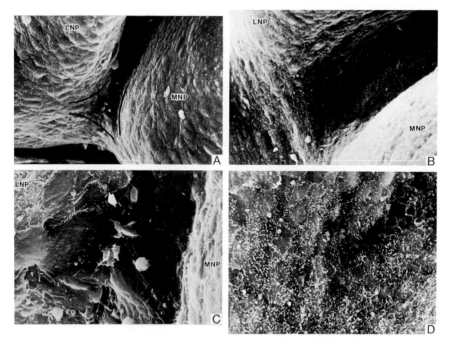

Fig. 6.15. Fusion of median nasal processes (MNP) and lateral nasal processes (LNP) in (A,B) C57BL and (C,D) CL/Fr mice. In C57BL 20–30 cells in the 'area of primary fusion' (triangle in A) lose their marginal microvilli (B). In CL/Fr the median and lateral nasal processes are widely separated (C) and microvilli are not lost (D). (From Millikovsky *et al.* 1982.)

some 64 per cent of CL/Fr to achieve a normal upper lip at the second attempt. Lateral nasal prominences were also smaller than normal in CL/Fr.

Millikovsky *et al.* considered that all these factors, facial geometry, lack of primary epithelial fusion, and hypoplasia of the lateral nasal processes, may contribute to the predisposition of CL/Fr to clefts. A/J mice can be made to show a lack of epithelial fusion by treatment with the anticoagulant phenytoin (Millikovsky and Johnson 1980, and see Chapter 9) and spontaneous clefts in CL/Fr and phenytoin-induced clefts in A/J can be abolished by increasing the concentration of maternal respiratory oxygen (Millikovsky and Johnson 1981*a,b,c*).

SPONTANEOUS CLEFT PALATE WITHOUT CLEFT LIP

Major genes

Many mutant genes produce cleft palate alone as part of their syndrome of abnormalities. As these are obviously heterogeneous in nature they are described below as separate entities. Many, however, seem to share with loci, previously described as major abnormalities of the pharyngeal arches, an effect on skull proportions. In this group it seems clear that this is often the result of a failure of the skull to undergo its normal elongation; the skulls of fetuses are normally broader and shorter in proportion than those of adult mice. The interruption in skull growth (often attributable here to one or other of the chondrodystrophies) may well be a factor predisposing to cleft palate. Conversely some genes included here (cn, ch, bh) clearly affect skull proportions but do not feature a cleft palate.

Chondrodysplasia (cho, Chr ?)

Seegmiller and Fraser (1977) worked on chondrodysplasia, which has both a regular cleft palate and marked micrognathia (Fig. 6.16), and suggested that the two might be linked. They sought to correlate mandibular growth retardation with palatal development by demonstrating that both occurred simultaneously: vitamin A teratogenesis (Shih *et al.* 1974) in contrast leads to both cleft palate and micrognathia, but the former precedes the latter, so that the micrognathia cannot be seen as a contributory factor to the palatal cleft.

Before palatal closure the tongue lies between that palatal shelves and behind the primary palate. After closure it is below both the fused palatal shelves and the primary palate. Displacement of the tongue may therefore be necessary for palatal closure (Larsson 1974), and this may in turn depend on the growth of the lower jaw. Seegmiller and Fraser showed that in cho, during the stages of palatal closure, Meckel's cartilage was growing much more slowly than in normal mice. The tongue was not carried forward, as it normally is, by the growth of the lower jaw and remained arched upwards between the palatal shelves, physically preventing palatal closure.

Cartilage matrix deficiency (cmd, Chr ?)

Rittenhouse *et al.* (1978) noted the regular occurrence of short snout, protruding tongue, and cleft palate in cmd/cmd individuals. The overall skull length is reduced at birth, but the width is normal.

Paddle (pad, Chr ?)

Johnson and Nash (1982) described the effects of paddle, a recessive with incomplete penetrance which produces both chondrodystrophy (see Chapter 3 and cleft palate, on the growth of the head using cephalometric measurements of the craniofacial complex.

They found that pad/pad embryos could be identified on day 14 of gestation. Normal littermates of this age had elevated palatal shelves with

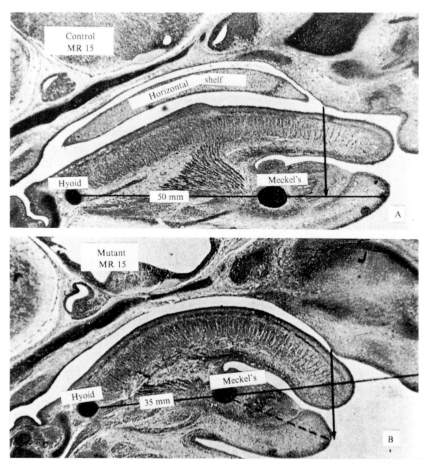

Fig. 6.16. Midsagittal sections through (A) normal and (B) cho/cho fetuses aged 15 days. (From Seegmiller and Fraser 1977.)

varying degrees of fusion. In paddle homozygotes the shelves were still ventrally directed at this age and remained so during days 15 and 16. Two methods of cephalometric measurement gave essentially the same results, that maxillary and mandibular growth were both retarded and the cranial base angle was reduced in the mutant. This did not allow the normal forward movement of the tongue under the primary palate and hence obstructed shelf movement.

Urinogenital (ur, Chr ?)

Urinogenital (Dunn and Gluecksohn-Schoenheimer 1947; Gluecksohn-Waelsch and Kamell 1955, and see Chapter 5) has a palatal cleft whose

258 The genetics of the skeleton

development was described in detail by Fitch (1957). Most ur mice have cleft palates, but a few do not; these live for several months, but are small, retarded, sterile, and have disturbed kidney function. The ur skull is reduced in length but not width, with short frontal and nasal bones, large frontal fontanelle, and an interfrontal bone. Ventrally all paired bones of the skull are widely separated and unpaired ones increased in width. The palatine processes of the maxilla appear normal in size but neither they nor the palatine bones meet in the midline. The rare individuals with complete palates showed a much less extreme phenotype with the skull not shortened by the usual amount and were considered by Fitch to show an incomplete manifestation of the syndrome.

Brain hernia (bh, probably Chr 7)

Bennett (1959) described a mutation whose main skeletal feature was the protrusion of part of the brain through a median opening in the skull. At birth bh/bh show a wide range of head anomalies ranging from a cerebral hernia (a barely noticeable blister in some cases, in others a bloody sac several millimeters across). A slight hydrocephaly may accompany the condition. The head is foreshortened (even in the absence of a frank hernia) and 75 per cent are microphthalmic or anophthalmic, often asymmetrically.

The brain hernia occurs between widely parted frontal and parietal bones. An interfrontal is universally present. The orbits are of normal size, even though they contain small eyes, or no eyes at all. Viability is normal *in utero* and at birth, but there is loss of individuals between birth and weaning; some survivors are sterile.

Polycystic kidneys develop later in life and bh/bh individuals have higher than normal levels of free amino acids in urine (Bennett 1961). As in ch/ch brain hernia changes facial proportions without affecting the development of the palate.

Hypophosphataemia (hyp, Chr X)

We have already seen the effect of hypophosphataemia on bone growth. Iorio *et al.* (1980) looked at the craniometry of normal and hypophosphataemic skulls on the same C57BL background to see if they in any way resembled those of vitamin-D-resistant rickets (VDRR) sufferers (Iorio *et al.* 1979a,b)

They found that the shorter, domed skulls of Hyp mice had a shorter neurocranium with frontal and occipital bossing, and a slight retardation in mandibular growth. The viscerocranium was also retarded.

Prominent bulges were present at the frontonasal suture and the premaxillary–maxillary suture, and there were small but consistent differences in the structure of the zygomatic arch. The effect on the heterozygous female was less than that seen in the hemizygous male, presumably due to the presence of a + gene in half the female cells after X-inactivation.

Mostafa *et al.* (1982) noted deformation and failure to close of the frontonasal and premaxillary–maxillary sutures in human VDRR rickets. The deficient growth of the nasal bone proved to be the most important measured variable contributing to the overall difference in skull shape. Both these groups of authors regarded the cranial changes in Hyp mice as a good model of the facial changes seen in human VDRR.

Sirenomelia (srn, Chr ?)

Orr *et al.* (1982) reported facial abnormalities in the sirenomelia mutant first described by Hoornbeek (1970). Hornbeek's original description and later work (Schreiner and Hornbeek 1973) suggested that the effect was primarily caudal; occasional exencephaly, cleft palate, and micrognathia were as frequent in normal as srn/srn mice.

Orr *et al.*'s colony of srn mice produced more consistent facial abnormalities. Micrognathia was present in 36 per cent of homozygotes, macroglossia in 26 per cent, and cleft palate in 21 per cent. Even when intact, siren palates were narrower than normal and high arched. Craniofacial anomaly was positively correlated with caudal abnormality. Analysis of uterine contents at 12–14 days suggested that although only 11 per cent of resorbtions were srn/srn the gene is fully penetrant but often lethal *in utero*.

Brachymorphic (bm, Chr 19), stubby (stb, Chr 2)

Lane and Dickie (1968) noted that brachymorphic and stubby mice had shortened, but not narrowed skulls. Pratt *et al.* (1980) and Brown (1978) looked at the susceptibility of bm to cortisone-induced cleft palate. bm is normally maintained on a C57BL background, and Pratt found that hydrocortisone on days 11–14 produced 20 per cent cleft palate in +/bm, +/+, or C57BL mice but 95 per cent in bm/bm. The median effective dose for bm was 45mg/kg (cf. 325mg/kg for C57BL). The time of palatal closure was delayed by the bm gene, with or without hydrocortisone treatment. The amount of cytoplasmic glucocorticoid receptor in the bm/bm palate on day 14 was similar to that of C57BL littermates, and not elevated as it is in A/J. Levels of cAMP were 30–70 per cent higher in bm than in controls, irrespective of cortisone treatment. These results suggest that both bm and A show delay in palatal shelf elevation and elevated levels of cAMP, which seems to predispose both to cortisone-induced cleft palate.

Palatial slit

Kusanagi (1983) reported a condition in control fetuses from a teratological study involving C57BL/6 mice. In 10 fetuses he noticed that the premaxilla and palatal shelves failed to fuse. This condition, with premaxilla not fused to the dorsal part of the palatal shelves was seen in 5.7 per cent of fetuses from 29 untreated dams. The identical condition was seen in 4.1 per cent of adult C57BL/6. In these mice the premaxilla was underdeveloped and seen in X-rays to be unfused with the palatine process of the maxilla. The condition seems not to be based on a developmental delay, as it is not repaired subsequently. The condition appears to be under genetic control, and not related to major palatal clefting.

Muscular dysgenesis (mdg, Chr ?)

Pai (1965*a,b*) reported that mice homozygous for the gene muscular dysgenesis usually (77 per cent) had a palatal cleft. Homozygotes appear shorter and broader than normal sibs and micrognathic. Alizarin clearance preparations show a slight shortening of the head with a minor enlargement of interparietal and occipital bones. Cervical vertebrae are broad, and clavicle, scapula, ribs, and sternebrae reduced. In individuals with a palatal cleft the tongue was lodged between the palatine processes. mdg/mdg embryos could be first recognized at 13.25–13.75 days by a generalized oedema. The mandibles were first seen to be subnormal at 15.5. days.

As the name suggests the primary abnormality in this gene affects skeletal musculature. The skeletal effects emphasize, however, how closely the skeleton is dependent for its shape on surrounding tissues.

Achondroplasia (cn, Chr 4)

Jolly and Moore (1975) looked at the growth of the skull in achondroplasia, which may be taken as representative of the disrupted growth of other chondrodystrophies although cleft palate is not, in fact, a feature of this gene. Since the gene interferes with cartilage formation, we should expect the main effect on skull growth to be on the cartilage replacement bones of the cranial base. Jolly and Moore found that the basicranial axis was reduced by 25 per cent, the viscerocranium by 18 per cent, and the condylar process by 11 per cent. This was attributed to the reduced growth of the spheno-occipital and midsphenoidal synchondroses, the nasal septal cartilage and the condylar cartilage. The disparity between these measurements suggest that synchondritic and septal cartilages are more affected by the gene than condylar cartilage, which grows by a different mechanism. Brewer *et al.* (1977) studied the development and growth of the basioccipital, sphenoid, and mandibular bones and were able to confirm that the percentage reduction in the sphenoid was twice that of the basioccipital and equal to that previously observed in the limbs, no doubt due to the fact that the sphenoid has contributions from two synchondroses, as do the long bones. The mandible was also affected, but less so; the authors were of the opinion that the mandibular changes were due to regulatory responses to the shortened cranium.

Congenital hydrocephalus (ch, Chr 13)

We have already seen (Chapter 2) that ch causes a systemic defect of the membranous skeleton. The abnormalities of the skull have been described and interpreted several times (Grüneberg 1953; Green 1970; Grüneberg and Wickramaratne 1974). Grüneberg originally believed that the hydrocephalus was secondary to a shortening of the skull base already present at the 12-day stage; absence of a subarachnoid space in the region of the foramen of Magendie was considered secondary. Green traced both hydrocephalus and lack of a subarachnoid space to the 11-day stage. If, as Green showed, the subarachnoid drainage system for cerebrospinal fluid (which normally drains from the brain to the subarachnoid sinuses and eventually to the venous sinuses of the dura mater) fails to function in the 11–12-day-old embryos, then this is sufficient explanation for the hydrocephalus. Even though the shortening of the basicranium is not now regarded as the cause of the hydrocephalus it is still present from 11–12 days onwards. The cartilages of the head region, like those elsewhere, appear late and remain small in ch. On the other hand blastemata giving rise to membrane bone tend to appear early and to be of excessive size. The mandible and the zygomatic process of the maxilla are examples of this. The zygomatic blastema is present at 13 days in ch and massively ossified by 14.5 days; no ossification is present in the normal at this stage. The mandible is in fact so large that on day 15 or 16 mandible and maxilla come together and undergo osseous fusion (see Chapter 7).

Palatal clefting in inbred strains

The above findings on palatal clefts in mutants also shed some light on the strain-dependent incidence of cleft palate. We have seen that face shape differs in various inbred strains as does the incidence of cleft lip and palate. Some inbred strains are susceptible to simple palatal clefting (Fraser and Fainstat 1951; Table 6.2). Particular emphasis has been placed on C57BL, which has a low incidence, and A which has a high incidence of cleft palate alone, as well as the previously discussed high incidence of CL(P). Since both CL(P) and simple cleft palate have high incidences in the same strains and both respond to the same teratogens, it sems likely that the abnormalities are in fact based on similar defects in development. Genetic investigation has shown that both maternal and fetal tissues are involved. Two or three independent loci were implicated by Biddle and Fraser (1977), one of which appears to be associated with the H-2 locus. Mice carrying H-2^a were highly susceptible, but those with H-2^b were resistant (Bonner and Slavkin 1975). However the H-2 linked factor accounted for only 80 per cent of the difference in susceptibility and it was shown (Goldman *et al.* 1977) that two strains (CBA/J and C3H/HeJ) both containing the same H-2^k haplotype differ in susceptibility, implicating

TABLE 6.2. *Cleft palate only in inbred strains*

Percentage	Stock	Reference
5	SW/Fr	Vekermans and Fraser (1977)
?	J/Glw	Staats (1972)
3.5	TlWh	Miller and Atnip (1977)
7	oel	Brown and Harne (1972)
95	ur/ur	Fitch (1957)
Most	p^{cp}/p^{cp}	Phillips (1973)
100?	mdg/mdg	Pai (1965a,b)
100	pc/pc	Fitch (1957)
100?	cho/cho	Seegmiller and Fraser (1977)
100	cmd/cmd	Rittenhouse *et al.* (1978)
100	sho/sho	Fitch (1961a,b)
35	Trisomy 19	White *et al.* (1972)

After Juriloff (1978).

other loci as well. Vekermans and Fraser (1982) investigated the hypothesis that the H-2^a and H-2^b genes might affect the time of palatal closure. They looked at two strains of C57BL differing only at this locus and found that the time of palatal closure did not differ. The strains do, however, differ in their cortisone response so the H-2 locus cannot be wholly responsible for slowing palatal closure. Vekermans and Fraser suggested that the H-2 contribution may be largely maternal (see also Biddle and Fraser 1976) and have elsewhere demonstrated (Vekermans *et al.* 1981) that the embryonic genetic component associated with cortisone-induced cleft palate in C57BL and DBA/2 is linked to a marker on chromosome 5 whereas H-2 is on chromosome 17. Thus the C57BL/A association with H-2 does not extend to other strains.

Long *et al.* (1973) looked at the growth of the cranial base in A/J mice, a strain susceptible to cleft palate. They thought, following Harris (1964, 1967) that cranial base growth might be an important factor in palatal closure. Harris suggested that the upward extension of the base would raise the nasal septum and primary palate so that the anterior ends of the palatine shelves could be brought above the tongue, and the increased size of the anterior oral cavity would then allow forward growth or movement of the tongue and lower jaw. Harris noted strain-dependent variation in this extension, and a decrease in cortisone-induced cleft palate fetuses. Verrusio (1970) used Harris's data as a basis of a model for palatal closure. Smiley (1967) and Babula *et al.* (1970) used basion, hypophysis cerebri, and crista galli to measure the anterior and posterior segments of the cranial base and cranial angle. Wragg *et al.* (1970) made similar measurements in rat. Long *et al.* (1973) extended these findings, looking for localized sites of rapid growth within the chondocranium, associated with increased mitotic rate identified by tritiated thymidine. They found that mice with vertical palatal shelves had a curved sphenoid cartilage (as seen in sagittal section) and that less curved sphenoids were seen in embryos whose palatal shelves had become horizontal but had not yet fused, while straight sphenoids were associated with complete fusion. In individuals treated wth the teratogen 6-aminonicotinamide (6-AN) the presphenoid was thinner than normal, and all had curved presphenoids and unfused, vertical palatal shelves.

Areas of active cartilaginous growth were seen in normal fetuses in the craniopharyngeal, presphenoid, and mesethmoid areas and low activity corresponded to the hypertrophied calcified cartilage of the basioccipital and sphenoid areas. 6-AN treatment specifically reduced the numbers of mitoses in the presphenoid and craniopharyngeal areas.

Miller *et al.* (1978) added evidence from allophenic mice. They used TlWh, susceptible to cortisone-induced cleft palate combined with C57BL. TlWh carries cytological markers enabling cells of this strain to be identified. TlWh and TlWh $< - - >$ TlWh chimeras were susceptible to cortisone. TlWh $< - - >$ C57BL chimeras less so. These authors were able to show that there was a positive correlation between the number of marked TlWh cells in the palate and the incidence of clefting. As TlWh had not been previously used in cleft palate research we do not know its status with regard to the H-2 locus, or if it shares abnormal cortisone receptor levels with A. This seems unlikely since the strain arose in a colony of mice random bred in a single large cage (Miller and Atnip 1977) and any relationship to A/J is thus accidental rather than by design.

Taken *in toto* these results lead to the very reasonable conclusion that normal development of the cranial base cartilages is a factor in palatal closure, a conclusion amply borne out by many other mutants which share both a shortened basicranium and a cleft palate. The persistence of the presphenoid curvature in cleft palate individuals suggests a possible causal relationship; the straightening of the cranial base seems to be connected with the movement of the palatal shelves in both normal and 6-AN-treated individuals, both processes being delayed by the latter.

The relationship between cleft lip and cleft palate now becomes a little clearer. If we allow that the rotation of the palatal shelves is based upon the straightening of the basicranial axis, then we must also allow that in some cases the basicranium would straighten and the shelves try to move but be impeded, perhaps by the tongue. This would produce a subgroup of palatal clefts with straight basicranial axis and incomplete palate. The cleft lip and cleft palate of A/J mice is a case in point. Smiley *et al.* (1971) noted that in A strain mice with a cleft lip, due to abnormal movement of the facial processes, the width of the maxillary arch increased faster than in normal fetuses. The palatal shelves, even had they rotated normally would not have met in the midline. In fact they are prevented from so doing by the tongue.

Trasler and Fraser (1963) investigated the role of the tongue in A/J cleft palates. They summarized clefting as due to one of the following:

1. Impairment of the intrinsic shelf force (cortisone).
2. Resistance of tongue increased (after amniotic puncture or as in sho mice).
3. Head too wide to allow palatine shelves to meet in the midline.
4. Shelves too narrow to meet in the midline.
5. Fusion inhibited

Trasler and Fraser looked at the implication that the tongue drops in the mouth before palatine fusion in mice with congenital cleft lip. It was postulated that here the same abnormality that leads to the cleft lip (i.e. rearrangement of facial processes) also causes cleft palate, i.e. that intrinsic shelf force is intact. They found no evidence that the onset of shelf rotation was delayed in cleft lip mice, and by inference that interference with the process started a little later, after palatal shelf movement had commenced. They found that the median nasal process was extra-large in cleft lip individuals; the tip of the tongue is closely applied to this process. The tongue, filling the space between palatal processes, is also enlarged. In the normal embryo the tongue moves forwards as the shelves start to move over it posteriorly; in cleft lip A strain mice this does not occur. The tongue remains applied to, and indented by, the median nasal process. The tongue therefore arches craniad as seen in longitudinal section. The misshapen tongue is then thought to obstruct further palatal closure, or at least to delay it until the shelves, on becoming horizontal, can no longer meet in the midline.

Trasler and Fraser thus united the idea of cleft lip and the commonly consequential cleft palate. They suggested that the process of shelf closure begins before the tongue drops clear; in A/J animals shelf force is normal (palatal rotation has been observed in cleft individuals when the tongue is removed—it is a common finding that shelf force is intact in cleft palate fetuses) and closure is impeded after its commencement by the tongue, misplaced due to its involvement with the over-large median processes.

Diewart (1982) developed a morphometric method for the comparison of craniofacial growth in strains of mice susceptible or resistant to cleft palate, bearing in mind that late palatal elevation is usually linked with susceptibility to palatal clefting (Fig. 6.17).

She used A/J, SWV, C3H, and C57BL/6J fetuses aged 13–16 days and found that essentially similar patterns of growth were present in all, with C57BL most advanced and A/J most retarded at a given chronological age.

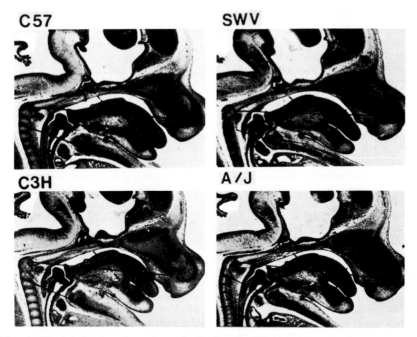

Fig. 6.17. Median sections through the head of four strains of mice at a fetal bodyweight of approximately 200 mg showing variation in head shape (From Diewert 1982.)

Diewart suggested that the timing of shelf elevation in different strains is directly related to changes in spatial relations that result from craniofacial growth. One key relationship seems to be that between the relative length of the mandible and the primary palate, which allows it to change the position of the tongue in the oronasal cavity. The change in position of the head relative to the thorax is also critical; head lifting increases the vertical dimension of the oronasal cavity and tends to decrease the effective length of the nasomaxillary complex by rotating the upper face. Again head lifting contributes to tongue-shelf relationships.

REFERENCES

Adelmann, H. B. (1934). A study of cyclopia in *Amblystoma punctatum* with special reference to the mesoderm. *Journal of experimental Zoology* **67**, 217–81.

Babula, W. J., Smiley, G. R., and Dixon, A. D. (1970). The role of the cartilaginous nasal septum in midfacial growth. *American Journal of Orthodontics* **58**, 250–63.

Bennett, D. (1959). Brain hernia, a new recessive mutation in the mouse. *Journal of Heredity* **50**, 264–8.

—— (1961). A chromatographic study of abnormal urinary amino acid excretion in mutant mice. *Annals of Human Genetics* **25**, 1–6.

Biddle, G. G., and Fraser, F. C. (1976). Genetics of cortisone-reduced cleft palate in the mouse—embryonic and maternal effects. *Genetics* **84**, 73–54.

—— —— (1977). Cortisone induced cleft palate in the mouse. A search for the genetic control of the embryonic response trait. *Genetics* **85**, 289–302.

Bonner, J. J. and Slavkin, H. C. (1975). Cleft palate susceptibility linked to histocompatibility-2 (H–2) in the mouse. *Immunogenetics* **2**, 213–18.

Bornstein, S., Trasler, D. G., and Fraser, F. C. (1970). Effects of the uterine environment on the frequency of spontaneous cleft lip in CL/Fr mice. *Teratology* **3**, 295–8.

Brewer, A. K., Johnson, D. R., and Moore, W. J. (1977). Further studies on skull growth in achondroplastic mice. *Journal of Embryology and experimental Morphology* **39**, 59–70.

Brown, K. S. (1978). Genetics of clefting in the mouse: a critique. In *Etiology of cleft lip and cleft palate* (ed. M. Melik, D. Bixler, and E. D. Shields), pp.77–89. Alan R. Liss, New York.

Brown, M. S. and Harne, L. C. (1972). Hereditary association of isolated cleft palate with open eye and cranioschisis in oel strain mice. *Teratology* **5**, 252.

Coccaro, P. J., D'Amico, R., and Chavoor, A. (1972). Craniofacial morphology of patients with and without cleft lip and palate children. *Cleft Palate Journal* **9**, 28–38.

Davidson, J. G. (1963). The genetic basis of spontaneous cleft lip in the inbred A/Jax mouse strain. PhD thesis, McGill University.

—— Fraser, F. C., and Schlayer, G. (1969). A maternal effect on the frequency of spontaneous cleft lip in the A/J mouse. *Teratology* **2**, 371–6.

Deol, M. S. (1961). Genetical studies on the skeleton of the mouse. XXVIII. Tail-short. *Proceedings of the Royal Society* **B155**, 78–95.

—— and Lane, P. W. (1966). A new gene affecting the morphogenesis of the vestibular part of the inner ear in the mouse. *Journal of Embryology and experimental Morphology* **16**, 543–58.

Diewart, V. M. (1982). A comparative study of craniofacial growth during secondary palate development in four strains of mice. *Journal of Craniofacial Genetics and Developmental Biology* **2**, 247–63.

Dunn, L. C. and Gluecksohn-Schoenheimer, S. (1947). A new complex of hereditary abnormalities in the house mouse. *Journal of experimental Zoology* **104**, 25–42.

Ede, D. A. and Kelly, W. A. (1964). Developmental abnormalities in the head region of the talpid³ mutant of the fowl. *Journal of Embryology and experimental Morphology* **12**, 161–82.

Erickson, C. A. and Weston, J. A. (1983). An SEM analysis of neural crest migration in the mouse. *Journal of Embryology and experimental Morphology* **74**, 97–118.

Erickson, J. D. (1974). Facial and oral form in sibs of children with cleft lip with or without cleft palate. *Annals of Human Genetics* **38**, 77–88.

Fitch, N. (1957). An embryological analysis of two mutants in the mouse, both producing cleft palates. *Journal of experimental Zoology* **136**, 329–57.

—— (1961*a*). A mutation in mice producing dwarfism, brachycephaly and micromelia. *Journal of Morphology* **109**, 141–9.

—— (1961*b*). Development of cleft palate in mice homozygous for the shorthead mutation. *Journal of Morphology* **109**, 141–9.

Flint, O. P. (1980). Cell behaviour and cleft palate in the mutant mouse amputated. *Journal of Embryology and experimental Morphology* **58**, 131–42.

—— and Ede, D. A. (1978). Facial development in the mouse: a comparison between normal and mutant (amputated) mouse embryos. *Journal of Embryology and experimental Morphology* **48**, 249–67.

Francis, B. M. (1973). Influence of sex-linked genes on embryonic sensitivity to cortisone in three strains of mice. *Teratology* **7**, 119–26.

Fraser, F. C. and Faintstat, T. D. (1951). Production of congenital defects in the offspring of pregnant mice treated with cortisone. *Pediatrics* **8**, 527–33.

—— and Pashagan, H. (1970). Relation of face shape to susceptibility to congenital cleft lip. A preliminary report. *Journal of Medical Genetics* **7**, 112–17.

Gluecksohn-Waelsch, S. Hagedora, S. D. and Sisken, B.F. (1956). Genetics and morphology of a recessive mutation in the house mouse affecting head and limb skeleton. *Journal of Morphology* **99**, 465–79.

—— and Kamell, S. (1955). Physiological investigation of a mutation in mice with pleiotropic effects. *Physiological Zoology* **28**, 68–73.

Goldman, A. S., Katsumata, M., Yaffe, S. J., and Gasser, D. L. (1977). Palatal cytosol cortisol binding protein associated with cleft palate susceptibility and H-2 genotype. *Nature* **265**, 643–5.

Green, M. C. (1970). The developmental effects of congenital hydrocephalus (ch) in the mouse. *Developmental Biology* **23**, 585–608.

Grüneberg, H. (1952). *The genetics of the mouse*, 2nd ed. Martinus Nijhoff, The Hague.

—— (1953). Genetical studies on the skeleton of the mouse. VII. Congential hydrocephalus. *Journal of Genetics* **51**, 327–58.

—— and Truslove, G. M. (1960). Two closely linked genes in the mouse. *Genetical Research* **1**, 69–90.

—— and Wickramaratne, G. A. de S. (1974). A re-examination of two skeletal mutants of the mouse vestigial tail (vt) and congenital hydrocephalus (ch). *Journal of Embryology and experimental Morphology* **31**, 207–22.

Harris, J. W. S. (1964). Oligohydramnios and cortisone induced cleft palate. *Nature* **203**, 533–4.

—— (1967). Experimental studies on closure and cleft palate formation in the secondary palate. *Scientific Basis of Medicine Annual Review*, 354–70.

Hoornbeek, F. K. (1970). A gene producing symmelia in the mouse. *Teratology* **3**, 7–10.

Iorio, R. J., Bell, W. A., Meyer, R. A. (1979a). Radiographic evidence of craniofacial and dental abnormalities in the X-linked hypophosphataemic mouse. *Annals of Dentistry* **38**, 31–7.

—— —— —— —— (1979b). Histologic evidence of calcification abnormalities in teeth and alveolar bone of mice with X linked dominant hypophosphataemia (VDRR). *Annals of Dentistry* **33**, 38–44.

—— Murray, G., and Meyer, R. A. (1980). Craniometric measurements of craniofacial malformations in mice with X linked dominant hypophosphataemia. *Teratology* **22**, 291–8.

Johnson, D. R. (1967). Extra-toes: a new mutant gene causing multiple abnormalities in the mouse. *Journal of Embryology and experimental Morphology* **17**, 543–81.

—— (1969). Brachyphalangy, an allele of extra-toes in the mouse. *Genetic Research* **13**, 257–80.

—— (1970). Trypan blue and the extra-toes locus in the mouse. *Teratology* **2**, 105–80.

—— (1971). Genes and genotypes affecting embryonic fluid relations in the mouse. *Genetical Research* **18**, 71–9.

—— (1976). The interfrontal bone and mutant genes in the mouse. *Journal of Anatomy* **121**, 507–13

Johnston, L. and Nash, D. J. (1982). Sagittal growth trends of the development of cleft palate in mice homozygous for the 'paddle' gene. *Journal of Craniofacial Genetics and Developmental Biology* **2**, 265–276.

Jolly, R. J. and Moore, W. J. (1975). Skull growth in achondroplastic (cn) mice; a craniometric study. *Journal of Embryology and experimental Morpholgy.* **33**, 1013–22.

Juriloff, D. M. (1977). Genesis of spontaneous and 6-aminonicotinamide induced cleft lip in mice. PhD thesis, McGill University.

—— (1978). The genetics of clefting in the mouse. In *Etiology of cleft lip and palate* (ed. M. Melnick, D. Bixler, and E. D. Shields), pp. 39–71. Alan J. Liss, New York.

—— and Harris, M. J. (1983). Abnormal facial development in the mouse mutant first arch. *Journal of Craniofacial Genetics and Developmental Biology* **3**, 317–37.

—— and Trasler, D. G. (1976). Test of the hypothesis that embryonic face shape is a causal factor in genetic predisposition to cleft lip in mice. *Teratology* **14**, 35–42.

Kalter, H. (1978). The structure and uses of genetically homogeneous lines of animals. In *Handbook of teratology*, Vol. 4 (ed. J. G. Wilson and F. C. Fraser), pp.155–90. Plenum, New York.

—— (1975). Prenatal epidemiology of spontaneous cleft lip and palate, open eyelid and embryonic death in A/J mice. *Teratology* **12**, 245–58.

Kurisu, K., Niswander, J. D. Johnston, M. C., and Mazaheri, M. (1974). Facial morphology as an indication of genetic predisposition to cleft lip and palate. *American Journal of Human Genetics* **26**, 702–14.

Kusanagi, T. (1983). Palatal slit: a new spontaneous defect of the palate of C57Bl/6 mice. *Teratology* **28**, 149–52.

Lane, P. W. and Dickie, M. M. (1968). Three recessive mutations producing disproportionate dwarfing in mice: achondroplasia, brachymorphic and stubby. *Journal of Heredity* **59**, 300–8.

Larsson, K. S. (1974). Studies on the closure of the secondary palate. V. Attempts to study the teratogenic action of cortisone in mice. *Acta odontologica scandanavica* **20**, 1–13.

Loevy, H. (1968). Cortisone induced teratogenic effects in mice. *Proceedings of the Society for experimental Biology and Medicine* **128**, 841–44.

Long, S. Y., Larsson, K. S., and Lohmander, S. (1973). Cell proliferation in the cranial base of A/J mice with 6-AN induced cleft palate. *Teratology* **8**, 127–38.

Lyon, M. F. (1958). Twirler: a mutant affecting the inner ear of the house mouse. *Journal of Embryology and experimental Morphology* **6**, 105–16.

Mangold, O. (1931). Das Determinationsproblem. III. Das Wirbeltierauge in der Entwicklung und Regeneration. *Ergebnisse der Biologie* **7**, 193–403.

Matta, C. A. (1981). Genetic background and the effects of the gene Tail short in the mouse. PhD Thesis, University of London.

McLeod, M. J., Harris, M. J., Chernoff, G. F., and Miller, J. R. (1980). First arch malformation: a new craniofacial mutant in the mouse. *Journal of Heredity* **71**, 331–5.

Miller, J. R. (1974). *Mouse News Letter* **51**, 12.

Miller, K. K. and Atnip, R. L. (1977). TlWh, a promising mouse strain with chromosomal markers for cleft palate research. *Teratology* **16**, 41–6.

—— Sulik, K. and Atnip, R. L. (1978). Allophenic mice in cleft palate investigations. *Journal of Embryology and experimental Morphology* **47**, 169–77.

Millikovsky, G., Ambrose, L. J. H., and Johnson, M. C. (1982). Developmental alterations associated with spontaneous cleft lip and palate in CL/Fr mice. *American Journal of Anatomy* **164**, 29–44.

—— and Johnson, M. C. (1980). Altered development in genetically and phenytoin-induced cleft lip. *Teratology* **21**, 56A–57A.

—— —— (1981*a*). Respiratory O$_2$ concentration alters expression of cleft lip in genetically predisposed CL/Fr mice. *Journal of Dental Research* **60, 306A.**

—— —— (1981*b*). Maternal hyperoxia greatly reduces the incidence of phenytoin-induced cleft lip and palate in A/J mice. *Science* **212**, 671–2.

—— —— (1981*c*). Respiratory O$_2$ concentration determines incidence of genetically-predisposed cleft palate in CL/Fr mice. *Proceedings of the National Academy of Sciences, USA* **78**, 5722–3.

Morgan, W. C. (1950). A new tail short mutation in the mouse whose lethal effects are conditioned by the residual genotypes. *Journal of Heredity* **41**, 208–15.

Mostafa. Y. A., El-Mangoury, N. H., Meyer, R. A., and Iorio, R. J. (1982). Deficient nasal bone growth in the X-linked hypophosphataemic mouse and its implication in craniofacial growth. *Archives of Oral Biology* **27**, 311–15.

Orr, B. Y., Long, S. Y., and Steffek, J. (1982). Craniofacial, caudal and visceral anomalies associated with mutant sirenomelic mice. *Teratology* **26**, 311–17.

Pai, A. C. (1965*a*). Developmental genetics of a lethal mutation, muscular dysgenesis (mdg) in the mouse. I. Genetic analysis and gross morphology. *Developmental Biology* **11**, 82–92.

—— (1965*b*). Developmental genetics of a lethal mutation, muscular disgenesis in the mouse. II. Developmental analysis. *Developmental Biology* **11**, 93–109.

Phillips, R. J. S. (1973). *Mouse News Letter* **48**, 30.

Pratt, R. M., Solomn, D. S., Diewart, U. M., Erickson, R. P., Burns, R. and Brown, K. S. (1980). Cortisone induced cleft palate in the brachymorphic mouse. *Teratogens, Carcinogenesis, Mutagenesis* **1**, 15–23.

Read, S. C. (1936*a*). Harelip in the house mouse. I. Effects of the external and internal environments. *Genetics* **21**, 333–60.

—— (1936*b*). Harelip in the house mouse. II. Mendelian units concerned with harelip and application of the data to the human harelip problem. *Genetics* **21**, 361–74.

Rittenhouse, E., Dunn, L. C., Cookingham, J., Calo, C., Spiegelman, M., Dooker, G. B., Bennett, D. (1978). Cartilage matrix deficiency (cmd) a new autosomal recessive lethal mutation in the mouse. *Journal of Embryology and experimental Morphology* **43**, 71–84.

Schreiner, C. A. and Hoornbeek, F. K. (1973). Developmental aspects of sirenomelia in the mouse. *Journal of Morphology* **141**, 345–58.

Scott, J. P. (1937). The embryology of the guinea pig. III. The development of the polydactylous monster. *Journal of experimental Zoology* **77**, 123–57.

—— (1938). The embryology of the guinea pig. II. The polydactylous monster. *Journal of Morphology* **62**, 299–321.

Seegmiller, R. E. and Fraser, F. C. (1977). Mandibular growth retardation as a cause of cleft palate in mice homozygous for the chondrodysplasia gene. *Journal of Embryology and experimental Morphology* **38**, 227–38.

Shih, L. Y., Trasler, D. G., and Fraser, F. C. (1974). Relation of mandible growth to palate closure. *Teratology* **9**, 191–202.

Sisken, B. T. and Gluecksohn-Waelsch, S. (1959). A developmental study of the mutation 'phocomelia' in the mouse. *Journal of experimental Zoology* **142**, 623–42.

Smiley, G. R. (1967). A profile of cephalometric appraisal of normal growth parameters in embryonic mice. *Anatomical Record* **157**, 323.

—— Vanek, R. J., and Dixon, A. D. (1971). Width of the craniofacial complex during formation of the secondary palate. *Cleft Palate Journal* **8**, 371–8.

Snell, G. D. and Stimpfling, J. H. (1966). Genetics of tissue transplantation. In

270   *The genetics of the skeleton*

Biology of the Laboratory mouse (ed. E. L. Green), 2nd edn, pp. 457–491. Dover, New York.

Staats, J. (1972). Standardised nomenclature for inbred strains of mice. 5th listing. *Cancer Research* **32**, 1609–46.

Trasler, D. G. (1960). Influence of uterine site on occurrence of spontaneous cleft lip in mice. *Science* **132**, 420–21.

—— (1968). Pathogenesis of cleft lip and its relation to embryonic face shape in A/J and C57BL mice. *Teratology* **1**, 33–50.

—— (1969). Differences in face shape of mouse with and without the gene dancer predisposing to cleft lip. *Teratology* **2**, 271.

—— and Fraser, F. C. (1963). Role of the tongue in producing cleft palate in mice with spontaneous cleft lip. *Developmental Biology* **6**, 45–60.

—— and Leong, S. (1976). Face shape and mitotic index in mice with 6-aminonicotinamide induced and inherited cleft lip. *Teratology* **9**, A39–40.

—— and Mackado, M. (1979). Newborn and adult face shapes related to mouse cleft lip predispostion. *Teratology* **19**, 197–206.

—— Rearden, C-A., and Rajchgot, H. (1978). A selection experiment for distinct types of 6-aminonicotinamide-induced cleft lip in mice. *Teratology* **18**, 49–54.

Truslove, G. M. (1952). Genetical studies on the skeleton of the mouse. V. Interfrontals and parted frontals. *Journal of Genetics* **51**, 115–22.

—— (1977). A new allele at the patch locus in the mouse. *Genetical Research* **29**, 183–6.

Vekermans, M. and Fraser, F.C. (1982). Susceptibility to cleft palate and the major histocompatibility complex (H-2) in the mouse. *Teratology* **25**, 267–70.

—— Taylor, B. M., and Fraser, F. C. (1981). The susceptibility to cortisone induced cleft palate of recombinant inbred strains of mice: lack of association with the H-2 genotype. *Genetical Research* **38**, 327–31.

Vehemans, M. and Fraser, F. C. (1977). Characteristics of a new strain of mice with unusually high sensitivity to cortisone-induced cleft palate. *Teratology* **15**, 18A.

Verrusio, A. C. (1970). A mechanism for closure of the secondary palate. *Teratology* **3**, 17–20.

White, B. J., Tjio, J-H., Van de Water, L. C., and Crandall, C. (1972). Trisomy for the smallest autosome of the mouse and identification of the TW1WL translocation chromosome. *Cytogenetics* **11**, 363–78.

Wragg, L. E., Klein M., Steinvorth, G. and Warpeha, R. (1970). Facial growth accommodating secondary palate closure in rat and man. *Archives of Oral Biology* **15**, 705–19.

7. The teeth

Teeth, as skeletal elements, are potentially just as prone to genetic variation and abnormality as bones, and many cases of abnormal dentition are recorded in man. The teeth are interesting in their own right as they develop so late as to be accessible as models of developmental processes postnatally in laboratory animals.

We may wish to consider here several properties of teeth: their number, their size, the arrangement of their cusps (which has considerable importance in paleontology, since the teeth of apes and men are often the only parts to survive), and the disposition of the roots. All these factors are known to be under genetic control and the structure of dentine or enamel may be affected by mutant genes.

The question must be asked as to the suitability of the teeth of laboratory rodents as a model for human dentition. The chief difficulties here are two in number. First the rodent incisor erupts continuously and is thus very specialized and not comparable with the incisors of other mammals. Matena (1972, 1973) sensibly suggested that the rodent molar is a better model for human tooth development than the rodent incisor. Second, the rodents do not have secondary dentition. This leads to certain difficulties in homology, discussed below. Basically the problem centres on whether the monophylodont dentition of the mouse and rat is equivalent to the milk teeth or permanent dentition of other species. Whilst we are considering the abnormal development of teeth of a single species, however, this controversy loses much of its relevance.

DEVELOPMENT OF TEETH

The early development of the teeth is an interesting example of the developmental interaction between ectoderm and mesoderm. In man the first sign of tooth development is the appearance in the sixth week of a C-shaped dental lamina, a ridge of ectoderm overlying the developing bones of upper and lower jaw. This gives rise to 10 buds per jaw which form the primordia of the ectodermal components of the teeth. The deep surface of each bud invaginates to form a cap of ectoderm over a mesenchymal thickening, the dental papilla (the cap stage). The cap consists of an outer layer of cells, the outer dental epithelium, and an inner layer, the inner dental epithelium, separated by a loosely packed stellate reticulum. As the cap grows the indentation of the dental papilla deepens, so that the tooth rudiment becomes bell-shaped (bell stage). The mesenchymal cells of the dental papilla adjacent to the inner dental lamina then differentiate into odontoblasts, which will later produce dentine. As the dentine layer thickens, the odontoblasts retreat into the dental papilla leaving behind thin processes, the dental processes, in the dentine. The odontoblast cells persist, producing predentine, throughout life. The remaining cells of the dental papilla form the pulp of the teeth.

The epithelial cells of the outer dental lamina meanwhile differentiate into ameloblasts which produce long enamel prisms which are deposited over the dentine. The enamel is first laid down at the apex of the tooth, then spreads towards its neck. As the enamel thickens the ameloblasts retreat into the stellate reticulum where they regress.

Root formation begins when the dental epithelial layers penetrate the underlying mesenchyme to form the epithelial root sheath. The odontoblasts of the dental papilla lay down a layer of dentine continuous with that of the crown, narrowing the pulp chamber to a mere root canal containing blood vessels and nerves supplying the tooth.

The mesenchymal cells in contact with root dentine produce a layer of cementum (specialized bone) and outside this the peridontal ligament, which holds the tooth in place.

During this process the mandible and maxilla have grown to surround the tooth, forming medial and lateral alveolar plates joined by transverese septa between the teeth. These walls unite above the crown of the developing tooth which must therefore erupt through a layer of overlying bone. The relative roles of root elongation and bone resorption in this process are discussed below.

In man the buds for the permanent teeth are located on the lingual aspect of the milk teeth. In mouse and rat there is only one generation of teeth and Gaunt (1966) addressed himself to the problem of homologies. He described the development of tooth buds from specific thickenings of the oral epithelium to form molar buds (at 13 days in the mouse the first lower molar is at the symmetrical bud stage) with a broad attachment to the overlying epidermis (Fig. 7.1). Soon the swelling bud becomes flask-shaped and asymmetrical with pronounced overgrowth away from the sagittal plane of the head, due to greater proliferation of cells on the buccal surface of the bud. Bud and cap stages are as described above but with the onset of the bell stage a prominent region of larger cells becomes apparent in the enamel epithelium of the lingual surface, which later folds on itself and becomes a double-layered fold with many mitoses. This fold may represent the dental laminae of the second generation of teeth of other mammals, and this phenomenon suggests that the dentition of rats and mice represents a long-standing primary dentition.

The pattern of the cusps in the developing cheek teeth is presumably a function of the shape of the developing crown. Ameloblasts and odontoblasts are separated initially only by the basement membrane of the ameloblasts (membrana praeformativa). The enamel–dentine junction

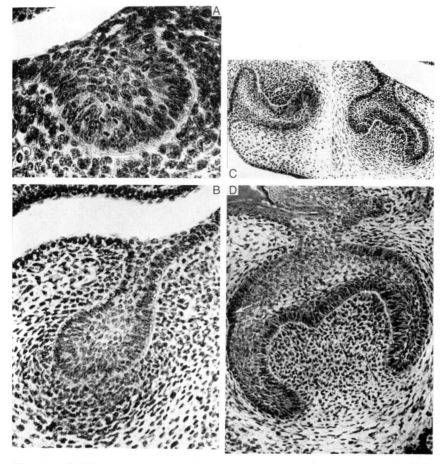

Fig. 7.1. (A) Section through tooth bud of the first lower molar (m_1) of 13–13.5 day mouse fetus (bud stage). (B) m_1 from 14.5-day-old fetus (cap stage). (C) m_1 (right) and m^1 (left) 15 day (bell stage). (D) m_1 at 16 days. External enamel epithelium thickened (arrow). (From Gaunt 1966.)

corresponds in shape to the final shape of this membrane, and the surface shape of the tooth differs from this only because of differing thicknesses of enamel at different points. At the bell stage the membrana praeformativa is smooth: the tips of cusps represent areas where mitosis has ceased; valleys and sulci between them represent continued mitotic activity. The rather sparse literature on this subject was reviewed by Butler (1956).

Gaunt (1955, 1956, 1961) gave an account of the development of the teeth of the mouse. He used an unspecified inbred line of mice and made graphical and three-dimensional reconstructions of developing teeth from sectioned material. He noted that the upper molar teeth of the adult mouse show a marked triserial arrangement of cusps in both longitudinal and transverse directions, the crown being divided by a pair of longitudinal and a pair of transverse valleys. On this basis he numbered the central cusps 1, 2, and 3 from anterior to posterior (Fig. 7.2). From each cusp Gaunt saw three ridges radiating anteriorly, buccally, and lingually. He characterized each ridge according to its cusp of origin, e.g. buccal 1, 2, 3. Cusps on these ridges were designated B1, B2, B3 on the buccal side, L1, L2, L3 on the lingual. How do these cusps develop? Gaunt (1961) made a comparative study of the cheek teeth of mouse and hamster and noted that the early stages of tooth cusp development were identical, with specific differences subvening later. In the mouse (Gaunt 1955) m^1 in the upper jaw was first described at 10–11 days as an elliptical cap in low relief. m^2 was simply a bulbous thickening of the dental lamina at this stage. The crown of m^1 bears two longitudinal ridges separated by a central valley in which lies an enamel knot (a centre of enamel formation). At 14 days m^2 has reached the stage seen in m^1 at 11 days; m^1 has progressed to show a large rounded cusp (cusp 2) halfway along its length. During the period 15–17 days this cusp increases in size and the medial ridge upon which cusp 1 will arise is first seen (Fig. 7.3). Gaunt went on to describe the development of all the adult cusps in both upper and lower molars in detail.

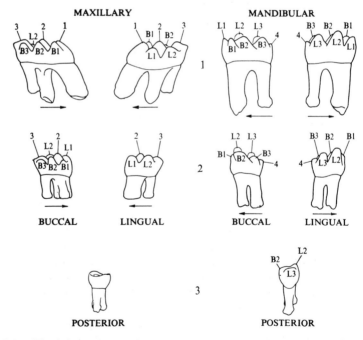

Fig. 7.2.   The left lower molars of a 60-day-old normal mouse. (From Grüneberg 1965.)

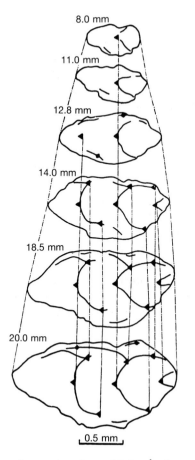

Fig. 7.3. Crown views of reconstructions of left m¹ of a mouse at various stages. (From Gaunt 1955.)

MINOR VARIATIONS OF MOLAR PATTERNS—THE INBRED STRAINS

Grüneberg (1965) and Wickramaratne (1974) studied the teeth of various inbred strains of mice. Grüneberg looked at CBA, C57BL, A, and BALB/c and Wickramaratne added NZB, CE, DBA, and AKR. They were able to show that, just as in the rest of the skeleton, minor variations of tooth shape, consistent enough to allow strain identification, were present (Fig. 7.4). Similar variants were also seen in examples of wild mice. These findings were not entirely unexpected, but interestingly the morphology of the roots of the teeth of inbred strains also varied. Certain root patterns were typical and consistent for any given strain. This is a surprising finding,

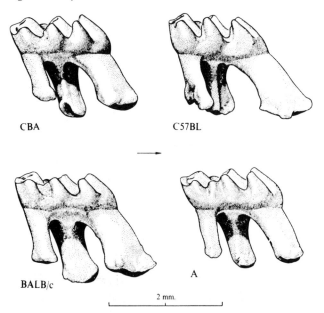

CBA   C57BL

BALB/c   A

2 mm.

Fig. 7.4. Buccal views of left m^1 from four inbred strains of mice. (From Grüneberg 1965.)

since the root morphology is not usually assumed to be under genetic control; it also provides an example of the utility of experimental animals, since it could have been revealed in man only by tedious twin studies.

TOOTH SIZE AND TOOTH REGRESSION

The small lower third molar in mouse is sometimes absent. The absence of these teeth is strain dependent, but with no parent–offspring correlation. The unit of variation is the mouse, or more accurately the litter to which it belongs. Maternal physiology often affects the offspring, but interstrain crosses show that the absence of third molars is a property of the embryo not the mother (Grüneberg 1963). The teeth of C57BL mice, where absent third molars are unknown, are large and uniform in size (Grüneberg 1951) whereas those of CBA, where the absence of third molars is common, are smaller and variable. Within CBA the teeth of litters containing no abnormals are larger than those of litters with absent third molar individuals. Clearly the absence of third molars is correlated with the size of the animal. Grewal (1962) demonstrated that on the the sixth postnatal day tooth germs for third molars are smaller than normal in CBA mice, and some are arrested in the cap stage—these fail to invaginate into a bell and subsequently regress (Fig. 7.5).

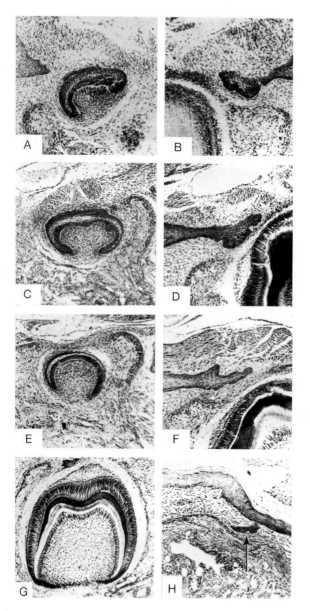

Fig. 7.5. Corresponding stages of normal (left) and regressing lower third molar aged 6–17 days. (From Grewal 1962.)

THE MUTANTS—MAJOR GENES AFFECTING THE TEETH

The tabby (Ta, Chr X), crinkled (cr, Chr 13), downless (dl, Chr 10), sleek (Dl^{Slk}, Chr 10) syndrome

Tabby, crinkled, and downless, and a more recently discovered mimic sleek (Sofaer 1977*a*) are a group of genes, tabby sex-linked and the others autosomal, which produce a syndrome involving both the coat and dentition of the mouse. Grüneberg (1965) provided the first description of tabby and crinkled teeth. These were first regarded as abnormalities of the incisors, which tended to be short and were sometimes lost, necessitating artificial trimming of the remaining incisors to counteract malocclusion. The abnormality was considerably affected by genetic background. Grüneberg's description was based on the Ta gene, cr being virtually identical. The incisors, as well as being of variable size, have a wider than normal covering of enamel, even when the tooth is no smaller that normal. The molars are also abnormal (Fig. 7.6). The crown of m^1 is reduced in size and has a simplified cusp pattern: the usual three roots are reduced to a single entity, grooved externally. The three central cusps (1, 2, 3) are present, as is B2, but B1 and B3 are not represented. The separation of L1 and L2 is indistinct. m^2 also has a single composite root and B1 is smaller than B2 and almost undetectable in some cases. m^3 is very variable and often absent. Lower molars are also variable: m$_1$ is reduced to a variable extent, with a root still slightly forked at its base. m$_1$ is often reduced to a similar size to m$_2$ which it then resembles in cusp pattern; the posterior end of both teeth is then formed by a composite cusp, L3 + B3 and cusp 4 is absent. Both these teeth have thus lost material anteriorly and posteriorly. m$_3$ is often enlarged, and if so bears cusps that are not normally seen on this tooth.

Grüneberg suggested that the size of m$_1$ is a limiting factor in the growth of m$_2$ and m$_3$. If m$_1$ is much reduced, then m$_2$ and m$_3$ can grow at its expense. He suggests the following scenario. Ta reduces the size of the molar teeth, most obviously m^1 and m$_1$ which develop first, but sometimes extending to m$_2$, m$_3$ and m^2, m^3. Perhaps because of differences in timing this process is sometimes not fulfilled and m$_2$,m^2 and m$_3$,m^3 are able to use material not appropriated by m$_1$ and m^1.

Sofaer (1969*a,b*) investigated the development of Tabby teeth. He pointed out that the areas which are abnormal in tabby teeth are those which develop late, and that the order of sensitivity, i.e. the areas most likely to be affected are L1, 4, B3–L3, B2–L2, almost exactly the reverse of the sequence in which these regions develop. Sofaer (1969*a*) showed that the essential defect in the 17-day m$_1$ Ta tooth germ was that the rudiment was shorter and wider than normal. This altered shape may well affect the distribution of forces within the crown which Butler (1956) considered to be a factor in the establishment of crown pattern.

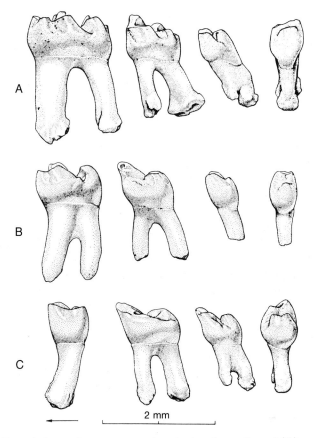

Fig. 7.6.   Buccal views of m_1, m_2, m_3 and posterior views of m_3 of (A) normal male (B) Ta/Ta female, (C) Ta/− male. (From Grüneberg 1965.)

Tabby teeth are also sometimes twinned; Grüneberg (1966) noted twinning in Tabby heterozygotes (Fig. 7.7) and the possibility of concealed twinning in crinkled. Sofaer (1969a,b) extended these findings to other genotypes. Grüneberg described cases in Tabby heterozygotes where four rather than three teeth made up the molar row, m_1 or m^1 being represented by two teeth, or in other cases by conjoined twin teeth. Sofaer (1969b) pointed out that the complexity of such supernumerary teeth is inversely related to the abnormality of the first molar proper. He also postulated (1969a) that the incidence of supernumeraries is a response to partial suppression of m^1 and hence the greater this suppression the higher the incidence of supernumeraries, Sofaer discussed this hypothesis in the light of studies on Ta, crinkled, and downless on various genetic backgrounds.

Another interesting feature of the dentition of heterozygotes for these

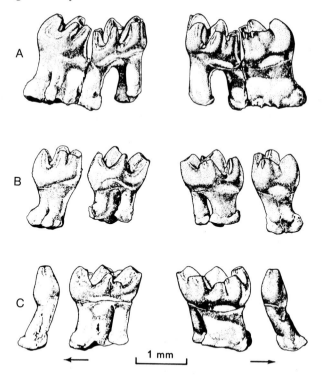

Fig. 7.7. Twinning of the right m¹ in Ta/+ females. In (A) the crowns are separate but the root (accidentally broken) is common. In (B) and (C) twin teeth are separate. (From Grüneberg 1966.)

genes is that the teeth are markedly mosaic. Grüneberg's arguments (1966) that the Ta mosaicism was not in line with the Lyon inactive X hypothesis (Lyon 1961) are not pertinent here, especially as three of the members of the series giving essentially similar phenotypes (cr, dl, Dl^{Slk}) are not sex-linked, Moutschen and Houbrechts (1976) did not agree that the Ta molar morphology invalidates the Lyon hypothesis. Grüneberg (1966) described the heterozygous dentition as a non-random mixture of normal and Tabby teeth (Fig. 7.8), or teeth which were part normal and part Tabby. He argued that this mixed dentition is the result of co-dominance, with normal and Ta, cr or dl alleles both active in all cells. The overall tooth morphology is thus determined by physiological factors at the multicellular level.

A more complete description of the non-dental aspects of the Tabby syndrome was given by Grüneberg (1971). These wider abnormalities affect exocrine glands, skin, and hair and so open up the possibility of finding the site of action of the gene(s). Sofaer (1973) used dermal–

epidermal recombinations of tail skin grown on chick chorio-allantoic membrane and showed that the dl gene acts on the epidermis. However (Sofaer 1974), Ta skin gave equivocal results, as a combination of Tabby epidermis + Tabby dermis produced hair when grafted on to the chorio-allantoic membrane; this combination is hairless when the dl gene is substituted. Mayer *et al.* (1977) used recombined skin made up of various combinations of + and cr dermis and epidermis and grown in the testes of histocompatible mice. In their experiments the cr locus was expressed through the epidermal component.

Now Ta, dl, and cr, although they produce very similar syndromes, must have distinct primary effects, since they show no interaction in double heterozygotes, although the effects are additive in double homozygotes (Sofaer 1979). It is possible that the site of these primary defects concerns ectodermal/mesodermal interaction, with the effects of cr and dl being on the epidermis and that of Ta still being undecided. Moutschen and Houbrechts (1976) discussed other aspects of the Tabby phenotype, including decreased fertility of the Tabby heterozygous female and the high incidence of abnormal spermatozoa in Ta males (Moutschen and Colizzi 1975). We now know that the X chromosome contains elements which may affect sex determination and sex ratio, and the results of Moutschen and Colizzi may reflect the presence of genes closely linked to Ta.

Hurley and Bell (1975) found that the effects of crinkled on skin, hair, and pigmentation could be ameliorated by supplementation of diet and

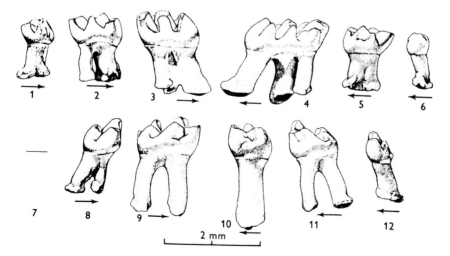

Fig. 7.8. Buccal views of the molars of a Ta/+ female. The dentition contains normal (4), tabby (3, 8, 10), and 'mixed' (2, 5, 9, 11) teeth. (From Grüneberg 1966.)

maternal diet with high levels (500 p.p.m.) of copper. Hairs produced by cr/cr mice on supplementary copper were normal and skin was thicker, whilst pigmentation was restored to normal (the development of pigmentation is delayed in crinkled mice). The morphology of the teeth was not assessed by these authors, but is of obvious interest.

The linking of crinkled and copper is especially interesting since another gene, the sex-linked brindled (Mo^{br}, Fraser *et al*. 1953) which, like the Tabby group, affects pigmentation and hair structure is also susceptible to copper supplementation (Hunt 1974) and has been shown to be based upon a copper transport defect in the gut. Reduced Cu levels affect many enzymes using a Cu cofactor including many involved in hair and skin metabolism, as well as some in the CNS. As far as I know brindled teeth are normal and Grüneberg (1965) found no abnormality in two individuals of the genotype $Mo^{br}/+$.

Sleek (Dl^{Slk}, Crocker and Cattanach 1979) maps very close to downless on chromosome 13 and Crocker considered that it may be an allele of dl, since sleek and downless interact in the double heterozygote, whereas sleek and crinkled do not. If this is so, then we have an interesting situation, a locus at which dominant and recessive alleles have similar phenotypes. Crocker proposed that sleek may be the regulator for the downless locus, and discussed two possible alternative mechanisms which might bring this about.

Fused and supernumerary molars in the rice rat *(Oryzomys palustris)*

Fused and supernumerary molars in the rice rat (Sofaer and Shaw 1971) are associated with reduced fertility and body size and all seem to be dependent on a single autosomal recessive gene subject to modification by genetic background. Molar fusion is preceded by stripping of the external enamel epithelium from the interdental lamina and may involve the first two, of all three molars of each jaw (Fig. 7.9). Supernumeraries are found posterior to the normal molar series and are usually, but not invariably associated with molar fusion. The mechanism of fusion seems to be similar to that proposed by Hitchin and Morris (1966) to account for fusion of teeth in the Lakeland terrier. Stripping of the external epithelium from the interdental lamina is thought to be due to rapid growth of the tooth rudiments and this leaves the internal enamel epithelia of adjacent germs free to come into contact and fuse. The supernumerary is distinct from similar teeth in Tabby (Sofaer 1969*a*,*b*) since it is posterior in position. In some cases it may develop as a response to the shortened antero-posterior dimension of the fused tooth row; however, this cannot be the only cause since fusion is not an inevitable prerequisite for supernumerary tooth development. Sofaer suggested that fusion and supernumerary formation are secondary to a condition of the dental lamina predisposing it to hyperactivity and stripping.

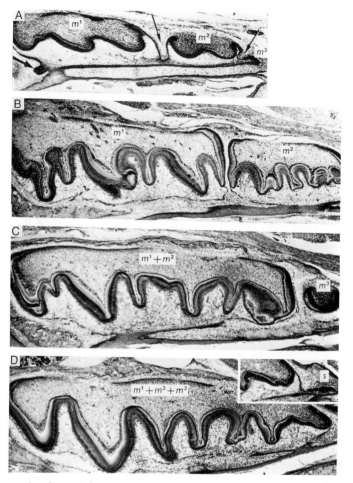

Fig. 7.9. m¹, m², and m³ from normal rice rat (A) at 3–4 days before birth and (B) at 1 day post partum. (C) Fused first and second molars and developmentally advanced third molar 1 day post partum. (D) Fused m¹m²m³ and rudiment of supplementary molar (S) 5 days post partum. (From Sofaer and Shaw 1971.)

Crooked-tail (Cd, Chr 6)

Crooked-tail (Morgan 1954; Grüneberg 1963) is a semidominant gene with characteristic abnormalities of the axial skeleton (see Chapter 5), a naked tail and anophthalmia when homozygous. The original description of crooked mice on the A genetic background noted the reduction of lower incisors which often failed to erupt. On other backgrounds the effect is less marked. Upper and lower third molars are often absent; the development of this condition was studied by Grewal (1962).

The molars also have a characteristic morphology (Grüneberg 1965). The crowns of m^1 and m^2 are reduced in size, with a narrow neck giving the crowns a bulbous appearance (Fig. 7.10). Of the main cusps 1 is more erect than normal, 2 and 3 and B1 and B3 reduced. L1 and L2 are almost united. The anterior roots of m_1 are more or less completely fused, and the lingual root wider than normal. The lower molars show a similar reduction in crown size and cusps.

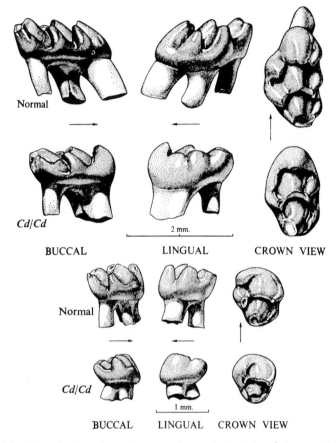

Fig. 7.10. Buccal, lingual, and crown views of the left m^1 (upper six) and m^2 (lower six) of normal and Cd/Cd mice. (From Grüneberg 1965.)

Sofaer (1977*b*) studied tooth development in Cd. Upper incisor germs did not differ from those of controls, but lower incisor germs were often abnormal, especially from day 16 *in utero* onwards. Some were reduced in size with abnormal odontoblasts and no ameloblasts, whilst others were fairly normal in size and histodifferentiation.

In the molar region Sofaer described tracts of cells in the dental lamina, first seen on day 14 and not present in controls. By day 17 these tracts, apparently caused by local cell proliferation and occupying areas of the dental lamina that would normally give rise to molar tooth germs, had regressed or had progressed further to form a tooth germ. In lower molars similar abnormal proliferations of the dental lamina were seen on day 15. The mean diameter of Cd/Cd molar tooth germs was always less than in equivalent normal controls, the difference being greatest on day 17. Sofaer suggested that, after an initial period when tooth germ size is normal in crooked, there is a divergence in tooth growth rates between normal and mutant first molars associated with abnormal proliferation of the dental lamina. After this initial divergence growth rates are similar, with crooked lagging approximately one day behind. It is interesting to note (see Chapter 9) that the failure of the crooked neural tube to close is ameliorated by various agents which interfere with DNA synthesis and hence slow cell division rate. Perhaps a similar localized lack of control over mitotic rate could be operating in crooked teeth.

As in Tabby the areas of crooked molar which are most affected are those appearing latest in development. Sofaer suggested that competition between the abnormal dental lamina and molar bud is responsible for the effects of the gene on the molars. Second and third molars are presumably affected via a knock-on effect. The lower incisors' abnormalities, however, favour the view that there is an intrinsic defect in the internal enamel epithelium or its interaction with pre-odontoblasts, and that this prevents the formation of ameloblasts. This finding is incidentally duplicated in Tabby homozygotes and hemizygotes and their mimics. The effect on molar crowns is also similar, with the distinction that second molars are always reduced in crooked, but may be larger than normal in the Tabby group.

The axial effects of crooked (Grüneberg 1963 and Chapter 5) suggest that the mesodermal component is to blame here, in contrast to the ectodermal or equivocal results seen in the Tabby group. Since we are dealing with ectodermal/mesodermal interactions, presumably a defect in either component could cause malformation.

Hairless (hr, Chr 14)

The dentition of hairless mice has variously been described as normal (Summer 1924; David 1932; Howard 1940) and 'abnormal' (Crew and Mirskaia 1931), although the latter authors gave no details. Payne *et al.* (1977) looked at the development of the cheek teeth and found no essential differences in morphology between hairless and normal, although there were chronological differences in staging compared to Cohn's (1957) findings in the 'albino' mouse of unspecified strain. Payne noted that the dental lamina of m_1 was initiated one day later than normal, but soon caught up by virtue of a shortened lamina stage. The late appositional stage of development occupied only one day in hr mice compared to three days in Cohn's stock. Mandibular and maxillary molars develop simultaneously in hairless, whereas the maxilla usually lags by 12–24 hours.

These differences are unsurprising. The comparison here is not between hairless mice and controls on the same genetic background, and differing at a minimum number of loci, raised under the same conditions and developing in the same uterus, but between mice which have the potential to differ at every locus in the genome, raised under different conditions, and divorced in space and time. We cannot, therefore, accept these findings as due to the hairless gene.

Payne *et al.* were unable to state whether the suggested interactions between ectomesenchyme and ectoderm are normal in hairless or not; this is not surprising in the light of their lack of controls. We still lack a convincing study of the development of hairless teeth, although by analogy with the tabby syndrome we might suspect abnormality. Grüneberg (1965) included two hairless and two rhino (hl^{rh}/hl^{rh}) individuals in his survey of mutant genotypes, but described no abnormality.

Screwtail (sc, Chr ?)

Bhasker *et al.* (1951) described dental abnormalities in screwtail mice. Molars are stunted and erupt after a delay and the third molars are often impacted. The incisors are reduced in size and their roots fail to orientate correctly in the maxilla and mandible, leading to malocclusion. They ascribed these defects to retardation of the growth of the connective tissue of the dental pulp. Dental anomalies affect the shape of the mandible which is heavy anteriorly but reduced in the region of the condylar process, presumably because of the unusual location of the lower incisor.

However there are other abnormalities, in the skull as well as in the axial skeleton (see Chapter 5) and Grüneberg (1963) suggested that both dental and axial defects are due to a systemic anomaly of the mesenchyme. The split vertebrae characteristic of screwtail homozygotes cannot have been formed later than the mesenchymal stage of development.

Pituitary dwarfing (dw, Chr 16)

This gene will serve as the first of a series of conditions in which relatively minor secondary abnormalities of teeth are consequent upon defects of jaw growth. The most obvious of these is malocclusion of the incisors, but cheek teeth may also be affected. Grüneberg (1965) noted that at the time when growth is most affected in the pituitary dwarf mouse (dw/dw), i.e. around seven days post partum, crown development is essentially complete whereas root development is not. By the time roots have fully formed dwarves are markedly smaller than normal, and the cheek teeth thus have normal crowns perched upon reduced roots.

Achondroplasia (cn, Chr 4) and stubby (stb, Chr 2)

Miller *et al.* (1974) looked at the teeth of cn and stb chondrodystrophic mice; reduced cusps and accessory cuspules were found in cn and the occlusal outline of m^2 was changed. In buccal and lingual outlines the cusps were more slanted than normal and had longer mesial slopes. The maxilla was retarded by 2 cusp units. Stubby (stb) was retarded 1 cusp unit, but the molars were normal.

Phocomelia (pc, Chr ?)

Fitch (1957) noted that at 14 days upper incisor tooth buds are smaller than normal in pc/pc. At 15 days 7/10 mutants were still at the bud stage, the other three at the cap stage. In normals four out of five were at the bell stage. Lower incisors and molars appeared normal. At 16 days no odontoblasts were yet visible in pc/pc. Lower incisors were a little abnormal at this stage with an absence of the normal increase in labial growth. All upper incisors seem to stop developing at the bud stage or soon after, or go on to form small but normal teeth. Lower incisors may be arrested at the bud stage, reduced, or completely normal. Development cannot be followed after birth as pc/pc dies at this time due to the effects of a palatal cleft.

Hypophosphataemia (Hyp, Chr X)

Iorio *et al.* (1979a,b) noted that the size of the incisors and molars was reduced in hypophosphataemic mice. Large pulp chambers were present and there was a wide predentine band which was not seen in normal mice. All mutant teeth examined had areas of poorly calcified (globular) dentine.

Extra incisors in the mouse

Fitzgerald (1973) described the presence of a vestigial deciduous incisor in the maxilla of four inbred mouse strains (AKR, Swiss Webster, Hale Stoner, and another unnamed strain). An occasional mandibular counterpart was seen in two of these strains. The supernumerary tooth is first seen at 12.5 days and reaches a maximum size at day 15.5. It persists until 15 days after birth. At its largest it consists of a cap-like shell of dentine with a maximum diameter of 40–60 μm. No enamel organ or enamel is formed. The odontoblasts degenerate at birth and the tooth sinks deeply into the connective tissue below rather than erupting. Fitzgerald considered that this strucure, from its morphology and position, may represent the deciduous counterpart of the mouse permanent incisor, rather than the suppressed second incisor.

Danforth (1958) described extra lower incisors behind and medial to the normal tooth in the descendants of mice treated with mustard gas. They were uni- or bilateral, and sometimes an accessory enamel organ produces only a tooth remnant. This sounds very similar to the condition described by Fitzgerald.

TOOTH SIZE AND ITS HERITABILITY

Leamy and Touchberry (1974) looked at molar width (m_2 and m_3) in seven inbred strains of mice and their crosses. They found significant differences between inbred strains and hybrids and were able to estimate heritabilities, which were high. High negative correlation was also found between tooth and litter size.

Leamy and Hrubant (1971) and Leamy (1981) extended these findings to the effect of single genes on coisogenic backgrounds. Alleles of the agouti series (aa, a^ta, $a^t a^t$, Aa, A^{VY}a and A^Ya) were all found to affect molar width. The first molar was most affected, the third the least. All dominant alleles tended to depress molar width from the standard aa genotype. It is interesting to note that chinchilla alleles at the albino locus (c^{ch}) had much smaller effects (Leamy and Touchberry 1974) which did not reach formal significance in any tooth. The agouti series affects pigmentation by an action on the dermis (Mayer and Fishbane 1972) at a site of ectodermal–mesodermal interaction and thus may have an effect on a similar site of interaction in the developing teeth.

THE OSTEOPETROSES

We now come to a group of mutants where the teeth may be abnormal due to a variety of causes. We know that the dentine of grey-lethal and microphthalmic mice is abnormal (Doykos *et al.* 1967; Al-Douri and Johnson 1983) and that root development and eruption in these mutants is also impeded by extra dense bone.

Tooth eruption

Tooth eruption, the migration of the forming tooth from its intraosseous location in the jaw to its functional position within the oral cavity, involves relative movement of tooth and jaw, mainly in the occlusal plane. The force necessary for this movement has been seen to lie in almost every adjacent tissue (Ten Cate 1969). Eruption has been divided into three phases, pre-eruptive, prefunctional, and functional. In the pre-eruptive phase the tooth moves prior to root formation; the prefunctional phase covers the time between the onset of root formation and occlusion; and the functional phase begins when the tooth reaches occlusion. In the late pre-eruptive phase the crown is fully formed and the tooth germ is surrounded by a loose mesh of connective tissue, the dental sac. The dental sac itself is surrounded in the cheek teeth by bone, except for a small opening near the lingual margin of the alveolar crest. This opening contains the gubernacular cord of fibrous connective tissue. During the eruption process the ameloblasts covering the crown are reduced in height to different extents at different locations. The prefunctional phase, in which the roots form, the hard tissue overlying the tooth is broken down, and the dental and oral epithelia are brought into contact, is not fully understood. Ten Cate (1980) suggested four possible eruptive mechanisms.

1. Root growth, the growing root being accommodated by occlusal movement of the crown.
2. Vascular pressure, whereby local increases in tissue fluid pressure in the periapical tissues push the tooth occlusally.
3. Selective deposition and resorption of bone around the tooth.
4. A pulling of the tooth into occlusion by contraction of the periodontal ligament.

The third, functional phase of eruption, is slow and prolonged accommodating growth of the jaws, occlusal wear, and interproximal wear of the teeth.

Since the growth of the teeth is eccentric, and varies from tooth to tooth the pattern of resorption of the wall of the bony crypt must also vary. This resorption must occur both above, to allow egress of the crown, and below, to allow growth of the roots.

Grey lethal (gl, Chr 10)

Grey lethal will serve as a model for tooth anomalies in osteopetrotic rats and mice. Grüneberg (1935, 1936a,b, 1938) discussed the retention of the teeth in the jaws and pointed out that, in the absence of resorption of bone, the dental crypts are too small and that the teeth are deformed by projecting lateral spicules of bone on which ameloblasts deposit a layer of enamel (Fig. 7.11). This leads to ankyloses of jaw and tooth, which, with the persistence of overlying bone above the tooth, make eruption impossible. The undercalcified nature of the dentine (Doykos *et al.* 1967; Fig 7.12) means that the roots are softer than normal and are bent and deformed. At the rear end of the lower incisor, which is also unable to erupt, the growing root pushes out of the mandible via the mental foramen where it forms a large irregular odontoma.

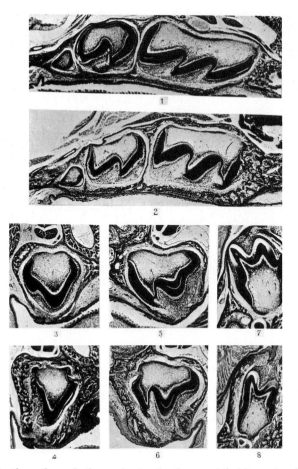

Fig. 7.11. Sections through the molar teeth of normal (odd numbers) and grey lethal (even numbers) teeth aged 9–10 days. Note the spicules of spongiosa pressing on the grey lethal tooth. (From Grüneberg 1936*b*.)

Doykos *et al.* (1967) found that gl and normal cheek teeth were identical until the eighth day post partum, when the dental follicle was impinged upon by surrounding unresorbed bone, as previously described by Grüneberg. This bone interrupts the epithelial root sheath. No radicular calcification is seen at this stage, but there is a belated attempt at root bifurcation at 20 days (normally this happens on day 12). Persisting alveolar bone causes interruptions in the tooth outline, and the cusps of the first molars appear compressed. The root sheath at 20 days is so compressed that it coils back on itself.

Hollinshead and Schneider (1975) were able to identify gl fetuses at 18 days of gestation. At this stage, as at birth, the posterior end of the short,

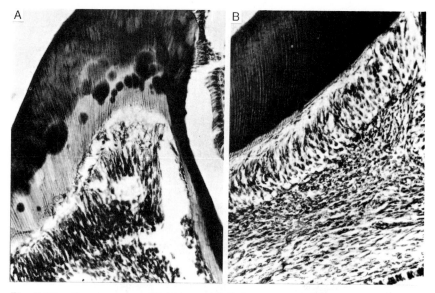

Fig. 7.12. Fourteen-day-old (A) grey lethal mouse and (B) normal littermate showing wide predentine area and interglobular dentine in the former. (From Doykos *et al.* 1967.)

stubby incisors did not extend caudally below the first lower molar tooth germ (Fig. 7.13). Spicules of thickened mandibular bone often impinged on the apex of the incisor even at this early stage of development. Molar follicles were also disrupted and invaded by bone. This process was more fully described by Schneider and Hollinshead (1976).

Microphthalmia (mi, Chr 6)

It is clear that the defects of gl teeth are secondary upon a defect of bone resorption. A similar situation is seen in microphthalmia; mi mice live longer than gl and teeth commonly erupt (Freye 1956), usually molars but occasionally incisors. The odontoma which is regularly present at the rear of the gl lower incisor is variable in mi (Grüneberg 1963). Freye (1956), using macerated specimens, found that even the incisors that erupt are retarded by at least six days, and are shorter than normal. Erupting molars are about 33 per cent smaller than normal. Konyukhov and Osipov (1966) gave details of improved eruption in older mice. Barnes *et al.* (1975) noted that parabiosis, although it resolved osteopetrosis elsewhere in the skeleton did not affect the jaw—teeth did not erupt—or the tail. Loutit and Sansom (1976) reported similar findings after intraperitoneal injections of haemopoietic cells.

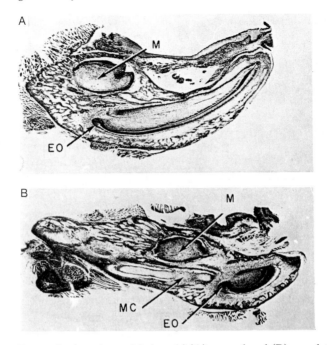

Fig. 7.13. Parasagittal sections of 1-day-old (A) normal and (B) grey lethal jaws. EO: enamel organ; M: molar; MC: Meckel's cartilage. (From Hollinshead and Schneider 1975.)

Al-Douri (1982) described the histology of mi teeth. The upper incisors of newborn mi did not resemble the classic segment of a circle (Addison and Appleton 1915) but looked more like an eagle's beak with a strongly downturned tip and a thick layer of dentine and narrowed pulp cavity. This resembles Grüneberg's (1936) description of gl upper incisors. At one-day-old the alveolar bone is closer to the tooth than normal, and the odontogenic epithelium has invaded the marrow spaces where ectopic dental tissue is found. With advancing age further deposits of enamel and dentine are laid down in marrow cavities leading to ankylosis. The first molar was regularly grossly affected by primary spicules of bone on the mesial and later on the distal surfaces. The cusps were more erect than usual. Second and third molars were usually unaffected. The molar roots never attained an appreciable length and were bordered by sclerosed bone from the outset. Walker (1966) was able to show that failure of molar eruption was due to excess production of bone matrix and not to failure of dentine production. Al-Douri and Johnson (1983) have shown, however, that mi dentine is abnormal. They also noted that the osteoclasts over the tip of the developing incisor at 10 days, but not at other times or in other places, appeared normal, and to be resorbing bone.

The incisors-absent (ia) rat

Incisors-absent rats have similar defects to the osteopetrotic mice described above, but show a spontaneous remission of the disease in later life. However incisors and first molars rarely erupt (Schour *et al.* 1949; O'Brien *et al.* 1958; Marks 1976) due to ankyloses of teeth to bone. As in osteopetrotic mice parabiosis (Marks 1976) cannot be used to rescue the teeth, presumably because it cannot be performed early enough. (But see Loutit and Sansom (1976) where suspensions of haemopoietic cells injected at birth failed to restore eruption in mi mice. We must remember that mi mice have abnormal bone as well as osteopetrosis.) Marks (1981) identified ia rats radiographically at birth, irradiated them so as to destroy mature haemopoietic tissue, then injected spleen cells from a normal littermate. In the ia rat the treatment was successful; in 14 out of 24 ia individuals all molars erupted. Odontoma at the base of the incisor was also absent or greatly reduced in size. Some molars erupted. Once again Johnson and Al-Douri (1982) were able to show that osteoclasts over the developing ia incisor were apparently normal and functional.

The teeth of these mutants are obviously secondarily affected by the osteopetrosis of the bone in which they find themselves. Work on mi shows that spicules of bone penetrate the teeth before birth, and this is presumably also true for the ia rat; prenatal relief of the osteopetrosis (which is recognizable in the 18-day gl fetus) might lead to reliable eruption of all teeth, and more normal root formation. Walker's work (1966) indicated that the abnormal dentine found in gl, mi, and ia is not a primary factor in the non-eruption of osteopetrotic teeth; we may surmise that the effects on the teeth in these mutants are purely secondary, and that the interaction of ectoderm and mesenchyme to produce the crown pattern is essentially normal.

Negative evidence from other genes

In the original paper describing both Tabby and crooked syndromes Grüneberg (1965) undertook systematic study of such material as was available to him, comprising 43 major loci and their normal alleles. With the exception of dw (see above) this survey was negative—it seems that cusp and root pattern is rarely upset by mutation in the mouse.

SUMMARY

If we subscribe to the idea that the defects in the teeth of osteopetrotic mice are secondary, then we are left with two major syndromes involving the teeth, and a series of minor defects.

The tabby–crinkled–downless syndrome, possibly controlled by the sleek locus is clearly a demonstration of the effects of a series of as yet unknown but disparate processes affecting mesodermal/ectodermal inter-actions and hence tooth development. The situation here is exactly comparable to that seen in chondrodysplastic mouse mutants and osteopetrosis, a variety of different first causes leading to a single well-marked developmental syndrome.

Crooked is a second major entity, with a basis (as far as its dental abnormalities are concerned; see Chapter 5 for other defects) in an abnormal dental lamina; so is the condition of molar fusion in the rice rat. In both cases the major interest lies in the fact that it is the maturation of the tooth that is affected rather than its cusp pattern—modifications to the cusp consist wholly of deletions of the later forming parts of the pattern which is itself basically unaffected although incomplete.

The effects of the agouti gene on the size of the molars probably reflect the action of this gene at another site of mesodermal–ectodermal interaction similar to that seen in the tooth crown; this is borne out by the much smaller effect of the albino locus on tooth size.

REFERENCES

Addison, W. H. T. and Appleton, J. L. (1915). The structure and growth of the incisor teeth of the albino rat. *Journal of Morphology* **26**, 43–96.

Al-Douri, S. (1982). Tooth eruption in osteopetrotic mice and rats. M.Phil Thesis, University of Leeds.

—— and Johnson, D. R. (1983). Ultrastructurally abnormal bone and dentine produced by microphthalmic mice. *Journal of Anatomy* **136**, 715–22.

Barnes, D. W. H., Loutit, J. F., and Sansom, J. M. (1975). Histocompatible cells for the resolution of osteopetrosis in microphthalmic mice. *Proceedings of the Royal Society* **B188**, 501–5.

Bhasker, S. N., Schour, I., McDowell, E. C., and Weinmann, J. P. (1951). The skull and dentition of screw tail mice. *Anatomical Record* **110**, 199–229.

Butler, P. M. (1956). The ontogeny of molar pattern. *Biological Reviews* **31**, 30–70.

Cohn, S. R. (1957). Development of the molar teeth in the albino mouse. *American Journal of Anatomy* **101**, 295–319.

Crew, F. A. and Mirskaia, L. (1931). The character hairless in the mouse. *Journal of Genetics* **25**, 17–24.

Crocker, M. and Cattanach, B. M. (1979). The genetics of sleek: a possible regulating mutation of the tabby–crinkled–downless syndrome. *Genetical Research* **34**, 231–8.

Danforth, C. H. (1958). The occurrence and genetic behaviour of duplicate lower incisors in the mouse. *Genetics* **43**, 139–48.

David, L. T. (1932). The external expression and comparative dermal histology of hereditary hairlessness in mammals. *Zeitschrift für Zellforschung und mikroscopische Anatomie* **14**, 616–719.

Doykos, J. D., Cohen, M., and Shklar, G. (1967). Physical, histological and roentgenographic characteristics of the grey lethal mouse. *American Journal of Anatomy* **121**, 29–40.

Fitch, N. (1957). An embryological analysis of two mutants in the house mouse, both producing cleft palate. *Journal of experimental Zoology* **136**, 329–57.

Fitzgerald, L. R. (1973). Deciduous incisor teeth of the mouse. (*Mus musculus*). *Archives of Oral Biology* **18**, 381–9.

Fraser, A. S., Sobey, S., and Spicer, C. C. (1953). Mottled, a sex modified lethal in the house mouse. *Journal of Genetics* **51**, 217–21.

Freye, H. (1956). Untersuchungen über die Zahnanomalie des Mikropthalmussyndrous de Hausmaus. *Zeitschrift für menschliche Vererbungs-und Konstitutionslehre* **33**, 492–505.

Gaunt, W. A. (1955). The development of the molar pattern of the mouse (*Mus musculus*). *Acta anatomica* **24**, 249–68.

—— (1956). The development of enamel and dentine on the molars of the mouse with an account of the enamel-free areas. *Acta anatomica* **28**, 111–34.

—— (1961). The development of the molar pattern of the golden hamster *(Mesocricetus auratus W.)* together with a re-assessment of the molar pattern of the mouse. *(Mus musculus)*. *Acta anatomica* **45**, 219–51.

—— (1966). The disposition of the developing cheek teeth in the albino mouse. *Acta anatomica* **64**, 572–85.

Grewal, M. S. (1962). The development of an inherited tooth defect in the mouse. *Journal of Embryology and experimental Morphology* **10**, 202–11.

Grüneberg, H. (1935). A new sub-lethal colour mutation in the house mouse. *Proceedings of the Royal Society* **B118**, 321–42.

—— (1936a). Grey lethal, a new mutation in the house mouse. *Journal of Heredity* **27**, 105–9.

—— (1936b). The relationship of endogenous and exogenous factors in bone and tooth development. *Journal of Anatomy* **71**, 236–44.

—— (1938). Some new data on the grey lethal mouse. *Journal of Genetics* **36**, 153–70.

—— (1951). The genetics of a tooth defect in the mouse. *Proceedings of the Royal Society* **B138**, 437–51.

—— (1963). *The pathology of development*. Blackwell, Oxford.

—— (1965). Genes and genotypes affecting the teeth of the mouse. *Journal of Embryology and experimental Morphology* **14**, 137–59.

—— (1966). The molars of the Tabby mouse, and a test of the 'single-active X chromosome' hypothesis. *Journal of Embryology and experimental Morphology* **15**, 223–44.

—— (1971). The Tabby syndrome in the mouse. *Proceedings of the Royal Society* **B179**, 139–56.

Hitchin, A. D. and Morris, I. (1966). Germinated odontome-connation of the incisors in the dog—its etiology and ontogeny. *Journal of Dental Research* **45**, 575–83.

Hollinshead, M. B. and Schneider, L. C. (1975). Prenatal development of the grey lethal mouse (teeth and jaws). *Anatomical Record* **182**, 305–20.

Howard, A. (1940). 'Rhino' an allele of hairless in the house mouse. *Journal of Heredity* **31**, 467–70.

Hunt, D. M. (1974). Primary defect in copper transport underlies mottled mutants in the mouse. *Nature* **249**, 852–4.

Hurley, L. S. and Bell, L. T. (1975). Amelioration by copper supplementation of mutant gene effects in the crinkled mouse. *Proceedings of the Society for experimental Biology and Medicine* **149**, 830–4.

Iorio, R. J., Bell, W. A., Meyer, M. H., and Meyer, R.A. (1979a). Radiographic evidence of craniofacial and dental abnormalities in the X-linked hypophosphataemic mouse. *Annals of Dentistry* **38**, 31–7.

—— —— —— —— (1979*b*). Histologic evidence of calcification abnormalities in teeth and alveolar bone of mice with X-linked dominant hypophosphataemia. *Annals of Dentistry* **38**, 38–44.

Johnson, D. R. and Al-Douri, S. (1982). Osteoclasts with ruffled borders from above the tip of the erupting incisors of osteopetrotic mice and rats. *Metabolic Bone Disease and related Research* **4**, 263–8.

Konyukhov, B. V. and Osipov, U. V. (1966). Study of maldevelopment of teeth in mice of the mutant strain microphthalmia. *Zhurnal Obschei biologii* **27**, 620–40.

Leamy, L. (1981). Effects of alleles at the albino locus on odontometric traits in coisogenic mice. *Journal of Heredity* **72**, 199–204.

—— and Hrubant, H. E. (1971). Effects of alleles at the agouti locus on odontometric traits in the C57BL/6 strain of house mice. *Genetics* **67**, 87–96.

—— and Touchberry, R. W. (1974). Additive and non-additive genetic variance in odontometric traits in crosses of seven inbred lines of house mice. *Genetical Research* **23**, 207–18.

Loutit, J. F. and Sansom, J. M. (1976). Ostopetrosis of microphthalmic mice. A defect of the haemopoietic stem cell? *Calcified Tissue Research* **20**, 251–9.

Lyon, M. F. (1961). Gene action in the X chromosome of the mouse *(Mus musculus L.)*. *Nature* **190**, 372–3.

Marks, S. C. (1976). Tooth eruption and bone resorption: experimental investigation of the ia (osteopetrotic) rat as a model for studying their relationships. *Journal of Oral Pathology* **5**, 149–63.

—— (1981). Tooth eruption depends on bone resorption: experimental evidence from osteopetrotic (ia) rats. *Metabolic Bone Disease and Related Research* **3**, 107–15.

Matena, V. (1972). The periodontum of the enamel aspects of the rat incisor tooth. *Journal of Periodontology* **44**, 311–15.

—— (1973). Periodontal ligament of the rat incisor tooth. *Journal of Periodontology* **44**, 629–35.

Mayer, T. C. and Fishbane, J. L. (1972). Mesoderm–ectoderm interaction in the production of the agouti pigmentation pattern in mice. *Genetics* **71**, 297–303.

—— Miller, C. K., and Green, M. C. (1977). Site of action of the crinkled (cr) locus in the mouse. *Developmental Biology* **55**, 397–401.

Miller, W. A., Flynn, K. L., and Drinnan, A. J. (1974). Dental and histological changes associated with chondrodystrophies in mice. *International Association for Dental Research Abstracts* **573**, 189

Morgan, W. C. (1954). A new crooked tail mutation involving distinct pleiotropism. *Journal of Genetics* **52**, 354–73.

Moutschen, J. and Colizzi, A. (1975). Absence of acrosome: an efficient tool in mammalian mutation research. *Mutation Research* **30**, 267–72.

—— and Houbrechts, N. (1976). Considerations sur le syndrome Tabby (gène lié au sexe) chez la souris. *Archives de Zoologie Expérimentale et générale* **117**, 359–81.

O'Brien, C., Bhasker, S. N., and Brodie, A. G. (1958). Eruptive mechanism and movement in the first molar of the rat. *Journal of Dental Research* **37**, 467–84.

Payne, T. M., Gartner, L. P., Hiatt, J. L., and Provenza, D. V. (1977). Molar odontogenesis in the hairless mouse. *Acta anatomica* **98**, 264–74.

Schneider, L. C., and Hollinshead, M. B. (1976). The role of alveolar bone in the noneruption of molar teeth in grey lethal mice. *Journal of Periodontology* **47**, 91–4.

Schour, I., Bhasker, S., Greep, P. R., and Weinmann, I. P. (1949). Odontomelike formations in a mutant strain of rats. *American Journal of Anatomy* **85**, 73–112.

Sofaer, J. A. (1969*a*). Aspects of the tabby–crinkled–downless syndrome. I. The development of tabby teeth. *Journal of Embryology and experimental Morphology* **22**, 181–205.

—— (1969*b*). Aspects of the tabby–crinkled–downless syndrome. II. Observations on the reaction to changes of genetic background. *Journal of Embryology and experimental Morphology* **22**, 207–27.

—— (1973). Hair follicle initiation in reciprocal combinations of downless homozygote and heterozygote mouse tail epidermis and dermis. *Developmental Biology* **34**, 289–96.

—— (1974). Differences between tabby and downless mouse epidermis and dermis in culture. *Genetical Research* **23**, 219–25.

—— (1977*a*). The teeth of the 'sleek' mouse. *Archives of Oral Biology* **22**, 299–301.

—— (1977*b*). Tooth development in the 'crooked' mouse. *Journal of Embryology and experimental Morphology* **41**, 279–87.

—— (1979). Additive effects of the genes tabby and crinkled on tooth size in the mouse. *Genetical Research* **33**, 169–74.

—— and Shaw, J. H. (1971). The genetics and development of fused and supernumerary molars in the rice rat. *Journal of Embryology and experimental Morphology* **26**, 99–110.

Summer, F. B. (1924). Hairless mice. *Journal of Heredity* **15**, 475–81.

Ten Cate, A. R. (1969). The mechanism of tooth eruption. In *Biology of the peridontium* (ed. A. H. Melchere and W. H. Bowen), pp. 91–104. Academic Press, New York.

—— (1980). *Oral histology, development, structure, and function.* C. V. Mosby, St. Louis Missouri.

Walker, D. G. (1966). Counteraction to parathyroid therapy in osteopetrotic mice as revealed in the plasma calcium levels and ability to incorporate ^{3}H-proline into bone. *Endocrinology* **79**, 836–42.

Wickramaratne, G. A.de S. (1974). The skeletal profile of some inbred strains of mice. *Journal of Anatomy* **117**, 565–73.

8. The limbs

The limbs arise from limb buds composed of a mass of mesoderm covered by a sheath of ectoderm. From this apparently simple stucture develops a limb which combines repetitive pattern (the five digits for example) with variation in pattern (none of the digits is identical to any other). In addition pentadactyl limbs embrace great specialization involving reductions, fusions, duplications, and differential growth (for reviews see Hinchliffe and Johnson 1980; Fallon and Caplan 1983) so that the basic pattern is modified to form the leg of a horse, the wing of a bird, or the flipper of a dolphin. All these limbs contain bone, cartilage, muscle, tendon nerves, and blood vessels. How does this diversity of pattern and histogenesis arise?

Much of the work on limb development has been performed on the chick, simply because the limb of an embryo bird *in ovo* is so much more accessible than that of an embryo mammal *in utero*. Recent developments in organ culture now allow us to repeat much of the classical work done on chick on mammalian limbs in organ culture, and so to establish whether or not mammals and birds do things in quite the same way. This is a valid question; another popular group of experimental animals, the urodele amphibians, form their limbs in a manner quite unlike that seen in chick.

Although this is not the place to review the development of the limb at great length (see Hinchliffe and Johnson 1980; Hinchliffe and Gumpel-Pinot 1983 for extensive reviews), we must at least consider the basis of development before we can hope to understand abnormalities. The following account is based upon the chick, and draws heavily upon the reviews already mentioned.

THE LIMB BUD

The limb bud, which arises from the Wolffian ridge, can first be identified as a mesodermal core covered by a thin ectodermal coat thickened distally into the apical epidermal ridge (AER) which runs in an antero-posterior direction. Perhaps the earliest significant experimental work (if we exclude the pioneering work of Harrison (1921) on the determination of the limb axes) depended upon the discovery that mesoderm and ectoderm can be separated and recombined. This led to the Saunders–Zwilling hypothesis (Zwilling 1961; Saunders 1977) of ectodermal–mesodermal interaction. The apical ectodermal ridge induces outgrowth of the underlying mesoderm but is dependent upon apical ectodermal maintenance factor

(AEMF) for its own continued health. Limb type (wing or leg in the chick) is determined by the mesoderm.

Inversion of the wing tip led to its duplication (Saunders and Gasseling 1968; Amprino 1965) in a mirror-image fashion. This was initially explained by a fancied asymmetry in the distribution of AEMF which was thought to be more concentrated posteriorly. Further experimentation led to the discovery of the zone of polarizing activity (ZPA) which was the real basis of the mirror-imaging (Fig. 8.1). The ZPA is a region of posterior mesoderm of the bud (or the flank near the posterior margin of the bud—polarizing activity varies in extent with time of development) which, if implanted preaxially, leads to the formation of a second wing tip whose polarity it controls. We shall return to the detailed effects of the ZPA later.

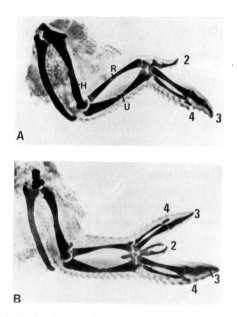

Fig. 8.1. Normal chick wing aged 10 days stained for cartilage. (B) Duplicated wing skeleton after the graft of zone of polarizing activity to anterior margin at the 3-day-stage. H: humerus; R: radius; U: ulna; 1–4, digits. (From Hinchliffe and Gumpel-Pinot 1983.)

The discovery of the ZPA together with recent work on amphibian and insect limb regeneration has produced a change of emphasis from the molecular biology of gene action to the reinvestigation of field phenomena in developmental biology. Harrison's (1921) pioneer experiments had suggested a series of biological fields at work within the developing limb; 50 years later the concept is once more fashionable. Mathematical models, using computers, and cell and organ culture have been added to classical grafting techniques as aids to the study of these phenomena.

Wolpert (1969, 1981) suggested that cells are in receipt of positional information—crudely that they know where they are within a field system. This knowledge is probably expressed with relation to the boundaries of the fields, and the ZPA is one boundary which can be neatly demonstrated experimentally.

Mapping skeletal areas

It is of obvious advantage to know what, in the primitive limb bud, is destined to become which particular part of the adult limb. Are all regions of the adult represented in the limb bud, or are new areas added, and if so where and what do they contain (Fig. 8.2). Saunders (1948) used carbon particles to map the developing dorso-ventral axis, and found that wrist and digit mesoderm was only identifiable from stage 20 onwards. Stark and Searls (1973) compiled a map based on the labelling of small blocks of limb mesoderm with tritiated thymidine and implanting these into the limb bud. Hinchliffe *et al.* (1981) used the nucleolar marker of quail cells. The prospective skeletal material seems to occupy a region approximately 2–3 somites wide; the anterior one-third of the limb bud makes no contribution to the digits.

Proximo-distal and antero-posterior regulation

The concept of an embryonic field implies regulation. If, as Wolpert (1981) suggested, a field contains a set of cells which have their position specified

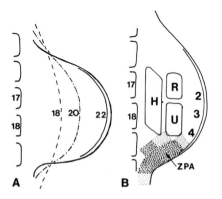

Fig. 8.2. (A) Outlines of chick wing buds at stages 18–22. (B) The prospective areas for humerus, H; radius, R; and ulna, U; digits 2, 3, 4; and the ZPA at stage 20. (From Hinchliffe and Gumpel-Pinot 1983.)

by the same boundary, then if the boundary is changed one would expect the field to change too—to regulate for excess or scarcity of tissue.

Regulation along the proximo-distal axis has been the subject of much discussion (see Hinchliffe and Johnson 1980 for review). We are dealing here with a region which encompasses many developmental stages at any one time; proximally blastemata may be cartilaginous or even starting to ossify while distally tissue is still apparently undifferentiated, and still being formed by mitosis of sub-ridge mesenchyme. Summerbell *et al.* (1973; review Wolpert 1978) explain the proximo-distal differentiation of the limb by reference to the concept of a 'progress zone', an area of mesoderm approximately 300 μm deep lying below, and maintained in a labile state by the overlying AER. Cells of the progress zone divide, and those which leave proximally have a positional value depending on how long they have spent in the zone, i.e. the first cells to leave are 'proximal' and later deserters become progressively more 'distal'. Once out of the progress zone the positional value of the cell is fixed; it is determined to form a particular structure, or at least to lie in a particular zone of the proximo-distal axis.

This hypothesis is in accord with the findings of Rubin and Saunders (1972) and Jorquera and Pugin (1971) that the age and even the species of AER does not influence the ability of the underlying mesoderm to form a complete skeleton. The model predicts that P–D differentiation is autonomous—once the cells have left the progress zone their fate is fixed; thus, if zeugopodal areas (radius + ulna or tibia + fibula) are duplicated or removed, the stump should not affect the differentiation of the operated tip. Summerbell's group (Summerbell *et al.* 1973; Summerbell and Lewis 1975; Summerbell 1977) demonstrated this lack of regulation by excision–duplication experiments. However, Hampe (1959) and Kieny (1964, 1977) found good regulation in apparently similar experiments carried out at similar developmental stages. If chick–quail grafts are used the contribution of host and graft to the final limb may be assessed and it is clear that stump tissues have had their fates shifted to produce structures more proximal than they would normally form.

This apparant contradiction is partly resolved by a close look at experimental technique. The French school use leg buds, which appear to regulate better than wing buds. Maximal graft–stump contact is necessary for regulation; some of Summerbell's experiments were carried out with stumps at stage 24, when cytodifferentiation of the zeugopod has already begun, and probably, therefore, too late for regulation. Hornbruch (1980) showed that the ability to regulate limb length is gradually lost between stage 19 (when it is perfect) and stage 25 (when it is nil).

Antero-posterior regulation seems to follow the same rules. Yallup and Hinchliffe (1983) produced limb buds in which the central 1.5-somite-wide section was either removed or duplicated. This section is believed to contain the prospective material of most of the skeleton. Good regulation was again seen over the stage 19–22 range but by stage 24 pattern regulation had virtually ceased and size regulation was poor.

These results suggest that regulation is in fact a property of limb buds of stages 19–23 and occurs in both AP and PD axes. After stage 23 the property is lost. This argues against the rigid application of the progress zone theory of differentiation.

The ZPA and antero-posterior differentiation of the limb

We have already noted that the rotation of the tip of the chick wing bud through 180° leads to duplication of the wing tip skeleton. Saunders and Gasseling (1968) showed that the same effect was achieved is a small piece of postaxial mesoderm, the ZPA, was transplanted to an anterior site. The AER over the graft thickened and elongated and excess mesoderm appeared. From this extra host mesoderm a second skeleton developed, always orientated so that its most posterior digit was adjacent to the graft (Fig. 8.1). Mesoderm from anterior or central areas of the limb bud failed to elicit this response. The pattern of muscles and tendons is also duplicated in this type of experiment (Shellswell and Wolpert 1977). The polarity of the newly initiated wingtip can be demonstrated to be determined by the transplanted mesoderm; the effect of any transplanted ectoderm is minimal. MacCabe et al. (1973) and Summerbell and Honig (1982) accurately mapped the extent of the ZPA at different stages of development. The ZPA appears to be a general feature of the amniote limb, having been demonstrated in reptiles, birds, rodents, and man. Any ZPA will duplicate the host tissue when transplanted into the chick wing bud. The ZPA does not contribute to the new structures, but initiates them; the new structures are of host origin and irradiated ZPA can affect a response although unable to make any cellular contribution (Smith 1979). MacCabe and co-workers (MacCabe et al. 1977; Calandra and MacCabe 1978) developed an assay for ZPA and identified two active substances, a low-molecular-weight (LMW) soluble component and a high-molecular-weight entity, probably a glycoprotein (McCabe and Richardson 1982) which is absent from anterior mesoderm. The high-molecular-weight fraction seems to consist of the LMW fraction bound to a large inactive carrier (MacCabe et al. 1982).

Cook and Summerbell (1980) demonstrated a significant increase in mitotic index in host cells after ZPA grafting. Honig (1981) showed, by an elegant grafting experiment, that the effect on mitotic index occurs throughout the host limb bud, rather than just locally.

The effect of the ZPA graft is thus twofold. First it imposes a new pattern and second it creates new mesoderm; the six or seven digits in a grafted limb are of normal size, not reduced. These two effects are linked.

Smith and Wolpert (1981) showed that the greater the widening of the host A–P axis the greater the degree of polydactyly.

The effect of the ZPA can also be shown without grafting by inserting a barrier in a normally developing limb. Summerbell (1979) isolated the anterior part of the developing chick wing by means of an impenetrable barrier in the centre of the bud, and found that usually no digits developed on its anterior side. Hinchliffe (1981) found greatly increased cell death anterior to a similar barrier 24 hours after its insertion. Anterior parts of the limb bud also regress if the posterior region containing the ZPA is removed (Hinchliffe and Gumpel-Pinot 1981).

These experiments suggested to several workers that the ZPA is a source of signal interpreted by the limb mesenchyme. Tickle *et al.* (1975) suggested that the ZPA is the source of a morphogen whose concentration declines anteriorly due to the presence of an anterior sink (Fig. 8.3). A morphogen gradient thus exists whose various levels specify the different digits. Wolpert (1981) believed that the signal in such a system is universal and the response depends upon the genome and developmental history of the responding tissue; certainly the signal is not species- or wing-leg-specific, although the response is. Hornbuch (in Hinchliffe and Gumpel-Pinot 1983) found that Hensen's node will substitute for ZPA. A second grafted ZPA would obviously alter the 'sag' of the morphogen curve (Fig. 8.4) and could affect the signal in various ways, changing the number, the width, and the identity of the digits specified (Fig. 8.5).

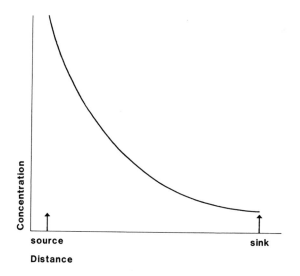

Fig. 8.3. According to Tickle *et al.* (1975) the ZPA is the source of a morphogen which is destroyed in an anterior sink. The concentration of such a morphogen would assume an exponential form.

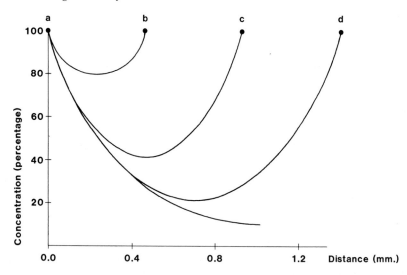

Fig. 8.4. The addition of a second anterior source at varying distances (b–d) from the ZPA (a) would alter the sag of the morphogen concentration curve and convert it to a parabolic shape.

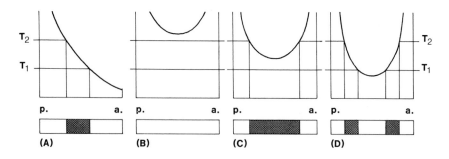

Fig. 8.5. If (A) we specify two thresholds (T_1, T_2) between which a cartilaginous element is specified by ZPA morphogen, then grafting a second ZPA gives the following possibilities: (B) The concentration may not fall between T_1 and T_2 and no digit will be formed. (C) It may fall below T_2 but not below T_1 specifying a single rather broad digit. (D) It may fall below T_1 specifying two digits. (After Slack 1977.)

The ZPA in normal development

The role of the ZPA has been specified largely by experiments creating an unusual set of circumstances; its role in normal development has been questioned by a number of workers (Saunders 1977, 1982; Saunders and Gasseling 1982; Rowe and Fallon 1981, 1982) since normal development does not always seem to depend on the presence of a ZPA, and ZPA-like properties can also be elicited from non-limb mesoderm.

MacCabe *et al.* (1973) and Fallon and Crosby (1975) were able to show normal development of wings after ZPA removal, a finding which appears to contradict the amputation work of Hinchliffe and Gumpel-Pinot (1983). The latter authors suggested that the former groups did not completely remove the ZPA activity, but only excised the area of its highest concentration, and pointed out that Tickle (1980) showed that the amount of active ZPA principle needed for normal development is in practice only a fraction of that normally contained within the limb.

Saunders (1977; also Saunders and Gasseling 1982) found ZPA activity in flank and tail tip mesoderm. This is hardly an argument against the normal role of the ZPA. Taken with Hornbuch's demonstration of the activity of Hensen's node it merely strengthens Wolpert's (1981) contention that the signal in morphogenic fields may be universal.

The concept of the ZPA is not, however, universally accepted. Iten and her co-workers do not accept the unique nature of the ZPA but suggest that limb development is governed by local differences in positional information in the limb mesoderm and therefore involves all of it. This view derives from another model of limb development, the polar coordinate model.

Polar coordinate model

French *et al.* (1976) and Bryant *et al.* (1981) proposed a rather different model to account for the presence of embryonic fields. The polar coordinate model was originally proposed in relation to insect and amphibian limb regeneration but Iten (Iten and Murphy 1980a,b, Iten 1982) suggested that this model is also applicable to the developing chick limb, and, by inference, to other amniotes. The polar coordinate model suggests that the cells around the circumference of a limb or limb bud have unique positional information values (Fig. 8.6); whenever an experiment, such as a graft or a reversal is performed, and tissues of inappropriate positional values find themselves next to each other, tissue is intercalated until the positional gradient is restored. Intercalated tissue will then produce structures appropriate to its positional value. This model accounts for amphibian and insect regeneration rather well; it differs from the ZPA model in suggesting that interactions are short-range and that positional values of graft and host do not change following experimentation. The ZPA on the polar coordinate model is thus not unique, but simply one of a series of positional values.

Iten concludes that the classical preaxial ZPA graft fits the polar coordinate model well. A growth zone will form between graft and anterior host tissue, which would intercalate positional values; these would be

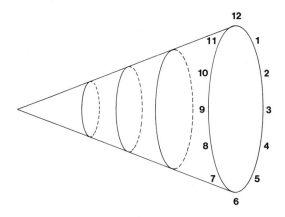

Fig. 8.6. In the model of Bryant *et al.* (1981) the limb is seen as a cone with a proximal base and distal apex. The position of a point on the limb surface is specified by its position on the proximo-distal axis and its position on the clock face of a section taken at that level.

appropriate for the formation of supernumerary digits. Iten and Murphy (1980*a*) grafted wedges of anterior tissue into slits made more posteriorly in the wing bud. Supernumerary digits were produced, and the greater the host–graft disparity the greater the degree of polydactyly. If ZPA tissue is removed and the remaining limb bud tissue pinned back to the stump in normal orientation, a normal wing is formed (Iten and Murphy 1980 *b*). This result is attributed to the growth of tissue intercalated at the wound surface.

In fact both these series of experiments can be interpreted by classic ZPA theory. A graft of anterior wing bud tissue inserted posteriorly will split the ZPA into two areas, one of which produces normal and the other supernumerary digits. Wedges inserted by Iten and Murphy away from the ZPA produced only a low incidence of polydactyly. If, as on the polar coordinate model, no area is special, then response should be equal in all experiments for grafts of the same size, no matter what their origin. ZPA removal experiments may not have been drastic enough to remove all activity; once more we cite Tickle's (1980) finding that only a fraction of the ZPA principle present in the limb is necessary to ensure normal differentiation. MacCabe *et al.* (1973) also showed high ZPA activity in the body wall posterior to the limb bud. Honig (1983) repeated Iten and Murphy's experiments using a graft of quail tissue. The resultant supernumerary digits were all quail, i.e. all graft, suggesting that they had been formed by the graft in response to the host ZPA. The coordinate model would predict that such digits should be of mixed graft and host origin.

The concept of immutable positional value is also difficult to reconcile with experimental results. For instance Yallup and Hinchliffe (1983) found that excesses formed by grafting extra tissue in the A–P axis were regulated away and a normal three-digit wing formed. On the polar model polydactyly should have resulted, with an extra digit or extra digits of mixed provenance intercalated.

The total excision of the ZPA (Hinchliffe and Gumpel-Pinot 1981) also leads to the death of anterior mesoderm, while the reciprocal experiment has little or no effect on the posterior ZPA-containing tissue. This also suggests a special role for the ZPA. The barrier experiments of Summerbell (1979) also suggest a long-range action for the ZPA.

Somites and limb buds in development

Limb bud mesenchyme uncontroversially contains somatopleural tissue. But the contribution of somites to the limb has aroused more controversy. Does the limb bud depend on the somite for its development? Does the somite make a cellular contribution?

Early work was contradictory on the first of these points. Hunt (1932) obtained good development of limb buds explanted with somites; Hamburger (1938) grew limbs from explants which he claimed were virtually somite-free. Murillo-Ferrol (1963, 1965) was able to block limb development by the insertion of a barrier, inorganic or organic, between somite and limb bud. A millipore filter, however, allowed development to proceed. Pinot (1970) grafted limbs with or without somites and found that the limb (at stage 15 or older) was able to develop autonomously. But before this stage (confirmed by Kieny 1971) the presence of somites was necesary. Dhouailly and Kieny (1972) found that wing or leg somites play an active role in specific further development; Pinot inclined to the view that somites provide a permissive influence.

The contribution of somites in terms of cells was demonstrated by chick–quail hybrids after some initial controversy (see Hinchliffe and Gumpel-Pinot 1983 for review). Similar experiments, involving the substitution of chick for quail somites and vice versa were performed independently by two groups of workers (Christ *et al.* 1974; Christ and Jacob 1980; Jacob *et al.* 1978; Jacob and Christ 1980; Chevallier *et al.* 1976, 1977*a,b*; Kieny 1980). The results were similar; limb skeletal muscle cells are somitic in origin but those of smooth muscle, cartilage, tendons, and connective tissue are somatopleural. Somites from any level of origin can produce limb muscles (Kieny and Chevallier 1980). Other authors (Zwilling 1966; Searls and Janners 1969; Nathanson *et al.* 1978) contested the view that the distinction is meaningful: they suggested that mesenchymal cells are determined by intrinsic conditions within the limb bud, such as position rather than origin. In abnormal conditions, such as a limb primordium grafted before somitic invasion (McLachlan and Hornbruch 1979), somatopleural cells can form skeletal muscle. Christ and Jacob (1980), however, claimed that the lineage is all, and this view is widely accepted.

Innervation

The limbs are innervated by several spinal segments which make up the brachial and lumbo-sacral plexuses. The actual segments involved are not critical and vary a little from strain to strain. A limb bud grafted on to a flank will acquire a normal nerve supply from quite inappropriate spinal ganglia (Hamburger 1939; Lewis 1980).

The main nerve trunk penetrates the chick limb but at around four days of incubation, and consists of motor and sensory axons. The pattern of innervation (Roncali 1970) is invariant in a particular strain, even though it is established at a stage when muscle and cartilage are just beginning to differentiate.

In a limb where the pattern of the musculo-skeletal system has been upset prior to nerve invasion by truncation, intersegmental duplication, or deletion either in the A–P or P–D axis, Stirling (1976) and Stirling and Summerbell (1977) showed that the route followed by the invading nerves is characteristic of the skeletal level they traverse. If a segment is missing, so is the corresponding branch of the nerve. The paths of nerve outgrowth are clearly defined by the intrinsic structure of the limb (Horder 1978).

In limbs experimentally produced without muscles (Lewis *et al.* 1981) the main mixed nerve trunk develops, but there are no muscular branches. Thus the motor nerve pattern appears to be controlled by the muscle pattern (Lewis 1980). The first nerve reaching a muscle innervates it and excludes all others. The muscles are thus innervated in an orderly sequence (Roncali 1970) and motor nerves not reaching a muscle die (Pettigrew *et al.* 1979). Nerveless wings have a normal muscle pattern, but the muscles gradually atrophy (Straznicky 1967; Shellswell 1977; Lewis 1980). Elimination of the sensory component of innervation by destroying part of the neural crest (Lewis 1980) does not affect motor neurones but the limbs which developed after operation lack cutaneous innervation. Thus motor axons do not rely on sensory axons or Schwann cells to reach their targets.

Innervation does not seem to govern the development of the skeleton. There is some evidence (reviewed by Hinchliffe and Gumpel-Pinot 1983) that spinal cord excision produces hememelia, but this is now thought to be due to operator effect in the difficult surgery involved. Lewis (1980) used focused ultraviolet irradiation to circumvent this problem and produced nerveless wings with perfect skeletons.

Cell death

Differential cell death (Zwilling 1964) helps to shape the limb bud. This topic was reviewed by Hinchliffe and Johnson (1980). Two zones of cell death, the anterior and posterior necrotic zones (ANZ, PNZ), seem to vary in size inversely as the number of digits in a particular species, being large in birds and virtually absent in mouse and rat (Milaire 1971, 1977a,b). Interdigital necrotic zones (INZ) help to separate and shape the developing digits. In web-footed birds the INZ is much reduced.

THE MUTANTS

Genetic effects on limb development may be divided into a number of groups purely for the sake of convenience. The first group concerns the size of the limb bud which may be decreased or increased by a mutation. The effect may be regulated either via the size (or length) of the apical epidermal ridge (AER), or the duration of the period over which it is active, or a combination of the two. The effects range from almost complete absence of the limb (if the bud is drastically reduced) through ectrodactyly (the absence of some digits) and syndactyly of various kinds (membranous, cartilaginous, osseous, soft tissue) and normality to polydactyly. The effects seen are largely distal; a deficiency or excess of mesenchyme seems to affect parts formed comparatively late in development, i.e. footplate rather than femur, and often preaxial rather than postaxial structures although these rules are by no means rigid. Often low-grade abnormality in a particular mutant is distal, and more grossly affected individuals show progressively more proximal defects. The preaxial rather than postaxial expression may be based on the posterior positioning of the ZPA, although morphological changes such as transition from mesenchyme to cartilage, cartilage to bone do not follow this sequence (Forsthoefel 1963a). In practice we may see both pre- and postaxial polydactyly, and often deletions and duplications in the same limb bud.

A second group, as yet poorly understood and poorly represented, is concerned with redistribution of the limb bud mesoderm between the segments of the developing limb; a third represents a heterogeneous group of conditions such as duplications of the whole limb bud or doubling of the AER. A fourth, included formally, comprises the chondrodystrophies, which affect the proportion of the limb but rarely the pattern.

REDUCTION OF THE LIMB BUD

Winglessness in the chick (wg, American wingless, ws)

Three distinct entities have been described under the name wingless in the chicken. Waters and Bywaters (1943) reported the effects of an autosomal recessive lethal, later given the symbol wg by Hutt (1949). This gene reduced the wings to small stumps or eliminated them altogether and had a somewhat less drastic effect on the legs. Affected individuals also lacked lungs and their associated air sacs, the metanephric kidney, and ureters. Zwilling (1949) reported that in this mutant the wing bud appeared but lacked an AER and did not progress beyond an early stage of development.

Zwilling's later (1956) classic experiments on the recombination of ectoderm and mesoderm were performed using mutant material from another wingless mutant, apparently never given a symbol and referred to here as American wingless. This arose in a flock at Storrs, Connecticut and closely resembled wg (which was by now extinct) in its effects on the limbs and kidneys, but lungs and air sacs were not affected. Saunders (in Hinchliffe and Ede 1973) found that in this mutant lateral-plate mesoderm in the region of the prospective wing was necrotic at stage 16–17 and that the wing AER did not form. Leg bud development is relatively normal until stage 19–20 when mesenchymal cell death and AER regression take place. The complete foot is never formed and a range of expression is found from complete absence of the hind limb to a reduction in the number of toes.

Hinchliffe and Ede (1973) and Hinchliffe (1976) studied a third wingless mutant (ws), a sex-linked gene discovered by Pease (1962) and christened by Lancaster (1968). Lancaster found a classic expression with absent wings and normal hindlimbs in 36 per cent of affected individuals (Fig. 8.7), rudimentary wings of various sizes, often with marked asymmetry in individuals with less than modal expression, and abnormality of the hind limbs in rather more severely affected birds. This included reduction in the number of digits and angulation of the tibia, which Lancaster considered to have 'broken and fused together again in an abnormal position before hatching'. Ede (1968) reported the absence of an apical ectodermal ridge in the wing of most embryos, that the wing mesenchyme was necrotic at four days, and that the angulation of the tibia appeared to be caused by weak cartilage matrix formation.

Hinchliffe and Ede (1973) showed that the wingless phenotype could first be seen at stage 19 (three days of incubation) as a region of cell death on the anterior side of the wing bud, which at this stage is merely a low ridge. At stage 20 wingless buds are smaller than normal and outgrowth soon ceases so that by stage 23–24 there is no wingbud (Fig. 8.8). Where the expression is mild the posterior part of the wingbud may form. Leg buds at stages 19–21 appear normal; in some cases anterior and distal parts seem to be missing by stage 22, and later the digital number can be seen to be reduced from 4 to 3, 2, or 1.

The enlarged area of cell death seen in wingless is thought to correspond to the normal anterior necrotic zone (ANZ) of the chick wing, but to be extended in both space and time. Other areas of cell death (opaque patch, PNZ) appear normal. The relationship between cell death and the AER is

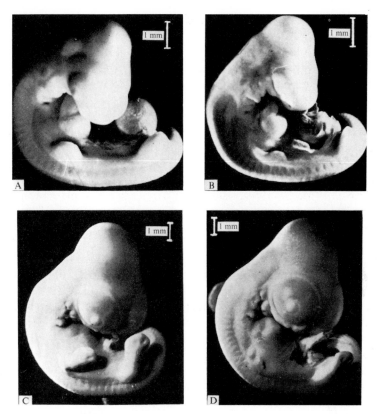

Fig. 8.7. (A,C) Normal and (B,D) wingless embryos at stages 22 and 26. (From Hinchliffe and Ede 1973.)

important. The AER normally appears at stage 19, but most wingless forelimbs have no ridge between stages 19 and 23, and after this any AER regresses. In lightly affected limb buds, however, the ectoderm overlying normal (non-degenerating) mesoderm does have a ridge in posterior parts of the bud. The ridge in ws hindlimbs is present posteriorly, but shortened anteriorly. The wingless leg bud is later missing anterior digits. The tibia may be absent or buckled but the fibula ia always present and well developed.

Hinchliffe and Ede (1967) suggested that the ANZ and PNZ in the normal chick limb act as 'end stops' to determine the extent of the AER, which is likely to be larger in pentadactyl animals than in birds, where limb skeletons have a reduced number of digits. They cite the example of talpid[3] (see below) where the absence of ANZ and PNZ is associated with extreme polydactyly. In wingless, extension of the ANZ limits the AER

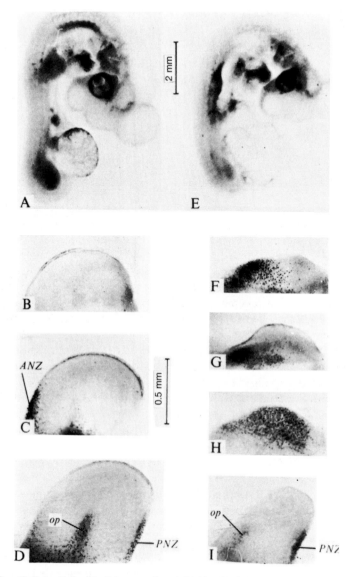

Fig. 8.8.   Cell death in (A–D) normal and (E–I) wingless forelimbs from stage 22–24. ANZ, PNZ, anterior and posterior necrotic zones; OP, opaque patch. (From Hinchliffe and Ede 1973.)

and reduces loss of anterior digits, and, in extreme cases, all or part of the tibia.

The anterior and posterior necrotic zones are clearly different from each other in wingless; this could be a manifestation of the phenomenon described by Cairns (in Ede 1971) that if the ectoderm is stripped from a developing limb the anterior mesoderm dies first. Also the anterior necrotic zone appears first in normal development. Is the anterior mesenchyme more susceptible to cell death?

Zwilling (1956), using the American wingless, found that when normal ectoderm was combined with wingless mesoderm the AER regressed after two or three days. Wingless ectoderm combined with normal mesoderm supported no further growth at all. He interpreted these results as showing a reciprocal dependence between AER and mesoderm—the mesoderm responding to the stimulus of the AER by growing distally and providing a factor (AEMF) necessary for the continued growth of the AER. In the mutant the latter appears to be deficient. However, Hinchliffe and Ede (1967) pointed out that the recombination was done at a fairly late stage, and a direct effect of the mutation on the ectoderm cannot be excluded.

If ws is a homologue of American wingless, then the phenotype could be explained on the basis of mesenchymal cell death interfering with the production of AEMF and consequent failure to maintain the ridge. Zwilling contended, however, that AEMF was essentially produced postaxially. It has been argued, not altogether convincingly, that the anterior increase in necrosis eventually spreads postaxially and would thus interfere with AEMF production and that postaxial ws cells may cease AEMF production some time before death. Since Hinchliffe and Ede (1973) found that failure of ridge initiation and cell death both occur at stage 19, they proposed two other hypotheses. First that mesenchymal cell death may prevent the initiation of the AER (which seems likely in Saunders' American wingless observations which noted necrotic mesoderm in the region of the wing from stage 16 onwards) or second that the initial lesion lies in the ectoderm.

Saunders (1972) reported experiments which suggest that mutant tissues lack the polarizing activity of normal wing buds. Sawyer (1982) looked at the ultrastructure of American wingless wing buds and described the reduced AER of the three-day-old mutant, where the cells are cuboidal and the nuclei irregular in shape. Below this defective AER there is no subectodermal space (Fig. 8.9) and the mesodermal cells are compact and rounded. At 4.5 days when the normal AER, is at its greatest extent, that of wingless is still cuboidal with less contact than normal between basal cells and less gap junctions. Underlying mesodermal cells are rounded with bulbous cytoplasmic extensions rather than stellate with thin filopodia. Sawyer describes the mutant mesoderm as tight-packed, and not oriented towards the ectoderm. Filopodia are reduced in number and length. This is thought to indicate that the gene is acting locally on the cell membrane in a way which could affect cell behaviour and association. No decrease in mitosis was noted, nor was excess cell death, as seen by Hinchliffe and Ede (1973) in ws.

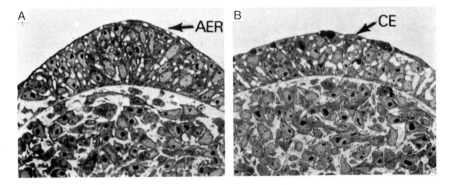

Fig. 8.9. Wing buds from (A) normal and (B) wingless embryos showing the compact ectoderm, CE, and lack of AER in the latter. (From Sawyer 1982.)

Ametapodia (Mp) in the chick

Cole (1967) described a dominant gene ametapodia in chicks which results in the absence of tarsometatarsus and carpometacarpus in the heterozygote but leaves toes intact (Fig. 8.10). The wing skeleton consists of two rows of two or three elements, the phalanges of digits II and III. The ulnare and radiale are present but metacarpals II and IV are absent. In the foot these distal elements present are rather broad, and there is some fusion, but in all they correspond well to the phalanges of digits II, III, and IV; there is no sign of digit I. Two spherical elements correspond to the proximal tarsals but metatarsals (fused in the normal to form the tarsometatarsus) are absent. Secondary fusions later tend to occur in the foot. Femur, tibiotarsus, humerus, radius, and ulna were all found to be slightly but significantly reduced in Mp. The homozygote is thought to be lethal during the first half of incubation and has been identified as having a 'distinct lateral projection from the distal end of each hindlimb'.

Ede (1968, 1977) showed that in this mutant the ANZ is suppressed, but the PNZ enlarged between stages 23 and 25. Later, at stage 27, there is cell death throughout the metapodal region which is thought to be responsible for the main effects of the gene.

Ectrodactyly (ec) in the chick

Abbot and MacCabe (1966) described a gene affecting the upper beak in the chick. When present in association with the scaleless gene this also produced defects of the feet, first recognizable at stage 17. The effects varied from reduction of the hallux to complete absence of legs. An average embryo has two digits per foot. No further details are available.

Disorganization (Ds, Chr 14)

Hummel (1958, 1959) reported limb defects in mice carrying the semidominant disorganization mutant. The syndrome produced by this gene is very complex indeed and may be summarized as follows.

In about 30 per cent of heterozygotes the limbs are abnormal with a characteristic presentation. Any limb may be affected, but usually not

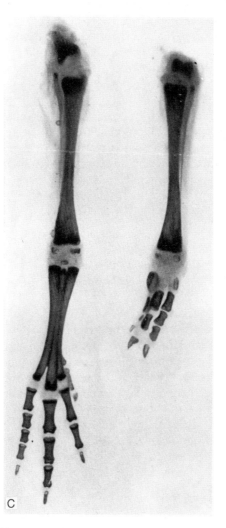

Fig. 8.10. (A,B) Skeleton of the foot and hand of an ametapodian chick aged six weeks. (C) Skeletons of the shank and foot of normal and ametapodian day-old chicks. (From Cole 1967.)

more than one. Polydactyly is present, but the extra digits produced are not recognizable as belonging to any particular part of the footplate. The number of digits may be almost doubled. Additional limbs are also found, usually close to normal limbs, and limb-like projections are found elsewhere in the fetus, particularly on the ventral abdomen. The gene also produces papillae of various sizes in almost any location on the body. The largest of these may merge in classification with the smaller supernumerary limbs since they contain bone and cartilage; others resemble urinogenital papillae. In addition almost half the embryos carrying disorganization have cranioschisis and exencephaly.

Ede (1980) recognized that neural ectoderm and epidermis are abnormal in this gene and suggested that the polydactyly may be ectoderm led—an outgrowth of ectoderm being filled with mesoderm and producing an attempt at a limb, better attempts being seen close to normal limb fields. However other abnormal structures reported by Hummel (1959) consist of nodules of embedded tissue with no ectodermal component. Disorganization resembles extra-toes (Johnson 1967) and talpid[3] (Ede and Kelly 1964*a,b*) in producing a non-organized type of extra digit. It is interesting to note that Johnson found abnormalities of other ectoderm–mesoderm structures (such as vibrissal papillae) in Xt.

Pupoid fetus (pf, Chr 4)

Meredith (1965) described a condition in the mouse, lethal at birth when homozygous, in which the whole fetus is cocooned in an epidermis which shows excessive proliferation. Within the cocoon internal organs are virtually normal, albeit mechanically constricted. The skeleton is a little distorted but otherwise normal.

The ectodermal abnormality is first seen in sections at 11–12 days of development as regions where the boundary between ectoderm and mesoderm is hard to define. These areas correspond to the areas first reached by peripheral sensory nerves. As development proceeds their association becomes more marked; beneath the growing and thickening areas of abnormal ectoderm, peripheral nerve endings become excessively branched and may penetrate the epidermis itself. Cells even break through the surface to form an additional layer (Ede and Watson, in Ede 1980, suggested that these are Schwann cells).

The interest in this gene from our point of view lies in the fact that the pattern of the limbs is not affected; presumably the overgrowth of the epidermis, stimulated by nerve endings, occurs after the pattern of the limbs has been laid down.

THE PREAXIAL HEMIMELIA–LUXATE GROUP

We now come to a large and interesting group of mutants in mouse, chick, and guinea pig which have been variously described as hemimelia, luxate, or luxoid. The group has two features of special fascination. First of all, abnormals in this group may display, within the same genotype, a phenotype which ranges from loss of preaxial digits and the tibia or ulna to a condition with extreme preaxial polydactyly where the tibia or ulna may be partially or completely duplicated. Second, the polydactyly which occurs is of a distinctive mirror-image type; I shall demonstrate later that this mirror imaging is in fact a feature of many preaxial polydactylys.

The hemimelic mutants are perhaps best arranged in a rough serial order beginning with those which resemble wingless in the chick in having a reduction in the preaxial and distal elements, but are rarely or never polydactylous, to those in which polydactyly is the dominant aspect of the syndrome, and preaxial loss is confined to the tibia or ulna.

Dominant hemimelia (Dh, Chr 1)

Dominant hemimelia (Searle 1964) has a series of effects on both the limb skeleton and the viscera. The spleen is absent and the alimentary canal shorter than normal. The gene is semidominant. Heterozygotes have external abnormalities confined to the preaxial side of the hindlimbs. These may include luxation and shortening of the limbs and loss of preaxial digits, loss of one or two digits or part of a digit, a slight thickening and lengthening of the hallux, triphalangy of the hallux, or addition of a digit or part of a digit ouside the hallux, i.e. preaxially. Soft tissue syndactylism is also common. Homozygotes usually die in the first few days of life but a few reach weaning and even breed. Their hindlimbs are short, twisted, and oligodactylous with the loss of up to three·preaxial digits: forelimbs are normal.

Clearance preparations show that skeletal effects are also varied (Fig. 8.11). The heterozygote may show a range of deletions or duplications ranging from loss of the tibia or fragmentation of the femur, inevitably associated with loss of the hallux (the pubis was also rarely involved), to the preaxial addition of up to three complete digits. Hemimelia of the tibia is seen in polydactylous feet, but not commonly. Polydactyly is usually expressed as a triphalangeous or thick hallux and the addition of an extra digit ahead of a triphalangeal hallux. Presacral vertebral number is slightly reduced. The homozygote is always oligodactylous, lacking up to four digits. The femur is often short, or distorted and fragmented. The tibia is usually absent. These defects are present in the cartilaginous stage where it is possible to see that most of the tarsus (save calcaneous and cuboideum) and metatarsals I and II and corresponding phalanges are also absent.

Precartilaginous abnormalities are seen in the limb skeleton at 12.5 days as absence of mesenchymal condensations.

Milaire (1965) found difficulty with the variable nature of the phenotype,

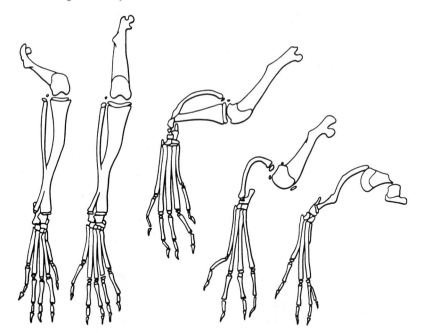

Fig. 8.11.   Right hindlimb skeletons of adult normal (left) and Dh/+ mice (right), showing various grades of abnormality from preaxial triphalangy to oligodactyly. (From Searle 1964.)

but was able to confirm Searle's findings. He also demonstrated that Dh/+ hindlimbs were deficient from the twelfth day (Fig. 8.12).

Rooze (1977) was able to confirm Searle's findings and was also able to show earlier defects. He found that the pattern of cell death was abnormal in Dh/+ *forelimbs* at the 10-day stage (surprising since the skeleton of the Dh forelimb is inevitably normal) with reduction in the preaxial necrotic zone. The amount of cell death in the preaxial part of the AER was reduced in Dh hindlimbs at 11 days (Fig. 8.13). The preaxial part of the mutant AER shows increased activity in response to histochemical stains for RNA and alkaline phosphatase.

Green (1967) was able to show that the visceral defects of Dh mice could be traced to a defect in splanchnic mesoderm visible from 9.5 days old onwards. At this stage normal splanchnic mesoderm is mainly epithelial in organization, especially anterior to the primordium of the dorsal pancreas (the anterior splanchnic mesodermal plate—ASMP). In Dh, the epithelial structure of the ASMP is diminished or absent behind the lung primordia. At the posterior end of the coelom, which seems not to elongate properly at this stage, mesoderm accumulates as an unorganized mass. It seems likely that abnormal splanchnic mesoderm may also give rise to abnormalities of the limb skeleton. Green pointed out that the hindlimb in Dh/+, Dh/Dh, and +/+ bears a constant relationship to the umbilical artery, and suggested that the latter may induce the former. We shall discuss the level of the pelvis in hemimelic mutants later.

Luxate (lx, Chr 5)

In the semidominant luxate (Carter 1951, 1953, 1954) anomalies are mainly confined to the hindlimbs of the homozygote, but lx/+ can also be affected. Like all genes in this group the manifestation is variable and sensitive to genetic background. The homozygote regularly shows a reduction or

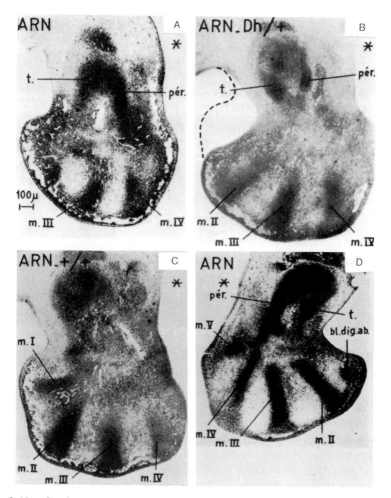

Fig. 8.12. Sections of the hindlimbs of (A,C,D) Dh/+ and (B) +/+ embryos aged 12 days. (From Milaire 1965.)

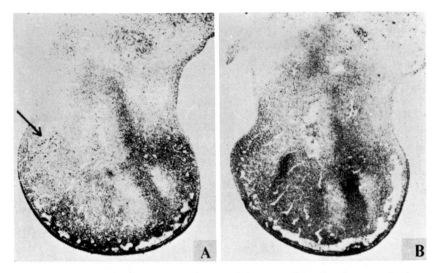

Fig. 8.13. 11-day hindlimb buds (A) +/+; (B) Dh/+. Note the absence of a deep preaxial necrotic site (arrowed in A) in the Dh/+ embryo. (From Rooze 1977.)

absence of the tibia allied to a decrease in digits or preaxial polydactyly. Femur and pelvis may again be involved in extreme cases. The lx/+ heterozygote often has a triphalangeal hallux or preaxial polydactyly, but the feet can equally be quite normal (Fig. 8.14). On a genetic background with six lumbar vertebrae lx reduces this number to five, a tendency stronger in homozygote than heterozygote. Horseshoe kidney and hydro-nephrosis are reported as aspects of the syndrome. The smallest feet found have three digits, missing I and II, and polydactyly is typically shown as the addition of a prehallux with three phalanges; sometimes two such are added giving seven toes in all. The tibia, when reduced, is more abnormal distally than proximally: in some cases the distal tibia is ligamentous. Carter (1954) was able to see a craniad shift of the hindlimbs at the 10-day stage of development, amounting to 0.5 of a somite in lx/+, and rather more in homozygotes. At 11 days the hindlimb buds are narrowed (3.5 somites rather than 4) and the AER reduced preaxially. At 12 days the tibial blastema is reduced or absent. If absent, then preaxial digital condensations are also reduced; if narrowed, then extra condensations are present in the region of the hallux. At 10 days the posterior end of the coelom is nearly a full segment ahead of its normal position in homozygotes, displacing the umbilical arteries and thus paving the way for kidney abnormalities. Long and Johnson (1968) were able to show a specific pattern of changed ontogeny of non-specific esterases and dehydrogenases of glucose-6-phosphate, malate, and lactate in lx, but the significance of these findings is not clear.

Green's luxoid (lu, Chr 9)

Luxoid mice (Green 1955) show abnormalities of both fore- and hindfeet, as well as some disturbance of the axial skeleton. Occasional females may breed, but males are sterile, with a complete absence of spermiogenesis (Elkins, in Johnson and Hunt 1971).

Forsthoefel (1958, 1959) described the anatomy and development of lu. lu shifts the hindlimbs caudally, increasing the number of presacral vertebrae. There is also a tendency to increase the total vertebral number, and the number of ribs. In the heterozygote the forelimbs are normal; in the homozygote they are regularly abnormal. The pollex (normally small in the mouse forelimb) is replaced by a triphalangeal structure, and further triphalangeous digits (to make a total of up to seven digits) may be added

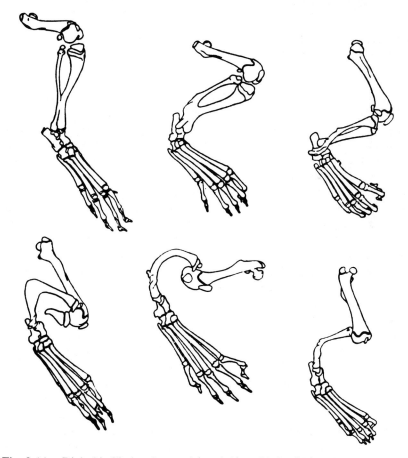

Fig. 8.14. Right hindlimbs of normal (top left) and lx/lx. (From Grüneberg 1963.)

preaxially. The radius is often shortened and there are minor defects in the carpus.

In the hindlimbs there is usually hypertrophy or preaxial polydactly in the heterozygote, again with up to seven digits in total, and the addition of a three-phalanged pre-pollex which may share a metatarsal with the pollex. In the homozygote the femur is shortened, the patella absent, the tibia slender, and the fibula heavy. Changing the genetic background produced a shortened or absent tibia and broader fibula. The homozygous foot has preaxial loss of digit I, or triphalangy of the hallux and addition of a prehallux. Splints representing incomplete metatarsals are present pre-axially. Forsthoefel (1959) described a development very similar to the type we have come to expect. The shift of the hindlimbs is seen at 10.5 days. At 11.5 days the forefeet show an excess of preaxial tissue and an elongated AER (Fig. 8.15). At 12.5 days the limb bud and AER may be either enlarged or deficient preaxially. The tail somites may be abnormal, and there are more of them than in controls. Cartilages from 13.5-day lu/lu embryos retain their defective morphology in culture (Burda and Center 1969).

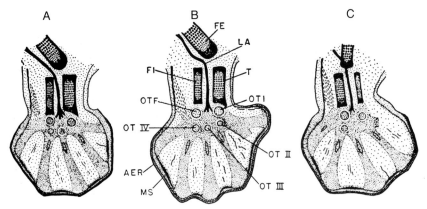

Fig. 8.15.   Left hindlimb buds of (A) +/+, (B) lu/+, and (C) lu/lu individuals aged 13.5 days. (From Forsthoefel 1959.)

Kobozieffs luxoid (Chr ?)

Kobozieff and Pomriaskinsky-Kobozieff (1959, 1960, 1961, 1962a,b,c) described in some detail a gene which is evidently virtually identical to lu, but which seems never to have been tested for allelism. No embryological data has been made available.

Strong's luxoid (lst, Chr 2)

Strong's luxoid (Strong and Hardy 1956; Strong 1961) also affects all four

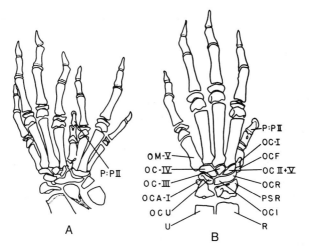

Fig. 8.16. Left forefoot of normal (right) and Strong's luxoid homozygote (left). (From Forsthoefel 1962.)

limbs. The position of the sacrum is unaffected. Forsthoefel (1962) described most lst/+ individuals as having preaxial polydactyly of the hindfeet. A complete triphalangeal digit is usually added on the medial side of an unchanged hallux. The forelimbs are usually normal.

In the homozygote there are abnormalities of the skull and limbs and their girdles. The forfeet have one to three digits added preaxially, the pollex remaining unchanged (Fig. 8.16). The radius is reduced but never absent. It may be duplicated, however, or represented by a compound bone, double distally or as two partly ossified ligamentous structures. In the hindlimb there is preaxial polydactyly, usually with six triphalangeal digits but sometimes up to eight (Fig. 8.17). The tibia is reduced, and occasionally double. The pubis is regularly reduced, but the femur essentially normal. All the bones of the braincase are shortened and the calvarium bulges dorsally; there is a large interfrontal bone.

The devlopmental picture (Forsthoefel 1963*b*) seems to be similar to that seen in other luxoid mutants, but a little more extreme (Fig. 8.18). Forsthoefel noted an outpushing of the preaxial margin of the forelimb bud at 11.5 days, which was prominent by 12 days and large by 12.5. In one case the limb bud was partly separated into two portions, and this individual had separate AERs on the limb proper and the excess swelling. Extra blastemata were present; reduced radii tended to lag behind in their development. In the hindlimb a similar picture was seen; the preaxial margin was swollen at 11.5 days. The swelling was larger by 12 days and the limb (as opposed to the footplate) narrower.

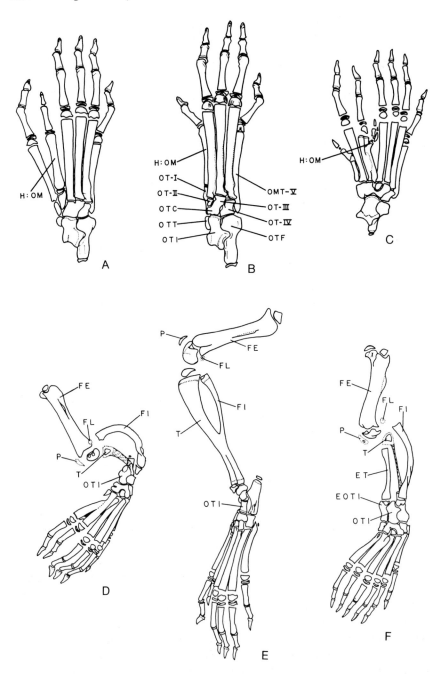

Fig. 8.17. Above: Right hindfoot of (B) normal mouse; (A) 1st/+; and (C) 1st/ 1st. Below: Whole hindlimbs, E: 1st/+, F,G: 1st/1st. (From Forsthoefel 1962.)

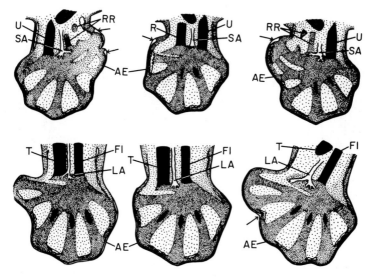

Fig. 8.18. Above: Forelimbs (above) and hindlimbs (below) of lst/lst (left); +/+ (centre); lst/+ (right); aged 13.5 days. T, ribia; Fi, fibula; U, ulna; LA, SA arterial supply. (From Forsthoefel 1963*b*.)

Interactions between mouse luxates

Forsthoefel (1958, 1959, 1962) painstakingly transferred Strong's luxoid, Green's luxoid, and Carter's luxate genes to the C57BL background and was thus able to make a true comparison of their effects and study their interactions. The reader is referred to Forsthoefel's original papers for further details.

Polydactyly Nagoya (Pdn, Chr ?)

Hayasaka *et al.* (1980) and Naruse and Kameyama (1982) described another mouse polydactyly. In this semidominant condition Pdn/Pdn mice have preaxial polydactyly of all four feet, with a slightly shortened tibia. The condition is lethal at two days post partum. About 20 per cent of these individuals have exencephaly and the remainder a varied spectrum of brain defects.

Heterozygotes have an extra preaxial digit in the hindlimb and a thickened first digit in the forelimb, often with a bifurcated terminal phalange to digit I. The effect of Pdn on the limb bud can be seen externally at 12 days when homozygotes have a wide footplate with an angular preaxial rim. The heterozygote can first be distinguished at 14 days. In sections presumed Pdn/Pdn (exencephalic) embryos had a thickened AER on day 11.5. In later embryos the ANZ was reduced or

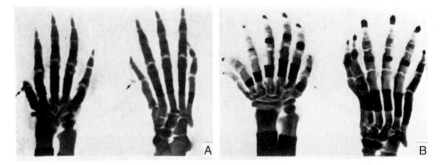

Fig. 8.19. Bone and cartilage preparations of the fore- and hindfeet of 18-day mouse embryos. (A) Pdn/+; (B) Pdn/Pdn. (From Naruse and Kameyama 1983.)

absent in Pdn/Pdn. Cell death was also locally disrupted in heterozygotes (Fig. 8.19).

Interestingly Naruse amd Kameyama (1983) showed that the effects of teratogens on Pdn is to produce hemimelia. The present manifestation of Pdn must therefore be considered that of a low-grade hemimelia. The condition thus forms a link between the hemimelias and the preaxial polydactylys, which are discussed below.

Scott's polydactylous guinea pig (Px) and the extra-toes (Xt, Chr 13) mouse

Scott's polydactylous monster in the guinea pig (Scott 1937, 1938) and the extra-toes mouse (Johnson 1967) form two very similar phenotypes which are probably best considered with the luxate series. Both genes cause widespread abnormalities throughout the developing embryo but only the effects on the limbs will be considered here.

In the normal guinea pig the pollex and hallux and little toe of the hindlimbs are absent. Heterozygotes for the gene pollex (Px/+, Wright 1934, 1935) tend to have the missing toes replaced: many have a total of 20 instead of the normal 14 toes. The arrangement of these pentadactyl feet is quite normal, so that they resemble the pentadactyl feet of other mammals. However, the homozygote (Scott 1937, 1938) has far from normal feet. They are paddle-shaped, bearing 9–12 short toes and clubbed, so that the palmar and plantar surfaces are directed inwards. Scott (1938) saw nine newborn Px/Px individuals (the condition being lethal at, or usually before, birth). The limbs were described as having three groups of digits, a lateral group resembling digits III–V, strongly flexed, a central group with four small toes, and an ulnar group with two straight toes and one free, weak medial toe. The outermost toe of the forefoot bore a small nubbin. Humerus, radius, and ulna were shortened and seven or eight metacarpals were present, often bearing two toes. The tibia was often missing, the fibula stout and curved, tarsals irregularly multiplied. Metatarsals were characteristically branched with splint metatarsals inserted between them. In the heterozygote extra digits were represented at least by a claw and a distal phalanx.

Scott (1937) described the embryology of the condition, noting first an increase in the width and thickness of the limb buds with delayed morphogenesis and the appearance of 'ectodermal thickenings' corresponding to extra digits (probably local thickenings of the AER).

Johnson (1967) described an essentially similar condition in the mouse. Extra-toes (and its allele brachyphalangy, Xt^{bph}, Johnson 1969a) is semidominant, and shows almost all of the complex syndrome of the Px guinea pig. Since the mouse normally has five digits on each foot, the heterozygote shows polydactyly rather than restoration of pentadactyly. The pollex is duplicated (phalanges plus metacarpal) and in extreme cases the phalanges of digit II are involved (Fig. 8.20). A supplementary postaxial digit at the base of digit V carries a small, rod-like skeletal element. In the hindlimb, duplication of the phalanges of the hallux is present, and a splint of bone between digits I and II forms the metatarsal of the extra digit. All these abnormalities can be seen in the cartilaginous skeleton. At birth the situation is a little complicated by osseous fusions of the separate cartilaginous elements, and in the adults the preaxial side of the forefoot is a mass of osseous elements, often of doubtful provenance. In the hindfoot a double first metatarsal may carry two digits, or a triple one three. Alternatively, metatarsals may be separate and the proximal phalange serving two digits Y-shaped. Splint metatarsals may be free, lying between digits II and III or fused to the tarsals, amongst which other fusions also occur. Two individuals showed tibial hemimelia.

In the homozygote (again lethal at birth) the number of cartilaginous digits may rise to eight or nine, with metatarsals and metacarpals split distally or represented only by their distal extremities (Fig. 8.21). The terminal phalange of any finger or toe may be duplicated or absent. Digits may be webbed by soft tissue. Carpals and tarsals are variously fused. One or more carpals or tarsals is usually united with the corresponding metatarsal by an isthmus of cartilage. The radius is occasionally represented only by a proximal rudiment but more usually present, and stout. The humerus is also wide and the scapula may be involved.

In the hindlimb bilateral defect of the tibia is the rule; the fibula is bowed and the femur short and stout. Ossification of this abnormal cartilaginous skeleton is retarded.

The Xt/Xt forelimb bud is first seen to be abnormal at 9.3 days when it is lobed, rather then smooth-edged, when viewed dorsally. No corresponding abnormality was seen in the hindlimb. At 10 days the AER of the forelimb is enlarged in cross-sectional area. The three genotypes can be distinguished by the external morphology of their limb buds at 12 days; heterozygotes

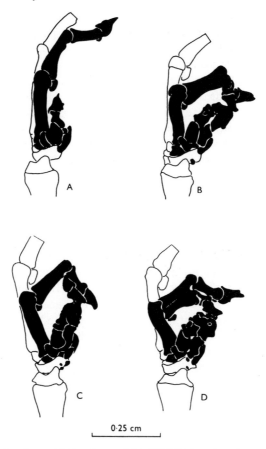

Fig. 8.20.   Left forefeet of (A) +/+, and (B,C,D) Xt/+ mice. Lateral view. (From Johnson 1967.)

have a distinct preaxial increase in size. In Xt/Xt there is both pre- and postaxial increase in size, which seems to occur at the expense of distal outgrowth.

The polydactyly of Px and Xt differs subtly from that seen in the luxate series of mice; it is postaxial as well as preaxial and the sequence of added digits is unclear. There is little or no evidence of mirror-imaging. The abnormality of the mesoderm on day 9 may suggest that the primary cause lies here, and that the enlarged AER is secondary.

Poldactyly (Po) and duplicate (Po^d) in the chick

Polydactyly in many breeds of chicken is based on a single gene Po. Landauer (1948) noted that the syndrome includes preaxial ectrodactyly.

The condition is usually noticed in the hindlimbs, but extra digits are also seen during the development of the wings (Baumann and Landauer 1944). A more extreme allele, duplicate (Po^d, Warren 1941, 1944) is also known, where preaxial additions may double the tarso-metatarsus and the adult wing is involved. The radius may be absent, normal, thickened, or duplicated (Landauer 1956*b*). There is no shift in sacral position (Zwilling and Ames 1958). Zwilling and Hansborough (1956) performed recombination experiments with ectoderm and mesoderm from duplicate and normal chicks and showed that duplicate mesoderm + normal AER gave abnormal limbs, whilst normal mesoderm + duplicate AER gave normal limbs.

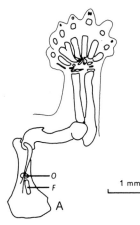
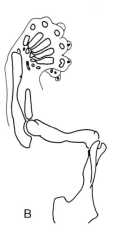
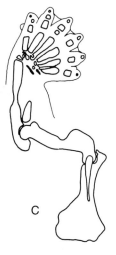
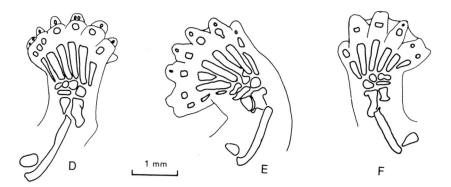

Fig. 8.21. (A,B,C) Forelimbs and (D,E,F) hindlimbs of 15-day-old Xt/Xt embryos. (From Johnson 1967.)

Pelvic level and the luxates

Several authors have stressed the importance of pelvic level in the luxate series (Forsthoefel 1959; Green 1967; Rooze 1977) and reference to Table 8.1 shows how this may be affected in various mutants. However the shift is craniad in some cases, caudad in others, and may be absent altogether. Grüneberg (1963) pointed out that pelvic shifts are also known which do not produce hemimelias and that this is inconsistent with the hypothesis that there is any causal relationship between hemimelia and pelvic level. If this objection is not seen as fatal, then the change of pelvic level may be seen as a means of uniting the axial and appendicular effects of some of these genes.

TABLE 8.1. *Pelvic levels in hemimelic mice*

Gene	Pelvic shift		
	Craniad	Nil	Caudad
Dh	x		
lx	x		
px	x		
lu		x	
lst			x

Diplopodia (dp) in the chick

Five mutations in the chick have been described under the umbrella heading of diplopodia (diplopodia dp-1, Taylor and Gunns 1947; dp-2 Landauer 1956*a*; dp-3 Taylor 1972; dp-4 Abbot and Kieny 1961; dp-5 Olympio *et al.* 1983). dp-4 is sex-linked and the other conditions are thought to be non-allelic (Landauer 1956*a*; Taylor 1972; Olympio *et al.* 1983). The diplopodias, therefore, although all given the same generic name because of similar phenotypes are no more alike than the various mouse luxoids. Olympio's description of dp-5 is typical of the group.

dp-5 individuals have extra digits preaxially on both hindlimb and wing, usually 2–4 on each leg, so that a typical individual has five or six toes on each foot, supernumerary metatarsals, and shortened limb bones, the tibia being stout but not duplicated. Similar changes are seen in the wing. The supernumerary digits have two or three phalanges, and there may be fusions so that a pair of digits share a basal phalanx. In rare cases the extra digits appear close to the hock joint, i.e. separated from the normal II, III, and IV. There is soft tissue webbing of the toes. Embryos have shortened

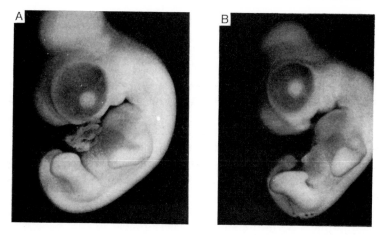

Fig. 8.22. (A) Normal and (B) dp-1 chick embryos at 5.5 days of incubation. Note the extreme preaxial thickening on both mutant limb buds. (From MacCabe and Abbott 1974.)

upper beaks, enlarged limb buds (Fig. 8.22), are small, and usually die at hatching.

Other diplopodias are a little more extreme. In the original diplopodia-1, Taylor and Gunns (1947) reported 10 per cent of individuals with an extra group of two toes, 75 per cent with three, and about 10 per cent with four. A few of these chicks hatched. Landauer described strain differences in dp-2 with a maximum of 7–8 toes. Taylor *et al* (1959) extended these findings. The tibia is again shortened, as is the femur, and the fibula bowed.

Abbott (1959) and Abbott and Kierney (1961) described the development of the condition. In early embryos two patterns of development were evident. In line 1, mesodermal condensations for digits IV, III, and II appeared, but none for the hallux; instead there was a group of preaxial condensations for extra digits, apparently not in mirror-image form. In another selected line the condensation for digit I appeared, followed almost immediately by another similar preaxial condensation differentiating ultimately into an extra very hallux-like digit. In either case the rate of cartilage formation and ossification is markedly retarded. The maximal digital number in dp-2 is seven, or in one exceptional case eight. Digits II, III, and IV are undisturbed and the hallux may be absent. If present, it is normal in size and position. Supernumerary digits again may be at the hock joint, or anywhere between it and the hallux. Low-grade duplication produces a prehallux only. Twenty-nine patterns of limb digits and 14 of wing digits were noted in a careful study, the wing being more variable. The most common pattern of extra toes was that the most proximal supernumerary digit, the largest, was equivalent in size to digit II; the most medial digit was the smallest, and the central one intermediate in size but smaller than the hallux. These extra digits 'did not correspond to the normal digits in size, number of phalanges or in general orientation'. Abbott summed up this detailed work as follows. The arrangements of extra digits are different in her selected lines. In one case the commonest abnormality was simply a duplicated hallux; in the others the duplicated hallux was accompanied by extra digits. If the preaxial region can obtain its independence from the postaxial region, it may follow its own pattern; if not, extra digits appear as additions to the main pattern. It seems evident that in none of the diplopod lines does mirror imaging form the basis of the duplication process.

MacCabe *et al.* (1975) looked at limb development in dp-4. They reported an increase in the length of the a-p axis and/or the size of the preaxial AER in the wing of stage-21 embryos. The leg bud at this stage has an AER thickened preaxially. By five days the apices of wing and leg are broadened and by seven days the tips have mushroomed into a broad limb paddle. The number of digits seen in cartilage preparations varied from five to seven per foot, six being commonest. Digits, II, III, and IV were short but otherwise normal with three, four, and five phalanges respectively. Digit I was always missing, being replaced by three to five extra digits preaxial to II and resembling it in having three phalanges. In the wings, digits III and IV were relatively normal, though shortened, and II missing—replaced by three to five extra digits preaxial to II and resembling it, or occasionally III. In two out of 22 embryos there was a duplicated radius.

Reciprocal recombinations of ectoderm and mesoderm show that, as in other mutants, limb type depends on the mesoderm. Polarizing activity, shown by grafting into normal hosts, was restricted to the postaxial region, as in normals and dp-1. Extra digits induced by the transplanted dp ZPA were normal, rather than dp in phenotype. McCabe *et al.* suggested the following scenario. The abnormality of the dp limb seems to be based on (1) an increase in the size of the preaxial ridge and (2) a lack of polarity of the anterior digits. The defect resides in the mesoderm and appears not to be due to any alteration or duplication of the ZPA. Transplants of normal ZPA were capable of eliciting response in most (10/15) unpolarized anterior regions of dp-4 limb buds, polarizing the digits to an extent not seen in control dp-4 limbs. In normal limb outgrowth the AER has a close temporal and spatial relationship to polarizing activity, suggesting that polarizing activity may enhance AEMF (Saunders and Gasseling 1968). In dp-4 the anterior mesoderm seems to have lost this necessity for close association with the polarizing zone and seems to maintain the ridge (by producing excess AEMF, or sufficient AEMF for longer). This, in turn, results in excess mesenchyme remote from polarizing activity—and producing non-polarized digits.

The talpid series (ta) in the chick

The talpid series are thought to represent another group of genetically independent recessive lethal mutations with similar phenotypes. ta[1] or Cole's talpid (Cole 1942; Inman 1946) and ta[3] (Ede and Kelly 1964a,b) are essentially similar whilst ta[2] (Abbott et al. 1960) survive rather longer and have a less abnormal head.

Abbott et al. (1960) described ta[2] wing buds as having five or six spatulate digits fused to form a hand-like structure and leg buds as having seven or eight. The limbs were short and broad. Inman (1946) noted that in her stock the limbs were polydactylous and retarded in development. Abbott reported that the whole limb bud was remodelled, so no extra digits *per se* are formed. The AER is increased in area apically, and remains thick and 'functional' over the digit-forming area long after its regression in normal sibs (Fig. 8.23); the retarded nature of the footplate must, however, be borne in mind in relation to this point.

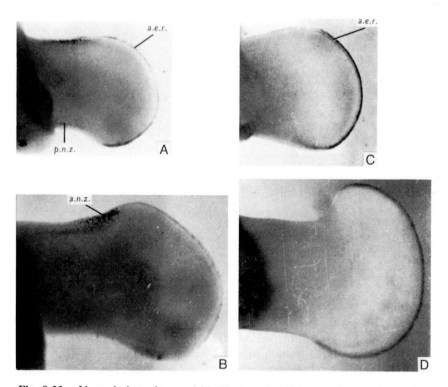

Fig. 8.23.   Ventral view of normal hindlimb bud: (A) stage 23 and (B) stage 25; and talpid[3] hindlimb bud: (C) stage 23 and (D) stage 25, showing the absence of necrosis of superficial mesenchyme but presence of necrosis in the AER of talpid[3]. *a.n.z*, *p.n.z*, necrotic zones. (From Hinchliffe and Ede 1967.)

Goetinck and Abbott (1964) stressed the fact that none of the eight to 10 digits of the talpid^2 leg can be recognized as a specific toe (Fig. 8.24). They found that the AER was indeed increased in size, as suggested by Abbott *et al.* (1960). At stages 18–22 no difference was seen in the area covered by the ridge, but from stage 23 on a significant excess was present; this persisted until the last stage measured, 25. Combinations of talpid^2 ectoderm with normal mesoderm and vice versa showed that limb buds with talpid mesoderm developed into typical talpid limbs, whereas those with talpid ectoderm and normal mesoderm were normal; the AER was of normal length and size despite its origin. These reciprocal grafts strongly suggest that the defect in ta^2 lies in the mesoderm.

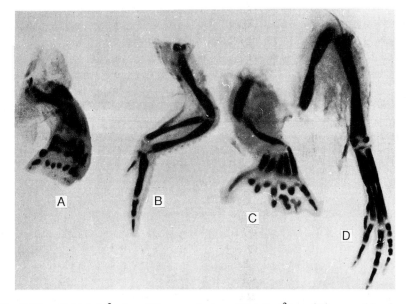

Fig. 8.24.   (A) talpid^2 wing; (B) normal wing; (C) talipd^2 leg; (D) normal leg from 9.5–day chick embryos. (From Goetinck and Abbott 1964.)

MacCabe and Abbott (1974) looked at the question of the unidentifiable digits of ta^2. They suspected that the ZPA in these mutants was not impressing the usual serial differentiation of the primordia of the digits. Grafts of the polarizing zone of talpid embryos (i.e. the posterior part of the limb bud) into the anterior wing bud of normal hosts showed polarizing activity by producing duplications. Grafts from other areas of the wing bud did not, although tissue from the tip of the bud induced outgrowths of donor tissue containing donor digits. Thus polarizing activity seems to be normal.

The AER in talpid seems to be developed symmetrically over the limb bud instead of in the usual asymmetrical fashion. This means that the preaxial part of the ridge is longer than usual. MacCabe and Abbott took this to suggest that maintenance activity is also uniformly distributed throughout the bud. The talpid wing tip, rotated and replaced on either a normal or its own stump, does not induce duplications. A normal wing tip does, because of its asymmetrical distribution of AEMF.

Cairns (1977) also studied the development of talpid^2. He noted that the observed increase in wing bud size over stages 18–21 was in fact minimal if accurate measurements were made, and that any differences were probably due to difficulty in accurately staging abnormal buds. The bud is wider than in normal sibs but outgrowth is not increased significantly. Neither is the rate of cell proliferation increased in ta^2 wing buds labelled with tritiated thymidine; in fact is is a little lower than normal. Cairns (1975) pointed out the importance of the distinctive distal mesoderm which runs beneath the AER in a band approximately 0.1 mm wide, and may be equated with the 'progress zone'. This looks normal in talpid^2 (cf. talpid^3 where Bell (in Cairns 1977) pointed out that all cells of the limb bud resemble the normal distal band). Cairns looked at cell adhesion in ta^2 and found that proximal ta cells adhered at the same rate as normal proximal cells and distal cells at the same rate as normal distal cells. This is in agreement with the findings of Niederman and Armstrong (1972) that normal and ta^2 cells did not resort after mixing. However, if all ta^3 are distal in type, it is not surprising that, as we saw in Chapter 2, cells from this mutant show changes in cell adhesion (Ede and Flint 1972, 1975*a,b*; Ede and Agerback 1968).

Talpid^2 also shows a marked lack of cell death (Fig. 8.25). Cairns (1977) argued that, as cell deaths in normal limb buds after removal of the AER are first seen anteriorly, and as the ridge is also at its least extent in its anterior part, a mesodermal control of both ridge and cell death is a possibility. Cairns transplanted central distal mesoderm to various sites within the limb bud. The cell death in this transplanted tissue varied according to location, being greatest anteriorly. Quail distal mesoderm showed less cell death when transplanted into talpid buds than into a corresponding position in the normal. Cairns suggested that this shows a decrease in the effect of proximal on distal mesoderm in ta^2.

The lack of cell death in ta^2 will, of course, lead to an increase in cell number through an increase in the size of the pool of proliferating cells. The cells of the distal mesoderm more nearly double in size at each cell division than cells elsewhere. They will therefore contribute a greater increase in volume per unit volume, i.e. distal parts of the bud will become relatively larger than proximal areas. The inhibited conversion of distal to proximal mesoderm will add to this effect especially preaxially; the increase in distal mesoderm will lead to an increase in the size of the AER.

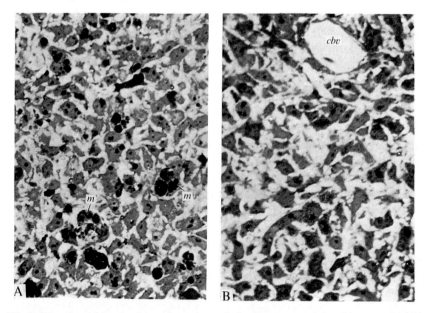

Fig. 8.25. Semithin plastic sections through the opaque patch level in stage 24 (A) normal and (B) talpid³ hindlimbs. Note the macrophages (m) in (A) and their absence in (B). (From Hinchliffe and Thorogood 1974.)

Ede and Kelly (1964*b*) looked at the limbs of talpid³. They found an essentially similar morphological picture to that seen in ta², with *c.* seven digits per foot, and, in the wing, carpals, and metacarpals forming a single block, as did radius and ulna. Syndactyly is also a common finding in talpid³. Hinchliffe and Ede (1967) studied the developing limbs in more detail. The talpid³ forelimb could be recognized as abnormal at stage 22, by abnormal extension on the cranial side and less than normal outgrowth. The AER is reported to be of normal extent at this stage, During stages 24–29 the 'mushroom' shape of the limb bud develops, and the AER is longer than normal at stage 24, attaining an increase of 65 per cent stage 28 and persisting until stage 29, when most of the normal AER has regressed. The hindlimb is less abnormal, but can be distinguished at stage 25, and gradually enlarges, being longer than that of the forelimb at stage 28. At stage 29 the talpid hindlimb ridge is still extensive. The ridge showed little histochemical change from normal as far as alkaline and acid phosphatase distribution was concerned and the histology of the ridge was normal. Talpid³ did, however, lack the massive areas of cell death normally seen in the superficial mesenchyme of the limb; there was no sign of ANZ or PNZ over the period 3.5–6 days. No interdigital zones of cell death were present in later embryos. The development of the limb skeleton was retarded, as shown by specific staining for acid mucopolysaccharides and the late-appearing central core of AMPS-positive material in the limb was not separated distally into radius and ulna as in the normal wing (Fig. 8.26). In later stages this core is divided, rather imprecisely, into a proximal block, representing the radius–ulna, and a distal one representing fused metacarpals. By stage 29 the talpid limb bud contains a series of blocks representing (in proximo-distal sequence) humerus, radius + ulna, carpals, metacarpals, and a poorly defined distal band which may represent the phalanges. The opaque patch of dead cells in the distal end of the initial humerus/radius/ulna condensation is small, and represented only by scattered cells.

Hinchliffe and Ede suggested that the excessive ridge size is linked to the

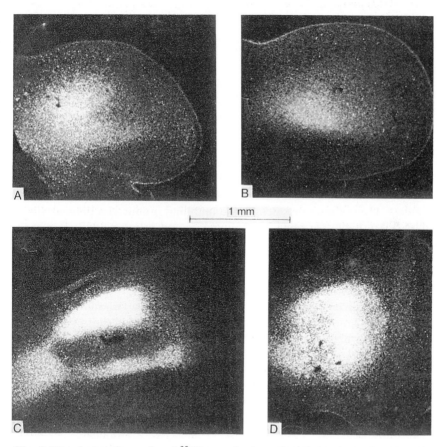

Fig. 8.26. Autoradiographs of $^{35}SO_4$ uptake into chondroitin sulphate in (A,C) normal and (B,D) talpid[3] hindlimbs at stages 24 and 26. (From Hinchliffe and Thorogood 1974.)

absence of cell death, and that the latter accounts for the failure of contouring of the limb, the extension of the AER (in the absence of ANZ and PNZ as end stops), and hence polydactyly and the lack of asymmetry in the limb.

Hinchliffe and Thorogood (1974) looked at the ta^3 limb in more detail and confirmed the lack of cell death in ANZ, PNZ, and the opaque patch. They also noted the absence of interdigital necrotic zones and the effect of this in producing syndactyly.

Talpid clearly has a series of alleles, or more likely, is composed of a series of unlinked loci. There is a suppression of cell death. There is an upset in the transition between proximal and distal mesodermal cells, less in ta^3 that in ta^2 where the proximal cell type is apparently abolished, and allied to a change in cell shape and cell adhesion. Ede and Law (1969) produced a computer model of limb bud shape. They started with a line of 40 cells representing flank mesoderm and instructed their computer to scan this, causing each rth cell to divide, and placing the two products of division in the space occupied by the original cell and in the nearest avaiable space. This produced a bee-hive-shaped mass whose shape could be altered by specifying a different value for r. However the mitotic rate in the normal limb bud (and indeed that rate in talpid) does not vary up to stage 22 in the chick (Janners and Searls 1970; Hornbruch and Wolpert 1970) so a second factor, rate of distal migration, was added. This produced shapes more akin to limb buds, and, if the rate of distal migration was reduced, concordant with the increase of cell adhesion in talpid^3 cells, more talpid-like outlines.

Wilby and Ede (1975, 1976) and Ede (1976) used more sophisticated models. Using these, broadening the limb bud produced a polydactylous, talpid-like shape (for review see Hinchliffe and Johnson 1980).

Polysyndactyly (Ps, Chr 4)

Johnson (1969*b*) described a polydactylous mouse mutant which also affects cell death. Polysyndactyly is a semidominant condition lethal at birth when homozygous. Ps/+ mice are about 10 per cent smaller than normal littermates. In the fore- and hindfoot claws are reduced and nail-like; Ps/+ are active at ground level but unable to climb well. Soft tissue syndactylism is common in the hindfoot, where the first toe is never free (Fig. 8.27), and rare in the forefoot. There are occasional fusions between carpals and tarsals. Terminal phalanges are poorly developed and rod-like. In the hindfoot the joint between terminal and subterminal phalanges is abnormal or fused. Twelve out of 30 mice had a swelling beneath the terminal pad of digit IV of the hindfoot which contained one or two small osseous elements. Accessory postaxial digits are rare in the hindfoot, but common in the forefoot. No abnormalities were found outside the skeleton

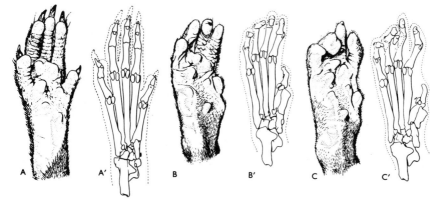

Fig. 8.27. Ventral views of the feet of (A,A') +/+ and (B,B',C,C') Ps/+ mice. (From Johnson 1969*b*).

of the feet. At birth Ps/+ feet are syndactylous and an extra digit is present between III and IV, bearing two, three or more peripheral nodules, perhaps rudimentary claws and corresponding to the structure seen beneath digit IV of the adult hindfoot(Fig. 8.28). The first digit is short and broad, sometimes clearly duplicated. The homozygote at birth has feet ending in a broad distal pad which extends pre- and postaxially beyond the rest of the foot. Fore- and hindfeet are oedematous, as far as the elbow and knee, respectively.

Ps/Ps mice can first be classified at 13 days of gestation. They lack the division of the footplate into digits normal at this stage. The AER is irregular and seen to be hypertrophied in transverse sections. The newly-appearing mesodermal condensations are irregularly spaced and the marginal blood sinus enlarged. At 14 days Ps/Ps has oedema of the limbs and trunk. The footplates are still not divided into digits and have a ragged appearance due to the presence of numerous irregular hillocks of mesoderm each capped with a thickened portion of AER. In the forefoot one central outgrowth predominates, becoming relatively enormous and protruding as a central cone. Beneath the AER is a continuous densely staining band of mesoderm. Chondrification of up to six metatarsals is poor with III and IV having a large included angle. The heterozygous footplate is more normal, but between digits III and IV are two indentations cutting off an intercalated digital region. There is preaxial enlargement of the hindfoot. The normal INZ–interdigital areas of cell death—are disrupted in Ps/+ and almost absent in Ps/Ps at this stage.

Milaire (1965) demonstrated that in normal 13-day limb buds the enzymic activity and RNA previously distributed along the length of the AER becomes localized in those portions of the AER which overlie a

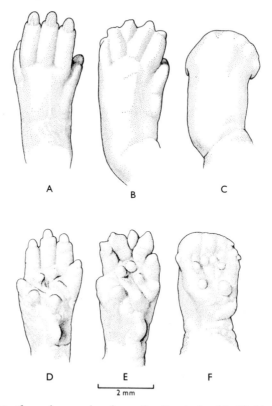

Fig. 8.28. Feet of newborn mice from the Ps stock. (A–C) hindfeet; (D–F) forefeet. (A,D) +/+; (B,E) Ps/+; (C,F) Ps/Ps. (From Johnson 1969*b*.)

developing digit and decreases elsewhere. This process seems to fail in Ps giving a uniform band dark-staining mesoderm beneath the AER and numerous small areas of active AER some of which may induce extra terminal phalanges, especially centrally where a large space is created by abnormally wide separation of metatarsals III and IV. The failure of the interdigital ridge to regress is presumably linked to the failure of cell death in interdigital areas, as well as in the AER. The erratic spacing of the blastemata of metacarpalia and tarsalia III and IV is presumably not due to the defect in the AER but indicates a probable underlying mesodermal defect.

Splitfoot in the chick

Abbott and Goetinck (1960) described the effects of the recessive splitfoot gene in the chick. The preaxial side of the foot was affected, leading to one of five different conditions: 1. The hallux was deleted; 2. The second toe was split in two, each digit with three phalanges; 3. As 2 but the lateral bifurcation had four phalanges; 4. Both parts of the second toe had four phalanges; and 5. The hallux was absent but two supernumerary digits were present. Additional toe nails and tarso-metatarsal bones were also noted, and the upper beak was reduced.

PREAXIAL POLYDACTYLY AS PART OF A WIDER SYNDROME

Preaxial polydactyly can occur as an inconstant part of a wider syndrome of abnormalities. Below are listed briefly a series of such syndromes where preaxial polydactyly is common.

Fidgit (fi, Chr 2)

This recessive gene (Grüneberg 1943; Truslove 1956) has widespread effects which include preaxial polydactyly of the hindlimbs. It is uncertain whether the effect is primarily one of the fi gene, or if fi is acting as a modifier (Bodmer 1960). The incidence of polydactyly in Truslove's stock was raised from 1.5–4 per cent in normal littermates to 20–65 per cent by the presence of fidgit. The condition appears to have a classic morphology (Fig. 8.29) with duplication of the hallux, often on a common metatarsal, and the production, in more abnormal individuals, of a prehallux with three phalanges. The gene has other effects on the skeleton, including dislocation of the hip based upon a very poorly defined acetabulum and a tendency of certain small bones to fuse.

Hop-sterile (hop, Chr ?)

Johnson and Hunt (1971) noted preaxial polydactyly of fore- and hindfeet in the hop-sterile mouse. This male sterile mutant has abnormal sperm tail development. Skeletal abnormalities are limited to the feet, which all show preaxial polydactyly with a doubling of the terminal phalanges of digit I in both fore- and hindfeet (Fig. 8.30). The animals walk with a characteristic hopping gait, but the researchers found no skeletal defect in the pelvic region to account for this. The association between male sterility and preaxial polydactyly (such as occurs in lst, Dh, px, lu, and hop) is probably fortuitous, since the abnormalities in the reproductive system do not tally; in many cases sperm production is normal and the sterility is probably due to difficulties in copulation. In lu, where spermiogenesis is abnormal the defect does not resemble that seen in hop, and in px the connection between vas and seminal vesicles is abnormal (Searle 1964).

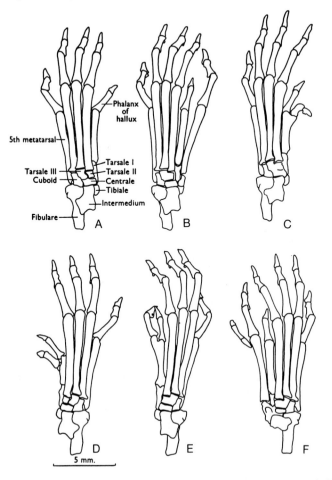

Fig. 8.29.   Dorsal views of the hindfeet of three fidgit mice. A: normal; B–F: fi/fi with increasing abnormality. (From Truslove 1956.)

Hydrocephalic polydactyly (hyp, Chr 6)

In this recessive condition Hollander (1976) described homozygous abnormals which had preaxial polydactyly of all feet, a hopping gait, and were male sterile. The sperm tail anomalies described by Bryan (1977) make it highly likely that hop and hpy are allelic or at least mimics; no linkage data with hop have been presented.

Myelencephalic blebs (my, Chr 3)

This complicated syndrome (reviewed by Grüneberg 1952) was reinvestigated fully by Carter (1956, 1959). It is named for fluid-filled blebs which

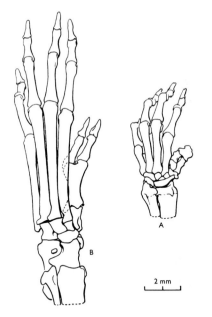

Fig. 8.30. (A) Left forefoot and (B) hindfoot of a hop/hop mouse. (From Johnson and Hunt 1971.)

are responsible for some of the defects, but clearly not for others, which include the effects of the gene on the limbs. Carter (1959) described a series of hindlimb abnormalities in my/my mice (Fig. 8.31). These included polydactyly, defined to mean hyperphalangy of the hallux in this context, although he also illustrated more severe types. This was preceeded by an

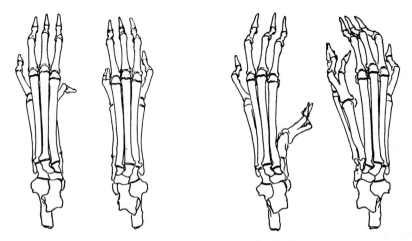

Fig. 8.31. Hindfeet of my/my mice. (From Carter 1959.)

excess of tissue on the preaxial margin of the 12.5-day-old footplate. Adult material was also described by Bean (1929) who added a higher grade of preaxial polydactyly, comprising a double hallux with separate metatarsal. Additionally a three-digit extra toe was seen, as was an occasional splint-like metatarsal. One of Carter's animals had tibial hemimelia and other defects of the feet included syndactyly, usually of the soft tissues but sometimes involving the phalanges. Sometimes this coexisted with tri-phalangy of the hallux.

Fetal haematoma (fh, Chr ?)

Center (1977) described a condition similar to my in many respects. Although fluid-filled blebs were present in the limbs, in this condition no mention of polydactyly is made. 'Luxoid' mice with 'shortened twisted limbs' were however sufficiently common in the fh stock to warrant a test for allelism, which proved negative. It seems that fh (which was also found to be non-allelic to my) embraces only part of the my syndrome, and not the part producing polydactyly. It is worth noting that eye-blebs (eb, Beasley and Cratchfield 1969) similarly produces blebs in various locations including the feet, but induces only degenerative lesions due to the presence of the blebs, which later becomes haemorrhagic. These lesions, common to my, fh, and eb limbs, are clearly nothing to do with limb pattern, and only affect the limbs secondarily.

Brachydactyly (br) in the rabbit

Brachydactyly in the rabbit (Greene and Saxton 1939; Jost *et al.* 1969) is an autosomal recessive characterized by congenital limb deformities resulting from haemorrhagic blebs which appear on the limbs between 16 and 21 days of development (Inman 1941).

Petter *et al.* (1977) investigated the defect and found that thrombosis in limb vessels was the probable cause of ischaemia and local hypoxia, leading to oedema, haemorrhage, and necrosis of the extremities. Polycystic blebs in the liver were also considered to be due to hypoxia. In the fetal liver erythropoiesis was also affected, but the cause-and-effect relationship between this and the hypoxia was unclear. Petter *et al.* speculated that the excessive number of large red blood corpuscles (rbcs) present in br may be due to intense haemopoietic activity by the yolk sac, whose products are normally larger in diameter than those of the liver. A similar process was described (Borghese 1952; Russell *et al.* (1957) in the W/W^v mouse. The large cells appear to obstruct small vessels, perhaps as a consequence of their size, number, poor deformability, or clumping properties (see Chapter 9).

Tail-short (Ts, Chr 11)

The axial effects of tail-short (Deol 1961) were dealt with in Chapter 5. Deol mentioned that Ts has a peculiar effect on the length of the limbs, with the left forelimb more reduced than the right (mainly due to shortening of the humerus) and the right hindlimb more than the left (mainly due to shortening of the tibia, although the femur is sometimes affected). Two cases of radial hememelia allied to absence of digit I were seen by Deol and 1 per cent of his mice were polydactylyous (Morgan (1950) found 5 per cent polydactyly in his Ts stock). Triphalangy of digit I (fore- and hindfeet) was accompanied by various fusions of carpals and tarsals.

SIMPLE PREAXIAL POLYDACTYLY AS A SINGLE ENTITY

As well as featuring in complex syndromes, polydactyly is common as an entity apparently unassociated with other defects. There is a finite occurrence of polydactyly in inbred strains (Staats 1976), probably due to minor genes whose additive effects were discussed by Fortuyn (1939) and Chase (1951). Other polydactylies which seem to be due to eitherone, or a small number of genes, are discussed below.

Holt's polydactyly (py, Chr 1)

Holt (1945, 1948) described a preaxial polydactyly chiefly of the hindfeet in the mouse. Fisher (1953) was later able to discern the activity of three contributory genes, one of which he was able to place in a linkage group. Expression is variable, but the usual range of abnormalities is described with up to seven digits.

Beck's polydactyly

Beck (1961, 1963) described preaxial polydactyly in the hind foot in mice selected for anophthalmia (ey). This was thought not to be pleiotropic but due to a linked gene or genes. Again the first digit of the hindfoot was doubled or tripled, sometimes with associated defects of the metatarsals.

Roberts's polydactyly

Roberts and Mendell (1975) discussed the nature of polydactyly and its genetic control at some length, in conjunction with studies on yet another genetically controlled polydactyly. This arose in a strain (CQ) selected for large body size. The new condition did not behave like a single gene when subjected to the usual back-cross regime, but appeared sporadically.

Roberts and Mendell undertook a programme of selection, originally to test again for segregation. However the results of this selection procedure were interesting in their own right. They found that selection for polydactyly: (a) increased the number of bilaterally affected individuals; (b) introduced an anomaly of the forelimbs; and (c) produced hemimelia. Polydactyly was eventually found in 80 per cent of the individuals in the stock. Skeletal deformity ranged from the initial polydactyly to a condition which could be exchanged for one of Carter's (1951) luxate figures 'without excessive misrepresentation'.

X-linked polydactyly (Xpl, Chr X)

Sweet and Lane (1980) described an X-linked polydactyly. Homozygous and heterozygous females are viable and fertile; hemizygous males mainly sterile with undescended testes and sometimes kidney abnormalities. All genotypes have preaxial polydactyly of the hindfeet varying from a thickened hallux to four extra toes per foot; tibial hemimelia is also seen.

PREAXIAL ABNORMALITIES—AN OVERVIEW

It should be clear from the above review that many of the abnormalities of the preaxial side of the footplate fall into a single series. The characteristics of this group are as follows:

1. The postaxial side of the foot is never affected.
2. The preaxial side of the foot may be either reduced or have additional preaxial digits.
3. The radius/tibia may be reduced/duplicated.

At first these abnormalities seem to be contradictory but it is clear that the same gene on different backgrounds, or even on a mixed background, can produce individuals covering a range from preaxial ectrodactyly to luxation and preaxial polydactyly, so all the effects must be part of the same syndrome. The key here lies in the amount of preaxial mesenchyme. It is clear that in conditions such as dominant hemimelia there is a defect in this mesenchyme which results in reduction of the anterior part of the limb. Yet it is also possible to find Dh/+ mice with a thickened or triphalangeal hallux. Surely this is an indication of rather poor timing? The mesenchyme may be not produced at all (giving ectrodactyly) or produced a little too late and in slight excess (giving polydactyly). If we extend this reasoning, it is easy to account for a limb with no tibia (mesenchyme reduced at a critical stage), frank polydactyly (mesenchymal excess present at a later stage), and even the situation seen in Strong's luxoid (lst) where there is excess mesenchyme throughout and polydactyly is combined with an extra radius or tibia which may ossify or remain ligamentous. Most of the hemimelic series show a combination of these effects. Another interesting feature is that polydactyly seen either as a simple entity (Holt 1945; Roberts and Mendell 1975) or as part of a complex syndrome (Ts, my) equates very well with the lower grades of hemimelia; in fact hemimelia may be a rare adjunct of these polydactylys or may be produced in them by teratogens (Pdn) or selection. Morphologically the entities are very similar.

At this stage we must ask ourselves exactly what is being duplicated in polydactylous mutants. Two classic schools of thought exist on this suggesting either that the limb bud field has been split into two parts, and is thus exhibiting mirror image duplication (Kaufmann-Wolfe 1908; Braus 1908; Harrison 1921; Gabriel 1946), or that extra digits have been added preaxially in no particular symmetry (Warren 1944).

We now know that splitting the limb field into two parts does not have the right effect. The insertion of a barrier into the midpoint of the developing wing bud has been described as causing loss of structures anterior to the barrier (Summerbell 1979). This finding is, however, contentious since Griffiths (in Hinchliffe and Gumpel-Pinot 1983) found that anterior wing AER survives even if the posterior AER is removed and Rowe and Fallon (1982) showed that insertion of a barrier in the leg bud does not result in the loss of anterior structures, but in normal development. No mirror image duplication, however, occurs in either case. But mirror image duplication can be caused by the grafting of a second ZPA anteriorly. Could this be what happens in polydactyly? In the chick every digit of the foot is identifiable morphologically and it is therefore possible to check the bona fides of extra digits and to ascertain that mirror imaging is genuine in some mutants. This type of polydactyly fits well with the theories of control of antero-posterior differentiation of Tickle *et al.* (1975), Wilby and Ede (1975), and Ede (1976). The formal argument is the same in each case, but I propose to use the model of Tickle *et al.* which deals only with A–P differentiation, and so is easier to follow. As we noted earlier, Tickle proposed a source of an unspecified morphogen (the ZPA) at or near the posterior border of the limb and a sink near the anterior border. The concentration of the morphogen at steady state is thus supposed to follow an exponential curve (Fig. 8.4). If we hypothesize that the tissue responds to this morphogen by the production of digits we can set up a two-threshold model (after Slack 1977). A digit is thus specified between two thresholds T_1 and T_2 (Fig. 8.5). Slack pointed out the effect of adding second ZPA anteriorly: the morphogen concentration may never fall to T_1 so that a given digit is not specified, it may fall to a level between T_1 and T_2 so that a digit is specified but remain between these values for an abnormal distance, so that the digit is broad, or it may fall below T_2 then rise again, so that two digits are specified.

A second variable is that of field width. We know that in polydactylous limbs (both mutant and induced by ZPA transplant) the width of the field is increased by the production of more mesenchyme. This will move the two ZPAs apart, and, if the morphogen works in absolute rather than relative units, will also upset the morphogen gradient.

Let us suppose that mirror image polydactyly is in fact formed in this way. We come to two conclusions. First, there are a surprising number of genes which add a second, anterior ZPA, and, second, in these genes the anterior border of the limbs should provide a higher than usual ZPA activity in bioassay.

The first of these points is a matter for conjecture. If we allow that low-grade polydactyly, as seen in the hemimelias, especially when heterozygous or on an unfavourable genetic background, is a variant of the more extreme polydactyly and hemimelia seen in the same genes when homozygous or on a more favourable genetic background (a reasonable assumption) and that many of the simple mirror-image polydactylys seen elsewhere are similar in aetiology, then we have a conundrum. Why should so many genes reiterate the ZPA in an anterior position?

One possibility is that the second ZPA (anterior ZPA or AZPA) is not *reiterated* but activated. Suppose that the AZPA regularly sits ahead of the limb bud, normally inactive because it is isolated from responsive mesoderm. The relationship is similar on the posterior side of the ZPA where no digits normally form in the unresponsive flank mesoderm. Saunders (1977) pointed out that ZPA activity is found elsewhere, for example in the flank mesoderm and tail tip which Saunders and Gasseling (1982) claimed is as morphologically active as the ZPA itself. Tissues frequently have this type of effect on other tissues with which they do not normally come into contract.

Any factor increasing the quantity of mesoderm in the anterior part of the bud would extend competent mesoderm into the influence of the AZPA. The concept of a second ZPA ahead of each limb is perhaps revolutionary; however, if we go back in phylogeny to fishes the fins are often very wide and symmetrical. Perhaps the territory of the ZPA signal is limited and has to be reiterated after so many digits. Perhaps the suppression of the AZPA allowed the development of the asymmetrical pendactyl limb.

The second point, that of PZA activity in the anterior part of the limb, is a matter for experiment. Such experiments have been performed, but only on diplopodia–4 and talpid (MacCabe and Abbott 1974). In neither case was there evidence to support the idea of a second ZPA: in diplopodia–4 the ZPA activity was normal posteriorly; in talpid no frank ZPA was found. Unfortunately neither of these mutants serves as a critical test. In all diplopodias a group of preaxial digits is found but many authors are careful to stress that the duplication is not a true mirror-image one: no extra ZPA would therefore be expected. In talpid the digits are all similar and not recognizable as specific entities—mirror-image duplication is again not evinced. Perhaps the best test of the theory would be the state of Po or Pd^d anterior AER.

Non-mirror-image polydactyly

MacCabe's experiments remind us that all preaxial polydactyly is not mirror-image in type. The diplopodias, the talpids, extra-toes, Scott's guinea pig, and Ps are clearly not part of the hemimelic series. We can perhaps speculate on the type of ZPA activity we might expect in these mutants. In talpid MacCabe and Abbott (1974) showed the lack of a specific ZPA and we can speculate that our hypothetical morphogen is more or less constant over the whole range of the limb field, producing a virtually continuous A–P mass: In diplopodia a number of anterior digits resembling II or III are formed. Perhaps the sink is blocked, and the concentration does not decay at its usual rate. Extra-toes may be like ta in some respects, producing many non-distinctive digits, although the heterozygote tends to produce non-mirror-image digits rather like diplopodia. In Ps the defect is clearly in local control of the AER; the main pattern production is not badly disarranged.

LIMB DEFICIENCIES AND DUPLICATIONS IN MAN

An extensive account of hemimelia and polydactyly in man was given by O'Rahilly (1951). O'Rahilly recognized two groups of abnormalities of interest to us.

1. *Absence of limb bones.* One major group presents as isolated defects of the limb bones. These can be either terminal, when the limb lacks a distal portion, the whole limb, the forearm and hand, or the digits only being affected, or intercalary, such as radial or ulnar hemimelia which may affect distal structures secondarily. Of these the terminal type may equate with the wingless chick and may well be due to defects in mesoderm/AER leading to non-formation of the affected parts. Later amputation *in utero* cannot, however, be ruled out.

More interesting are the intercalary types. Although no good estimate of frequency has been made, O'Rahilly listed congenital absence of the radius as most common; ulnar deficiency was rarer. Warkany (1971) gave an estimate of 1 in 30 000 births for the radius and one in 100 000 for the ulna. In the leg fibular hemimelia is commoner than tibial. Interestingly, amongst many hundreds of tabulated cases O'Rahilly does not mention polydactyly associated with radial (1193 cases) or tibial (639 cases) hemimelia although he lists losses amongst the distal bones. Warkany (1971) noted however that hexodactyly is associated with radial absence and that double thumb has been reported repeatedly as an additional anomaly.

2. *Polydactyly*. O'Rahilly gave many examples of postaxial, i.e. ulnar and fibular, dimelia but was unable to find an example of preaxial duplication. He listed many examples of mirror imaging, often with clear duplication of carpals or tarsals as well as a mirror-image sequence of digits. In some cases two ulnars flank a single radius. A similar condition was described in disorganization in the mouse by Hummel (1958). Warkany (1971) noted that in these cases 'all elements are duplicated except the radius, scaphoid, trapezium, metacarpal I and digit I. Digital formula is then V, IV, III, II, (II), III, IV, V.'

O'Rahilly also listed a series of cases described as multipollicalism and multihallucasism where there is no clear evidence for mirror-image duplication. These two types clearly correspond to the mirror-image and non-mirror-image polydactylism seen in mouse and chick.

In man polydactyly not associated with hemimelia or duplications usually involves the first or fifth finger, although cases of intercalation of an extra finger are known. A combination of pre- and postaxial polydactyly occurs rarely. Polydactyly of a higher degree is rarer.

ABSENCE OF THE DISTAL PART OF THE FOOTPLATE

Reduction of the distal limb

Dachyplasia (Dac, Chr ?)

Chai (1981) described a mouse mutation resembling lobster claw in man. The typical condition is the absence of the three middle digits of each foot. Phalanges are missing and metacarpals/metatarsals intact, incomplete, or absent. If present, they are often fused. The forefeet are generally more affected than the hindfeet. In severely affected feet, only the fifth digit may survive. The defect is visible at 13–14 days of development as a reduction of the margin of the developing footplate. Chai considered that the data on Dac suggest a 2-locus mode of inheritance with the structural gene Dac protected by an epistatic gene Mdac at another locus.

Split hand (Sh) in the cat

Searle (1953) described this condition in a single female cat, and her offspring. Seven of 26 offspring, by at least three different males, were affected and the condition can thus be considered a dominant (Fig. 8.32). The propositus had a central cleft in each manus with a reduction of the central ray (digit III). In the left hand a characteristic 'cross-bone' linked

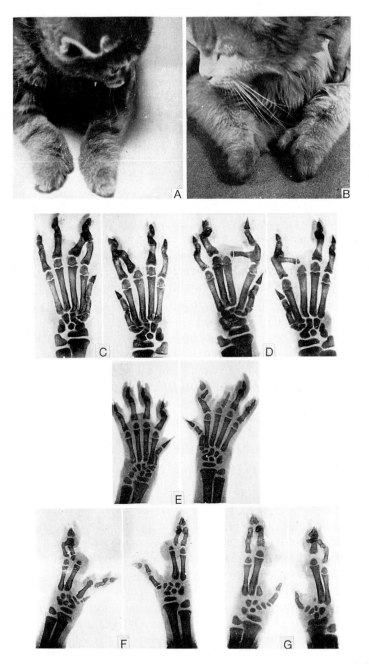

Fig. 8.32. (A,B) External appearance of split hand in the cat and (C–G) radiographs of the paws of affected litters at two months. (From Searle 1953.)

metacarpal III and the proximal phalanx of the second digit. In the right the proximal phalanx of digit II was atrophic distally and no distal phalanges were present.

Amongst the kittens deep clefts of the forepaws were seen together with absence of digits and syndactyly or absence of phalanges in digit II and III. 'Cross-bones' were again seen. In a more severe manifestation the whole of the third ray was absent with syndactyly of digits IV and V. In the contralateral paw digit II was also missing and digit I reduced. The carpus was also affected. Hindfeet were never seen to be abnormal.

Hypodactyly (Hd, Chr 6)

Hummel (1970) described a semidominant hypodactyly in the mouse. The homozygote, which rarely survives to term, has only one digit on each foot. The heterozygote fares better, with a shortening of the hallux involving the deletion of the terminal phalanx and claw and variable shortening of the basal phalanx; heterozygotes are usually fully viable and fertile.

The homozygotes, and some heterozygotes, die before 16 days of gestation. Alizarin clearance preparations of the few homozygotes surviving to birth show gross defects of carpals, metacarpals, tarsals, and metatarsals as well as ectrodactyly. Each forepaw has a single digit with two phalanges, probably the second and third, with a terminal claw. The proximal phalanx articulates with a single metacarpal which in turn articulates with a carpal distal to the ulna. Hummel suggests that this identifies the surviving digit as the fifth. The hindfoot is more variably and less selectively affected. A short fifth metatarsal and digit may be present, often with a trace of a phalanx of digit IV. The condition thus appears to be a preaxial reduction of the limb bud.

Hypodactyly (hd) in the rat

Moutier *et al.* (1973) described a male sterile recessive leading to hypodactyly in the rat. The usual condition in affected homozygotes is a tridactyl forefoot (two digits and a pollex) and tetradactyl hindfeet. No further details are given.

This group of mutants form a reasonable model of the split-hand and split-foot condition seen in man, but must be considered heterogeneous. In the cat digit III is the first to be affected, with spread to digit II. In Dac II, III, IV are the first to go, followed by I. In the hd rat the middle digits are similarly removed, but in the Hd mouse, where the highest degree of expression leaves only digit V, the hallux is first affected in the heterozygote.

POSTAXIAL POLYDACTYLY

Postaxial polydactyly is much less common in mouse than man, and never has extensive effects. We have already seen that there is a suggestion of postaxial polydactyly in Xt with the extra digits represented only by a sliver of cartilage. A similar situation obtains in Ps and fidgit. Apart from these, only one or two genes have been described which regularly produce postaxial polydactyly not allied to extensive foot abnormalities.

Postaxial polydactyly (toe-ulnar tu, Chr ?)

Center (1955) described a condition, genetically complex, which produced postaxial polydactyly of the forelimbs (Fig. 8.33). The extra digit may have one or two phalanges and an atypical nail. In other cases it is vestigial and regresses. The condition is first seen as an excess of tissue on the ulnar border of the forelimb bud at 13–14 days which is not subsequently resorbed as in normal mice. Center suggested that the defect is thus not one of excess mesoderm, but of failure to remove normally occurring tissue.

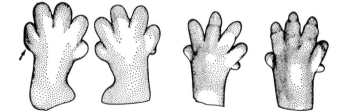

Fig. 8.33. Postaxial polydactyly in forelimbs of embryos aged 14 days (left) and 15 days (right). (From Grüneberg 1963.)

Postaxial polydactyly (Po, Chr ?)

Nakamura *et al.* (1962) described a similar condition, again limited to the postaxial side of the forefeet, and again dependent upon a complex genetic situation involving the dominant Po and modifying genes.

POSTAXIAL HEMIMELIA

Reduction of the posterior border of the limb bud, like postaxial polydactyly, is less common than effects on the preaxial border. However two conditions where posterior reduction occurs have been described in the mouse.

Oligodactylism (ol, Chr 7)

Oligodactyly (Hertwig 1939, 1942) is a recessive. Homozygotes show reductions of the fifth digit of all four feet. In more extreme cases the ulna and/or fibula may also be reduced or absent. The tail may be kinked or shortened, and the last rib reduced in size. There are associated abnormalities of the spleen and kidneys (Freye 1954) which usually lead to death within one month; any survivors are runted and do not breed.

The forelimbs are more affected than the hind. Digital loss starts with V and may spread to IV with associated loss or fusion of carpalia and tarsalia. In more extreme cases ulna and fibula are reduced, the latter sometimes being represented by only its proximal end, the olecranon process. Extreme reduction of the hindlimb may also involve the femur. The abnormality has been traced back to the 12–13-day stage where reductions are visible as soon as the blastemata form. Freye (1954) suggested that insufficient mesenchyme migrates into the limb bud. With present-day knowledge of the development of the limb we would no doubt phrase this rather differently, and suggest that the same effect is seen here as in px, loss of mesenchyme perhaps due to a breakdown of interaction between AER and mesenchyme in the posterior part of the limb bud.

Postaxial hemimelia (px)

Searle (1964) described the recessive postaxial hemimelia. The sterile homozygote is characterized by loss of the postaxial part of the forelimb, so that digits V, IV + V, or III, IV, and V are missing (Fig. 8.34). Sometimes digit V is present as a mere rudiment and there is rare syndactyly (1–2 per cent). About 20 per cent of affected individuals also lack digit V of the hindlimb. On the dorsum of the px foot small papillae are seen at birth, usually three to each foot, which later develop dark pigmentation and sometimes a rudimentary claw. These black pigmented excrescences consist of a thick cornified layer with a pyramidal cone of soft tissue. The proper claws of px tend to be malformed. Skeletal preparations show postaxial abnormalities of the whole forelimb, from the postaxial margin of the scapula, which has a large foramen in the infraspinous fossa which sometimes includes the posterior margin of the bone, through distal reduction of the humerus, ulnar hemimelia, and absence of proximal carpals metacarpals and phalanges. There were extra sesamoid bones on the extensor surface of the manus which may, from Searle's illustrations, correspond to the position of the ectodermal 'claws'.

The postaxial quarter of the fore footplate was absent at 11.5 days in

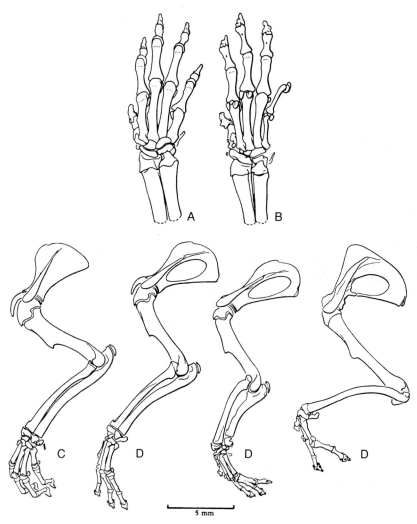

Fig. 8.34. Upper aspect of (A) normal and (B) px/px forefeet. Below, left forelimbs of (C) normal and (D) px/px mice showing increasing severity of the action of px. (From Searle 1964.)

some presumed px/px embryos. In some 10.5-day embryos the footplate occupied only three to four somites (normally five). Searle suggested that the production of abnormal sesamoids, associated tendons (not attached to muscles), and epidermal claws (equivalent to ventral footpads) indicates a failure of the mutant to discriminate between dorsal and ventral axes. However, if so, this must be either selective or local, as the bones of the px foot have distinct dorsal and ventral morphology. We now know that tendons are derived from lateral plate mesoderm and limb muscles from somites, so it is no surprise that abnormally placed tendons are not associated with muscle masses. One effect of this gene is clearly to abolish part of the posterior footplate, just like wingless in its excessive form. If we believe in the ZPA then this abolition must either not include the ZPA or act at a late stage, since the pattern of the limb is intact, even if not fully expressed (the digits remaining have a clear identity—we can tell that the most postaxial surviving digit is a IV or a II). The claw-like excrescences and sesamoids and the malformation of the phalanges and claws suggest a defect in the AER—the AER persists in digital regions and contributes to the phalanges and claws; it could be that abnormal ectoderm over the dorsal sesamoid bones is able to form some response to their presence, just as the ectoderm over the terminal phalanx does.

EUDIPLOPODIA (EU) IN THE CHICK

Eudiplopodia (Rosenblatt *et al.* 1959) produces a very characteristic foot which may have something in common with px and Ps. The eudiplopod foot typically has two rows of digits set one above the other (Fig. 8.35). Goetinck (1964) noted that the eudiplopod limb bud first became noticeably abnormal at stage 22 or 23 when an ectodermal thickening is seen on the dorsal side of some limbs. By stage 24 all have this extra AER, with columnar and later pseudostratified epithelium (Fig. 8.36). At stages 25–7 the mesoderm beneath the extra ridge forms an outgrowth. There is marked variation in how closely the dorsal AER resembles the normal

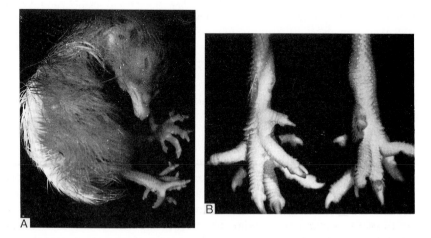

Fig. 8.35. (A) Eudiplopod embryo aged 15 days. (B) Legs from a eudiplopod embryo showing a full expression of the gene. (From Fraser and Abbott 1971*a*.)

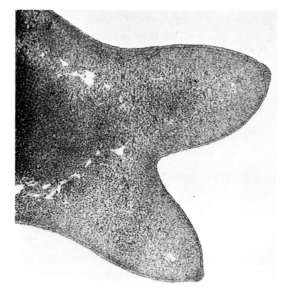

Fig. 8.36. Sagittal section of the left of a eudiplopod embryo (stage 27). Note the symmetrical duplication and the presence of two AERs. (From Goetinck 1964.)

one, even in the same individual where right and left buds are often markedly asymmetrical. In some individuals the 'extra' AER is discontinuous; in others it forms an X with the normal AER, united in the region of digit three. Goetinck found that recombinations between eudiplopod and normal tissue at stage 18–19 produce abnormal feet when eudiplopod ectoderm is combined with normal mesoderm. The reverse combination produces normal feet. The eudiplopod gene is thus unusual in affecting ectoderm rather than mesoderm. Fraser and Abbott (1971a,b) showed that the AER will fail to regenerate in normal limb buds from which it is removed at stage 18. Eudiplopod AER will however, regenerate as late as stage 20. The accessory ridge will still form at stage 22–3 on limbs from which dorsal ectoderm has been previously removed. This suggests that the two ridges are independent. Eudiplopod limb ectoderm grafted to the flank of a normal host also produces two ridges. Eudiplopod flank ectoderm grafted to the limb bud produces a similar response. The capacity of the ectoderm to form a ridge is clearly extended in eudiplopod from the normal stage 19 to 23 when the second ridge appears. The activity of the second ridge could be decreased by exposing mutant embryos to a low temperature during development. This could either interfere with epithelial mitotic rates or mesenchymal stimulus transmission, the former being more likely.

Eudiplopodia thus offers a fascinating glimpse of the mechanisms of normal development. The limb pattern produced by both ridges is normal, and presumably governed by a normal mesodermal ZPA. The extension of competence of the ectoderm to form a ridge to later stages seemingly leads to a complete duplication, the new ridge forming with orientation corresponding to the existing one. Is this an extension of the process seen in Ps and px, where small areas of ectoderm seem to retain their potential longer than normal?

SYNDACTYLY

The syndactylous union of toes may occur at various stages of development; the membranous blastemata for two units may arise as a single entity, or arise separately and then fuse. The cartilaginous blastemata may unite, the osseous elements may fuse or two digits may be united by soft tissue. Two or more of these processes may be seen in a single individual. The following group is therefore a somewhat heterogeneous one.

Oligosyndactylism (Os, Chr 8)

Oligosyndactylism is a semidominant condition which is an early lethal when homozygous. Van Valen (1966) described abnormalities of the seventh and eighth mitotic divisions leading to death at the 64-cell stage or shortly afterwards, on the fifth day of development.

Heterozygotes (Grüneberg 1956) have mild-to-severe diabetes insipidus (Stewart and Stewart 1969; Falconer *et al.* 1964) and abnormalities of all four feet. The forefeet are more affected than the hind. Syndactylism between digits II and III may involve soft tissues only or osseous fusion of basal, basal and middle, or all phalanges (Fig. 8.37). Ultimately the fusion may be so intimate that there is no hint of the compound nature of the digit so formed. A similar effect can also be achieved by another means: the cartilaginous *anlagen* making up digit II may be reduced in calibre so that they are too small to ossify and therefore regress. In rare cases digit II is said to be absent and digit I doubled, thus restoring a pentadactyl foot. Secondary fusions are found amongst that tarsalia and carpalia, through cuboideum and cuneiform 3 are primarily fused from the outset. Carpals are also fused secondarily.

Grüneberg (1961) described abnormalities in the Os/+ foot dating from day 11 or 12 (Fig. 8.38). At this stage there is a reduction of (mainly) the preaxial border of the limb bud. Milaire (1962, 1967) showed an increase in cell death in the anterior part of the AER at this stage and that a small area of mesoderm between digits II and III also underwent a 'cytolytic injury' responsible for blastemal fusion. As the blastemata are formed those of digits II and III are seen to be closer together than normal and less

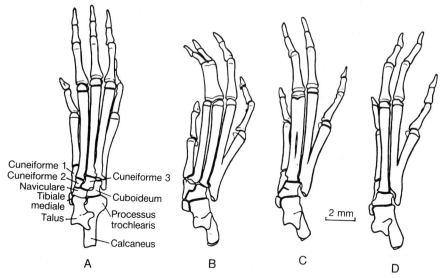

Fig. 8.37. Right hindfeet of (A) normal and (B–D) progressively more affected Os/+ mice. (From Grüneberg 1963.)

divergent. This propinquity leads to soft tissue or more intimate fusion of the digits. The calibre of the blastemata for digit II is seen to be reduced in some individuals and in others an 'anomalous shift of the blastema for digit II towards digit I' (Grüneberg 1963) is supposed to produce the type of foot with a duplicated hallux but no digit II. This is probably less of a polydactyly than a rearrangement of material; it makes sense to suppose that the least medial of the two halluces in fact represents the missing digit II rather than suppression and duplication having taken place. Kadam (1962) investigated the musculature of Os/+ feet and found anomalous arrangements in both foot and lower leg which are not necessarily correlated with skeletal defects. In the hindlimbs, for example, the most striking differences are in the postaxial musculature whilst skeletal defects are based on digit II.

Shaker with syndactylism (sy, Chr 18)

We have already discussed (Chapter 2) the abnormality of the membranous skeleton of shaker with syndactylism (Hertwig 1942) which appears to affect mainly the limbs and the labyrinth. In this recessive condition surviving homozygotes, which are small, do not breed and have disturbed behaviour and variable involvement of the fore- and hindfeet. Grüneberg (1956) showed that syndactylism is confined to digits II and III, III and IV, or all three. Fusions may involve soft tissue or be osseous; in the latter case they tend to involve all, or merely the two distal phalanges. Metatarsals

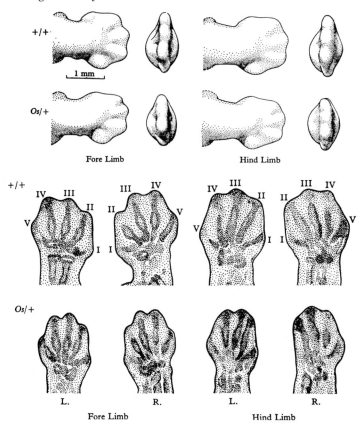

Fig. 8.38. Right fore- and hindlimbs of +/+ and Os/+ embryos aged 13 days. Drawings below based on micrographs of sections through limbs of embryos of the same age. (From Grüneberg 1961.)

and metacarpals are never involved but tarsals and carpals show extensive fusions. The fusions between digits are primary, those between carpals and tarsals secondary.

These abnormalities can first be seen at 12.5 days of gestation (Grüneberg 1962) when the footplates are seen to be narrowed both pre- and postaxially. This narrowing forces together the blastemata of the middle digits which tend to coalesce distally. There are small changes in the long bones, girdles, vertebrae, and scapula none of which is frankly pathological.

A similar allele, fused phalanges sy^{fp} reported by Lane and Hummel (1973) has fusions amongst the middle three digits of the hindfeet, and often the forefeet; behaviour in this allele is normal.

Syndactylism (sm, Chr ?)

Syndactylism (Grüneberg 1956) was described as a systemic disturbance with the skeleton involved as a secondary consequence of a disturbance of the embryonic epidermis. Many animals homozygous for sm reach adulthood and breed. The forefeet usually exhibit soft tissue syndactyly between digits, the hindfeet osseous union. The primary fusion is between digits III and IV; digit II and even digit I in the hindfoot is occasionally included. Again the familiar pattern of fusion between phalanges and carpals or tarsals but not metacarpals or metatarsals is described. Sixty per cent of sm tails show kinks or twists. Kadam (1962) found that the tendons of muscles were more affected than the muscles themselves. This is of interest in the light of subsequent work which shows that leg musculature is derived from somitic, but tendons from somatic mesoderm.

The osseous defects seen in sm are preformed in cartilage, and once again the fusions between phalanges are primary and those of tarsals and carpals secondary. Limb buds can be seen to be abnormal at 12 days when their dorsal surfaces are more strongly curved than normal, and palmar and plantar surfaces more flattened or, later, actually concave. In sections this can be seen to be correlated with an increase in thickness of the AER, first visible at 10–10.5 days. Grüneberg (1963) curiously stated that this excessive size of the AER may be causally related to the *overgrowth* of the footplates and thus syndactylism. This appears to be contradictory; the sy footplate is not overgrown but deformed and overgrowth of the AER would lead to polydactyly. It is clear from Grüneberg's illustration that the AER, in common with ectoderm elsewhere, is thicker than usual and that this appears to be correlated with abnormalities of limb and tail—it is not however an overgrowth in the sense seen in polydactylous mutants since there is no corresponding outgrowth of limb bud mesoderm. The thickened AER later becomes prematurely keratinized. Milaire (1967) found no evidence for overgrowth of the footplate, and noted that hyperplasia was confined to the anterior part of the AER; he postulated that syndactyly was due to abnormal mechanical rigidity of the limb bud due to the excessively thick ectoderm.

Syndactylism (jt, Chr ?)

Center (1966) described a recessive syndactylism which is viable and usually expressed as soft tissue union of digits II and III, III and IV, or all three. The hindfeet are affected more than the forefeet. In more extreme cases there is osseous fusion, again of the phalanges from terminal to proximal, and Center's illustrations also show fusions amongst tarsalia.

The jt/jt feet can be seen to be abnormal as early as 11.5 days by the loss of postaxial and a little preaxial tissue from the developing footplate.

There is no hyperplasia of the AER (cf. sm), and no dorso-ventral swelling. On day 13 the digits are compressed in the same way as in other syndactylous mutants.

Bovine syndactylism (sy)

Grüneberg and Huston (1968) described the development of the embryos of syndactylous cattle. This condition is caused by a recessive gene and is included here to show the effect of syndactylism genes on the highly modified bovine foot.

In cattle blastemata for digits II, III, and IV are visible at 37 days; the blastema of digit II does not chondrify. Metatarsalia III and IV move closer at their distal extremities and ultimately fuse as cartilaginous elements. In sy/sy cattle the blastemata are much closer together from the time when they can first be seen and fuse distally before chondrification sets in (Fig. 8.39). The material destined to form the phalanges of the two

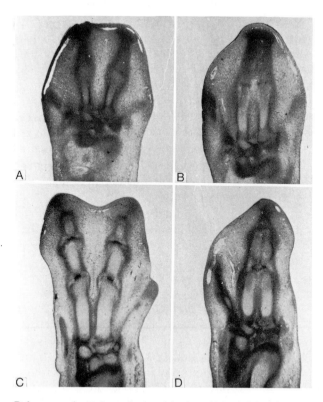

Fig. 8.39. Palmar sections through the right forelimb of (A,C) normal and (B,D) syndactylous cattle embryos aged 37–41 days. (From Grüneberg and Huston 1968.)

extant toes is thus brought close together and forms a single extensive toe
seen only to be composite, in some individuals, at the proximal end of its
basal phalanx.

The syndactylism of the digits in the cow is thus an exact counterpart of
that seen in the mouse—blastemata that are too close together fuse.
Secondary fusion of the metatarsals is a normal event in cattle: in sy cattle
it becomes a primary fusion.

SYNDACTYLISM—AN OVERVIEW

The above group of syndactylous mutants of mice provide a remarkably
consistent picture. All the genes in this group act on digits II, III, and IV,
and feature a small reduction in the limb bud, pre- and postaxially. Fusion
between precartilaginous blastemata of the phalanges originates distally
and may spread proximally. The metacarpals or metatarsals are spared,
but the carpals and tarsals in the narrower wrist region (formed, and
normal, before the fusion of the phalanges) show secondary fusion as a
result of the crowding of the digital rays. In sy we can see that the whole
process is due to hyperplasia of the ectoderm; in other mutations the cause
is less obvious. In all cases the whole of the limb pattern is present and
correct—the changes leading to syndactyly occur at a much later stage than
those causing polydactyly.

The sole exception to all this seems to be in Os where digit II is formed
as a rather narrow blastema which may never form an osseous element.
The key to this condition perhaps lies in the carpals and tarsals where
Grüneberg (1963) noted that, amongst many secondary fusions, primary
fusion of the cuboideum and cuneiforme 3 is always present; this perhaps
indicates an earlier defect at the time of the formation of the precartilaginous blastemata of these elements.

Brachypodism (bp, Chr 2)

We have already dealt in earlier chapters (2 and 3) with the cellular basis of
brachypodism. Here we should merely consider its overall effect, which
seems to be different from any of the mutants described above.

Grüneberg and Lee (1973) described the way in which various segments
of the limb skeleton are shortened by various amounts in brachypod, the
ulna/tibia being slightly reduced, the humerus and femur rather more, and
the manus and pes more still (Fig. 8.40). The brachypod limb bud is normal
in size and form until the digital condensations can be seen at 13 days (Fig.
8.41). Grüneberg and Lee suggested that the defect is due to a
misallocation of material amongst the various proximo-distal divisions of

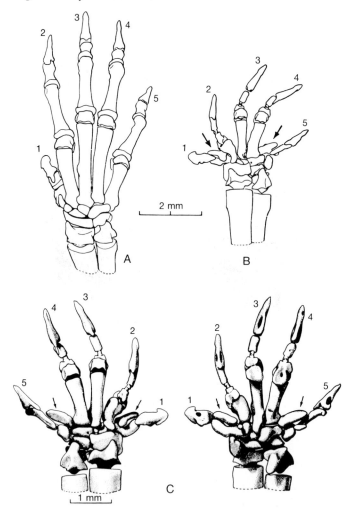

Fig. 8.40.   Dorsal views of the right manus of (A) normal and (B) a brachypod mouse aged 19 days. (C) The left manus of the same brachypod individual. (From Grüneberg and Lee 1973.)

the limb. At chondrification the metacarpal is normally longer than the sum of the two basal phalanges but in brachypod the reverse obtains, with a single element representing both phalanges being longer than the metcarpal (Fig. 8.42). At 13 days the brachypod digital condensations are thinner than normal and the metacarpal–phalangeal joint is moved proximally. Perhaps the primary defect in cell adhesion known to be the basis of the brachypod syndrome interferes with the allocation of overall blastemal pattern; we have already noted the importance of cell adhesion in blastemal formation.

Brachypod is thus important in suggesting that changes in adult limb bone proportion may be under genetic control; this must be carefully distinguished from changes brought about by the chondrodystrophies which operate at a totally different stage in development, and by different mechanisms.

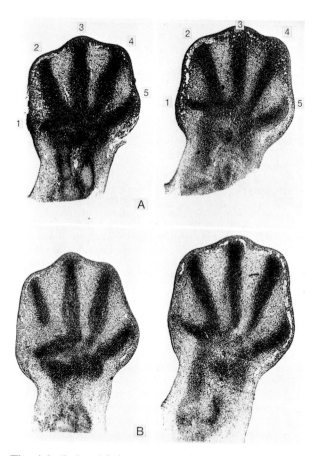

Fig. 8.41. The right limbs of (A) normal and (B) brachypod embryos at 13 days. Forelimbs on the left. (From Grüneberg and Lee 1973.)

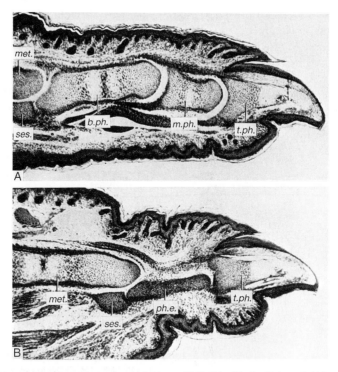

Fig. 8.42. Sagittal sections through the middle hindlimb digits of (A) a normal and (B) a brachypod mouse, 1 day old. met, metatarsal; ses, sesamoid; b.ph, m.ph, t.ph, phalanges; ph.e, phalangeal element equivalent to basal and middle phalanges. (From Grüneberg and Lee 1973.)

REFERENCES

Abbott, U. K. (1959). Further studies on diplopodia. II Embryological features. *Journal of Genetics* **56**, 179–96.

—— and Goetinck, P. F. (1960). Splitfoot, a lethal mutation affecting the foot structure of the fowl. *Journal of Heredity* **51**, 161–6.

—— and Kieny, M. (1961). Sur la croissance in vitro du tibiotarse et du péroné de l'embryon de poulet 'diplopode'. *Comptes rendus de l'Académie des Sciences* **252**, 1863–5.

—— and MacCabe, J. A. (1966). Ectrodactyly: a new embryonic lethal mutation in the chicken. *Journal of Heredity* **57**, 207–11.

—— Taylor, L. W., and Abplanalp, H. (1960). Studies with talpid^2, an embryonic lethal of the fowl. *Journal of Heredity* **51**, 195–202.

Amprino, R. (1965). Aspects of limb morphogenesis in the chicken. In *Organogenesis* (ed. R. L. De Haan and H. Ursprung). Holt Reinhart & Winston, New York.

Baumann, L. and Landauer, W. (1944). On the expression of polydactylism in the wings of the fowl. *Anatomical Record* **90**, 377–98.

Bean, A. M. (1929). A morphological analysis of the foot abnormalities occurring in the descendants of X-rayed mice. *American Journal of Anatomy* **43**, 221–46.

Beasley, A. B. and Cratchfield, F. L. (1969). Development of a mutant with abnormalities of the eye and extremities. *Anatomical Record* **63**, 293.

Beck, S. L. (1961). A system of polydactyly in the mouse. *American Zoologist* **1**, 183.

—— (1963). The anophthalmic mutant of the mouse. II. An association of anophthalmia and polydactyly. *Journal of Heredity* **54**, 79–83.

Boder, W. F. (1960). Interaction of modifiers: the effect of pallid and fidgit on polydactyly in the mouse. *Heredity* **14**, 455–8.

Borghese, E. (1952). Richerche embriologiche ed istologische sui mutanti del locus W mei topi. *Genetics* **3**, 86–118.

Braus, H. (1908). Entwicklungsgeschichtlich Analyse der Hyperdaktylie. *München Medizinische Wochenschrift* **55**, 386–90.

Bryan, J. H. D. (1977). Spermatogenesis revisited. IV. Abnormal spermiogenesis in mice homozygous for another male sterilty inducing mutation hpy. *Cell and Tissue Research* **180**, 187–201.

Bryant, S. V., French, V., and Bryant, P. J. (1981). Distal regeneration and symmetry. *Science* **212**, 993–1002.

Burda, D. J. and Center, E. M. (1969). Development of luxoid (lu) skeletal defects in vitro. *Journal of Embryology and experimental Morphology* **21**, 347–60.

Cairns, J. M. (1975). The function of the apical ectodermal ridge and distinctive characteristics of adjacent distal mesoderm in the avian wing bud. *Journal of Embryology and experimental Morphology* **34**, 155–69.

—— (1977). Growth of normal and talpid^2 chick wing buds: an experimental analysis. In *Vertebrate limb and somite morphogenesis* (ed. D. A. Ede, J. R. Hinchliffe, and M. Balls), pp.123–37. Cambridge University Press.

Calandra, A. J. and MacCabe, J. A. (1978). The in vitro maintenance of the limb bud apical ridge by cell free preparations. *Developmental Biology* **62**, 258–69.

Carter, T. C. (1951). The genetics of luxate mice. I. Morphological abnormalities of heterozygotes and homozygotes. *Journal of Genetics* **50**, 277–99.

—— (1953). The genetics of luxate mice. III. Horsehoe kidney hydronephrosis and lumbar reduction. *Journal of Genetics* **51**, 441–57.

—— (1954). The genetics of luxate mice. *Journal of Genetics* **52**, 1–35.

—— (1956). Genetics of the Little and Bagg X-rayed mouse stock. *Journal of Genetics* **54**, 311–26.

—— (1959). Embryology of the Little and Bagg X-rayed mouse stock. *Journal of Genetics* **56**, 401–35.

Center, E. M. (1955). Postaxial polydactyly in the mouse. *Journal of Heredity* **46**, 144–8.

—— (1966). Genetical and embryological studies of the jt form of syndactylism in the mouse. *Genetical Research* **8**, 33–40.

—— (1977). Genetical and embryological comparison of two mutants which cause foetal blebs in mice. *Genetical Research* **29**, 147–58.

Chai, C. K. (1981). Dactylaplasia in mice: a two locus model for developmental anomalies. *Journal of Heredity* **72**, 234–7.

Chase, H. B. (1951). Inheritance of polydactyly in the mouse. *Genetics* **86**, 697–710.

Chevallier, A., Kieny, M., and Mauger, A. (1976). Sur l'origine de la musculature de l'aile chez les oiseaux. *Comptes rendus de l'Academie des Sciences* **282**, 309–11.

—— —— —— (1977a). Limb–somite relationship: origin of the limb musculature.

Journal of Embryology and experimental Morphology **41**, 245–58.

—— —— —— and Sengel, P. (1977*b*). Developmental fate of the somitic mesoderm in the chick embryo. In *Vertebrate limb and somite morphogenesis* (ed. D. A. Ede, J. R. Hinchliffe, and M. Balls), pp.421–32. Cambridge University Press.

Christ, B., Jacob, H-J., and Jacob, M. (1974). Uber den Ursprung der Flügel-muskulatur. Experimentelle Untersuchungen mit Wachtel- und Hühner-embryonen. *Experientia* **30**, 1446–8.

—— and Jacob, H. T. (1980). Origin, distribution, and determination of chick mesenchymal cells. In *Teratology of the limbs* (ed. H-J. Merker, H. Nau, and D. Neubert), pp. 67–77. de Gruyter, Berlin.

Cole, R. K. (1942). The talpid lethal in the domestic fowl. *Journal of Heredity* **33**, 82–6.

—— (1967). Ametapodia, a dominant mutation in the fowl. *Journal of Heredity* **58**, 141–6.

Cook, J. and Summerbell, D. (1980). Growth control early in embryonic development: the cell cycle during experimental pattern duplication in the chick wing. *Nature, London* **287**, 687–701.

Deol, M. S. (1961). Genetical studies on the skeleton of the mouse. XXVIII. Tail-short. *Proceedings of the Royal Society* **B155**, 78–95.

Dhouailly, D. and Kieny, M. (1972). The capacity of the flank mesoderm of early bird embryos to participate in limb development. *Developmental Biology* **28**, 162–75.

Ede, D. A. (1968). Abnormal development at the cellular level in talpid and other mutants. In *The fertility and hatchability of the hen's egg* (ed. T. C. Carter and B. M. Freeman), pp. 71–83. Oliver & Boyd, Edinburgh.

—— (1971). Control of form and pattern in the vertebrate limb. *Symposia of the Society for Experimental Biology* **25**, 235–54.

—— (1976). Cell interactions in vertebrate limb development. In *The cell surface in animal embryogenesis and development* (ed. G. Poste and G. L. Nicolson), pp. 493–543. Elsevier, Amsterdam.

—— (1977). Limb skeleton deficiencies in some mutants of the chick embryo. In *Mécanismes de la rudimentation des organes chez les embryons de vertèbres*, pp.187–91. Coilloques CNRS 266, Paris.

—— (1980). Role of the ectoderm in limb development of normal and mutant mouse (Disorganisation, pupoid foetus) and fowl (talpid³ embryos. In *Teratology of the limbs* (ed. H. J. Merker, H. Nau, and D. Neubert), pp. 53–7. de Gruyter, Berlin.

—— and Agerback, G. S. (1968). Cell adhesion and movement in relation to the developing limb pattern in normal and talpid³ mutant chick embryos. *Journal of Embryology and experimental Morphology* **20**, 81–100.

—— and Flint, O. (1972). Patterns of cell division, cell death and chondrogenesis in cultured aggregates of normal and talpid³ mutant chick limb mesenchyme cells. *Journal of Embryology and experimental Morphology* **27**, 245–60.

—— —— (1975*a*). Intercellular adhesion and formation of aggregates in normal and talpid³ mutant chick limb mesenchyme. *Journal of Cell Science* 18, 97–111.

—— —— (1975*b*). Cell movement and adhesion in the developing chick wing bud: studies of cultured mesenchyme cells from normal and talpid³ mutant embryos. *Journal of Cell Science* **18**, 301–13.

—— and Kelly, W. A. (1964*a*). Developmental abnormalities in the head region of the talpid³ mutant of the fowl. *Journal of Embryology and experimental Morphology* **12**, 161–82.

—— —— (1964*b*). Developmental abnormalities in the trunk and limbs of the talpid[3] mutant of the fowl. *Journal of Embryology and experimental Morphology* **12**, 339–56.

—— and Law, J. T. (1969). Computer simulation of vertebrate limb morphogenesis. *Nature* **221**, 244–8.

Falconer, D. S., Latyszewski, M., and Isaacson, J. H. (1964). Diabetes insipidus associated with oligosyndactylism in the mouse. *Genetical Research* **5**, 473–88.

Fallon, J. F. and Caplan, A. I. (1983). *Limb development and regeneration.* Alan B. Liss, New York.

—— and Crosby, G. M. (1975). Normal development of the chick wing following removal of the polarising zone. *Journal of experimental Zoology* **193**, 449–55.

Fisher, R. A. (1953). The linkage of polydactyly with leaden in the house mouse. *Heredity* **7**, 91–5.

Forsthoefel, P. F. (1958). The skeletal effects of the luxoid gene in the mouse, including its interactions with the luxate gene. *Journal of Morphology* **102**, 247–87.

—— (1959). The embryological development of the skeletal effects of the luxoid gene in the mouse including its interaction with the luxate gene. *Journal of Morphology* **104**, 89–141.

—— (1962). Genetic and manifold effects of Strong's luxoid gene in the mouse including its interactions with Green's luxoid and Carter's luxate genes. *Journal of Morphology* **110**, 391–420.

—— (1963a). Observations on the sequence of blastemal condensations in the limbs of the mouse embryo. *Anatomical Record* **147**, 129–38.

—— (1963b). The embryological development of the effect of Strong's luxoid gene in the mouse. *Journal of Morphology* **113**, 427–52.

Fortuyn, A. B. D. (1939). A polydactylous strain of mice. *Genetica* **21**, 97–108.

Fraser, R. A. and Abbott, U. K. (1971a). Studies on limb morphogenesis. V. The expression of eudiplopodia and its experimental modification. *Journal of experimental Zoology* **176**, 219–36.

—— —— (1971b). Studies in limb morphogenesis. VI. Experiments with early stages of the polydactylous mutant eudiplopodia. *Journal of experimental Zoology* **176**, 237–48.

French, V., Bryant, P. J., and Bryant, S. V. (1976). Pattern regulation in epimorphic fields. *Science* **193**, 969–81.

Freye, H. (1954). Anatomische und entwicklungsgeschichtliche Untersuchungen am Skelett normaler und oligodactyler Mäuse. *Wissenschaftliche Zeitschrifte der Martin Luther Universität halle-Wittenberg* **3**, 801–24.

Gabriel, M. L. (1946). The effect of local applications of colchicine on leghorn and polydactylous chick embryos. *Journal of experimental Zoology* **101**, 339–50.

Goetinck, P. F. (1964). Studies on limb morphogenesis. II. Experiments with the polydactylous mutant eudiplopodia. *Developmental Biology* **10**, 71–91

—— and Abbott, U. K. (1964). Studies in limb morphogenesis. I. Experiments with the polydactylous mutant talpid[2]. *Journal of experimental Zoology* **155**, 161–70.

Green, M. C. (1955). Luxoid, a new hereditary leg and foot abnormality in the house mouse. *Journal of Heredity* **46**, 91–9.

—— (1967). A defect of the splanchnic mesoderm caused by the mutant gene dominant hemimelia in the mouse. *Developmental Biology* **15**, 62–89.

Greene, H. S. N. and Saxton, J. A. (1939). Hereditary brachydactyly and allied anomalies in the rabbit. *Journal of experimental Medicine* **69**, 301–14.

Grüneberg, H. (1943). Two new mutant genes in the house mouse. *Journal of Genetics* **45**, 22–8.

—— (1952). *The genetics of the mouse*, 2nd edn. Martinus Nijhoff, The Hague.

—— (1956). Genetical studies on the skeleton of the mouse. XVIII. Three genes for syndactylism. *Journal of Genetics* **54**, 113–45.

—— (1961). Genetical studies on the skeleton of the mouse. XXVII. The development of oligosyndactylism. *Genetical Research* **2**, 33–42.

—— (1962). Genetical studies on the skeleton of the mouse. XXXII. The development of shaker with syndactylism. *Genetical Research* **3**, 157–66.

—— (1963). *The pathology of development.* Blackwell, Oxford.

—— and Huston, K. (1968). The development of bovine sundactylism. *Journal of Embryology and experimental Morphology* **19**, 251–9.

—— and Lee, A. J. (1973). The antomy and development of brachypodism in the mouse. *Journal of Embryology and experimental Morph* syndactylism.'–41.

Hamburger, V. (1938). Morphogenetic and axial self-differentiation of transplanted limb primordium of 2 day old chick embryos. *Journal of experimental Zoology* **77**, 379–400.

—— (1939). The development and innervation of transplanted limb primordia of chick embryos. *Journal of experimental Zoology* **80**, 347–89.

Hampe, A. (1959). Contribution a l'étude du dévelopment et de la régulation des déficiences et des excédents dans la patte de l'embryon de poulet. *Archives d'Anatomie microscopique et de Morphologie expérimentale* **48**, 345–78.

Harrison, R. G. (1921). On relations of symmetry in transplanted limbs. *Journal of experimental Zoology* **32**, 1–136.

Hayasaka, I., Nakatsuka, T., Fujii, T., and Naruse, I. (1980). Polydactyly Nagoya *Pdn*: a new mutant gene in the mouse. *Experimental Animals* **29**, 391–5.

Hertwig, P. (1939). Zwei subletale recessive Mutationen in der Nachkommenschaft von röntgenbestrahlten Mäusen *Erbarzt* **6**, 41–43.

—— (1942). Neue mutationen und Koppelungsgruppen bei der Hausmaus. *Zeitschrift für inducktive Abstammungs- und Vererbungslehre* **80**, 220–46.

Hinchliffe, J. R. (1976). The development of winglessness in the chick. In *Mecanismes de la rudimentation des organes chez les embryons de vertebrès*, pp.173–85. Editions du CNRS **266**, Paris.

—— (1981). Cell death in embryogenesis. In *Cell death* (ed. I. D. Bowen and R. A. Lockshin). Chapman & Hall, London.

—— and Ede, D. A. (1967). Limb development in the polydactylous talpid[3] mutant of the fowl. *Journal of Embryology and experimental Morphology* **17**, 385–404.

—— —— (1973). Cell death and the development of form and skeletal pattern in normal and wingless (ws) chick embryos. *Journal of Embryology and experimental Morphology* **30**, 753–72.

—— Garcia-Porrero, J. A., and Gumpel-Pinot, M. (1981). The role of the zone of polarising activity in controlling the maintenance and anteroposterior differentiation of the apical mesoderm of the chick wing bud: histochemical techniques in the analysis of a developmental problem. *Journal of Histochemistry* **13**, 643–58.

—— and Gumpel-Pinot, M. (1981). Control of maintenance and anteroposterior skeletal differentiation of the anterior mesenchyme of the chick wing bud by its posterior margin (the ZPA). *Journal of Embryology and experimental Morphology* **62**, 63–82.

—— —— (1983). Experimental analysis of avian limb morphogenesis. *Current Ornithology* **1**, 293–327.

—— and Johnson, D. R. (1980). *The development of the vertebrate limb.* Clarendon Press, Oxford.

—— and Thorogood, P. V. (1974). Genetic inhibition of mesenchymal cell death and the development of form and skeletal pattern in the limbs of talpid[3] mutant chick embryos. *Journal of Embryology and experimental Morphology* **31**, 747–60.

Hollander, W. F. (1976). Hydrocephalic-polydactyly, a recessive pleiotropic mutant in the mouse and its location on chromosome 6. *Iowa State Journal of Research* **51**, 13–23.

Holt, S. B. (1945). A polydactyly gene in mice capable of nearly regular manifestation. *Annals of Eugenics* **12**, 220–49.

—— (1948). The effect of maternal age on the manifestation of a polydactyly gene in mice. *Annals of Eugenics* **14**, 144–57.

Honig, L. S. (1981). Positional signal transmission in the developing chick limb. *Nature* **291**, 72–83.

—— (1985). Does anterior (non-polarising region) tissue signal in the developing chick limb? *Developmental Biology* (in press).

Horder, T. J. (1978). Functional adaptability and morphogenetic opportunism, the only rules for limb development. *Zoon* **6**, 181–92.

Hornbruch, A. (1980). Abnormalities of the proximo-distal axis of the chick wing bud: the effect of surgical intervention. In *Teratology of the limbs* (ed. H-J. Merker, H. Nau, and D. Neubert), pp. 191–8. De Gruyter, Berlin.

—— and Wolpert, L. (1970). Cell division in the early growth and morphogenesis of the chick limb bud. *Nature* **226**, 764–6.

Hummel, K. P. (1958). The inheritance and expression of disorganisation, an unusual mutation in the mouse. *Journal of experimental Zoology* **137**, 389–424.

—— (1959). Developmental anomalies in mice resulting from action of the gene disorganisation, a semi-dominant lethal. *Pediatrics* **23**, 212–21.

—— (1970). Hypodactyly, a semi-dominant lethal mutation in mice. *Journal of Heredity* **61**, 219–20.

Hunt, E. A. (1932). The differentiation of chick limb buds in chorio-allantoic grafts, with special reference to muscles. *Journal of experimental Zoology* **62**, 57–91.

Hutt, F. B. (1949). *Genetics of the fowl*. McGraw Hill, New York, London.

Inman, O. R. (1941). Embryology of hereditary brachydactyly in the rabbit. *Anatomical Record* **79**, 483–505.

—— (1946). Developmental abnormalities in the talpid lethal (tata) chick. PhD. Thesis, Cornell University.

Iten, L. E. (1982). Pattern specification and pattern regulation in the embryonic chicken limb bud. *American Zoologist* **22**, 117–29.

—— and Murphy, D. J. (1980a). Pattern regulation in the embryonic chick limb: supernumerary limb formation with anterior (non ZPA) limb bud tissue. *Developmental Biology* **75**, 373–85.

—— —— (1980b). Supernumerary limb structures with regenerated posterior wing bud tissues. *Journal of experimental Zoology* **213**, 327–35.

Jacob, H. J. and Christ, B. (1980). On the formation of the muscular pattern in the chick limb. In *Teratology of the limbs* (ed. H-J. Merker, H. Nau, and D. Neubert). pp. 89–98. de Gruyter, Berlin.

Jacob, M., Christ, B., and Jacob, H. J. (1978). On the migration of myogenic stem cells into the prospective wing region of chick embryos. A scanning and transmission electronmicroscopy study. *Anatomy and Embryology* **153**, 179–93.

Janners, M. Y. and Searls, R. L. (1970). Changes in the rate of cellular proliferation during the differentiation of cartilage and muscle in the mesenchyme of the embryonic chick wing. *Developmental Biology* **23**, 136–65.

Johnson, D. R. (1967). Extra-toes: a new mutant gene causing multiple abnormalities in the mouse. *Journal of Embryology and experimental Morphology* **17**, 543–82.

—— (1969*a*). Brachyphalangy, an allele of extra-toes in the mouse. *Genetical Research* **13**, 275–80.

—— (1969*b*). Polysyndactyly, a new mutant gene in the mouse. *Journal of Embryology and experimental Morphology* **21**, 285–94.

—— and Hunt, D. M. (1971). Hop-sterile, a mutant gene affecting sperm-tail development in the mouse. *Journal of Embryology and experimental morphology* **25**, 223–36.

Jorquera, B. and Pugin, E. (1971). Sur le comportement du mésoderme et de l'ectoderme du bourgeon de membre dans les éschanges entre le poulet et la rat. *Comptes rendus de l'Académie des Sciences* **272**, 522–5.

Jost, A., Roffi, J., and Courtat, M. (1969). Congenital amputations determined by the br gene and those induced by adrenalin injection in the rabbit fetus. In *Limb development and deformity*, pp. 187–99, (ed. C. A. Swingard). Thomas, Springfield, Illinois.

Kadam, K. M. (1962). Genetical studies on the skeleton of the mouse. XXXI. The muscular anatomy of syndactylism and oligosyndactylism. *Genetical Research* **3**, 139–56.

Kaufmann-Wolf, M. (1908). Embryologische und anatomische Beiträge zur Hypodactylie (Houdanhühn). *Morphologisches Jahrbuch* **38**, 471–531.

Kieny, M. (1964). Etude du mécanisme de la regulation dans le dévelopment du bourgeon de membre de l'embryon de poulet. *Journal of Embryology and experimental Morphology* **12**, 357–71.

—— (1971). Les phases d'activité morphogènes du mésoderme somatopleural pendant le dévelopment précoce du membre chez l'embryon de poulet. *Annales d'Embryologie et de Morphogenèse* **4**, 281–8.

—— (1977). Proximo-distal pattern formation in avian limb development. In *Vertebrate limb and somite morphogenesis* (ed. D. A. Ede, J. R. Hinchliffe, and M. Balls), pp.87–103. Cambridge University Press.

—— (1980). The concept of the myogenic cell line in developing avian limb buds. In *Teratology of the limbs* (ed. H-J. Merker, H. Nau, and D. Neubert), pp. 79–88. de Gruyter, Berlin.

—— and Chevallier, A. (1980). Exite-t-il une relation spatiale entre le niveau d'origine des cellules somitiques myogenes et leur colonisation terminale dans l'aile? *Archives d'Anatomie microscopique et de Morphologie expérimentale* **69**, 35–46.

Kobozieff, N. and Pomriaskinsky-Kobozieff, N. A. (1959). Hémimélie longitudinale chez la souris. I. Étude morphologique des hétérozygotes atteints de différentes anomalies du squelette du segment distal. A. Hyperphalangie et polydactylie squelettique. *Recueil de médecine vétérinaire* **135**, 877–902.

—— —— (1960). Hémimélie longitudinale chez la souris. I. Étude morphologique des hétérozygotes atteints de différéntes anomalies du squelette du segment distal. B. Polydactylie intégral. *Recueil de médicine vétérinaire* **136**, 189–218.

—— —— avec l'assistance de E. Gemahling (1961). Hémimélie longitudinale chez la souris. II. Étude morphologique des homozygotes atteints de différentes anomalies du squelette. A. Membres antérieurs. *Recueil de Médecine Vétérinaire* **137**, 965–96.

—— —— (1962*a*). Hémimélie longitudinale chez la souris. II. Étude morphologique des homozygotes atteints de différentes anomalies du squelette. B. Membres postérieurs: hyperphalangie oligodactylie et polydactylie squelettique. *Recueil de médicine vétérinaire* **138**, 271–303.

—— —— (1962*b*). Hémimélie chez la souris. II. Étude morphologique des homozygotes atteints de différentes anomalies du squelette. C. Membres postérieurs: polydactylie intégrale et ceinture pelvienne. *Recueil de Médecine vétérinaire* **138**, 485–505.

—— —— avec l'assistance de E. Gemahling (1962c). Hémimélie chez la souris. II. Étude morphologique des homozygotes atteints de différentes anomalies du squelette. D. Squelette axial. *Recueil de Médicine vétérinaire* **138**, 671–86.

Lancaster, T. M. (1968). Sex-linked winglessness in the fowl. *Heredity* **23**, 257–62.

Landauer, W. (1948). The phenotypic modification of hereditary polydactylism of fowl by selection and insulin. *Genetics* **33**, 133–57.

—— (1956a). A second diplopod mutation of the fowl. *Journal of Heredity* **47**, 57–63.

—— (1956b). Rudimentation and duplication of the radius in the duplicate mutant of the fowl. *Journal of Genetics* **54**, 199–218.

Lane, P. W. and Hummell, K. P. (1973). *Mouse News Letter* **49**, 32.

Lewis, J. (1980). Defective innervation and defective limbs: causes and effects in the developing chick wing. In *Teratology of limbs* (ed. H-J. Merker, H. Nau, and D. Neubold), pp. 235–92. de Gruyter, Berlin.

—— Chevallier, A., Kieny, M., and Wolpert, L. (1981). Muscle nerve branches do not develop in chick wings devoid of muscle. *Journal of Embryology and experimental Morphology* **64**, 211–32

Long, S. Y. and Johnson, E. M. (1968). Enzyme ontogeny in normal and hemimelic limbs of mice. *Journal of Embryology and experimental Morphology* **20**, 415–30.

MacCabe, A. B., Gasseling, N. T., and Saunders, J. W. (1973). Spatiotemporal distribution of mechanisms that control outgrowth and antero-posterior polarisation of the limb bud in chick embryos. *Mechanisms of Ageing and Development* **2**,1–12.

MacCabe, J. A. and Abbott, U. K. (1974). Polarizing and maintenance activity in two polydactylous mutants of the fowl: diplopodia[1] and talpid[2]. *Journal of Embryology and experimental Morphology* **31**, 735–46.

—— Calendra, A. J., and Parker, B. W. (1977). In vitro analysis of the distribution and nature of a morphogenetic factor in the developing chick wing. In *Vertebrate limb and somite morphogenesis* (ed. D. A. Ede, J. R. Hinchliffe, and M. Balls), pp.25–40. Cambridge University Press.

—— Leal, K. W., and Leal, C. W. (1982). The control of axial polarity. I. A low molecular weight morphogen affecting the ectodermal ridge. II. Ectodermal control of the dorso-ventral axis. In *Limb development and regeneration* (ed. J. F. Fallon and A. I. Caplan), pp. 237–44. Alan R. Liss, New York.

—— MacCabe, A. B., and Abbott, U. K. (1975). Limb development in diplopodia[4] a polydactylous mutation in the chicken. *Journal of experimental Zoology* **191**, 383–94.

—— and Richardson, K. E. Y. (1982). Partial characterisation of a morphogenetic factor in the developing chick limb. *Journal of Embryology and experimental Morphology* **67**, 1–12.

McLachlan, J. C. and Hornbruch, A. (1979). Muscle forming potential of the non-somitic cells of the early avian limb bud. *Journal of Embryology and experimental Morphology* **54**, 209–17.

Meredith, R. (1965). Genetical and embryological studies of some congenital malformations in the mouse. Institute of Biology, Thesis for Membership.

Milaire, J. (1962). Détection histochimique de modifications des ébauches dans les membres en formation chez la souris oligosyndactyle. *Bulletin de l'Académie royale de médicine de Belgique* **48**, 505–28.

—— (1965). Étude morphogénétique de trois malformations congénitales de l'autopode chez la souris (syndactylisme—brachypodisme—hémimélie dominante) par des methodes cytochemique.*Académie Royale de Belgique, Classe de Science. Mémoires* **16**, 1–120.

—— (1967). Histochemical observations on the developing foot of the normal oligosyndactylous (Os/+) and syndactylous (sm/sm) mouse embryos. *Archives de biologie, Liège* **78**, 223–88.

—— (1971). Évolution et déterminisme des dégénérescences cellulaires au cours de la morphogenèse des membres et leurs modifications tératogeniques. In *Malformations Congéitales des Mammifères* (ed. H. Tuchmann-Duplessis), pp. 131–49. Colloques Pfizer Amboise.

—— (1977*a*). Histochemical expression of morphogenetic gradients during limb morphogenesis (with particular reference to mammalian embryos). *Birth Defects* **13**, 37–67.

—— (1977*b*). Rudimentation digitale au cours du dévelopment normale de l'autopode chez les mammifères. In *Mécanismes de la rudimentation des organes chez les embryos de vertèbres*. Editions du CNRS **266**, Paris.

Morgan, W. C. (1950). A new short tailed mutation in the mouse. *Journal of Heredity* **41**, 208–15.

Moutier, R., Toyania, K., and Charrier, M. F. (1973). Hypodactyly, a new recessive mutation in the Norway rat. *Journal of Heredity* **64**, 99–100.

Murillo-Ferrol, N. L. (1963). Analysis expérimentale de la participation del mesoblasto paraxial sobre la morfogenesis de los miembros en el embrion de las aves. *Annales Descarrollo* **11**, 63–76.

—— (1965). Étude causale de la différentiation la plus précoce de l'ébauche morphologique des membres. Analyse expérimental chez les embryons d'oiseaux. *Acta anatomica* **62**, 80–103.

Nakamura, A., Sakamoto, H., and Moriwaki, K. (1962). Genetical studies of postaxial polydactyly in the house mouse. *Annual Report of the National Institute of Genetics of Japan* **13**, 31.

Naruse, I. and Kameyama, Y. (1982). Morphogenesis of genetic preaxial polydactyly, polydactyly Nagoya *Pdn* in mice. *Congenital Anomalies* **22**, 137–44.

—— —— (1983). Effects of 5-fluororacil and cytosine arabinoside on the manifestation of digital malformations in Pdn mice. *Congenital Anomalies* **23**, 211–21.

Nathanson, M. A., Hilfer, S. R., and Searls, R. L. (1978). Formation of cartilage by non-chondrogenic cell types. *Developmental Biology* **64**, 99–117.

Niederman, R. and Armstrong, P. B. (1972). Is abnormal limb bud morphology in the mutant talpid² chick embryo a result of altered cellular adhesion? Studies employing cell sorting and fragment fusion. *Journal of experimental Zoology* **181**, 17–32.

Olympio, O. S., Crawford, R. D., and Classen, H. L. (1983). Genetics of the diplopodia⁵ mutation in domestic fowl. *Journal of Heredity* **74**, 341–3.

O'Rahilly, R. (1951). Morphological patterns in limb deficiences and duplications. *American Journal of Anatomy* **89**, 135–87.

Pease, M. S. (1962). Wingless poultry. *Journal of Heredity* **53**, 109–10.

Petter, C., Bourbon, J., Maltier, J. P., and Jost, A. (1977). Simultaneous prevention of blood abnormalities and hereditary congenital amputations in brachydactylous rabbit stock. *Teratology* **15**, 149–58.

Pettigrew, A. G., Lindeman, R., and Bennett, M. R. (1979). Development of the segmental innervation of the chick forelimb. *Journal of Embryology and experimental Morphology* **49**, 115–37.

Pinot, M. (1970). Relations entre le mesenchyme somitique et la plaque des membres chez le poulet. *Année Biologique* **9**, 277–84.

Roberts, R. C. and Mendell, N. R. (1975). A case of polydactyly with multiple thresholds in the mouse. *Proceedings of the Royal Society London* **B191**, 427–44.

Ronacali, L. (1970). The brachial plexus and the wing nerve pattern during early developmental phases in chick embryos. *Monitore zoologico italiano N.S.* **4**, 81–98.

Rooze, M. A. (1977). The effects of the Dh gene on limb morphogenesis in the mouse. *Birth Defects* **13**, 69–95.

Rosenblatt, L. S. Kreutziger, G. O., and Taylor, L. W. (1959). Eudiplopodia. *Poultry Science* **38**, 1242.

Rowe, D. A. and Fallon J. F. (1981). The effect of removing posterior apical ectodermal ridge of the chick wing and leg on pattern formation. *Journal of Embryology and experimental Morphology (suppl.)* **65**, 309–25.

—— —— (1982). Normal anterior pattern formation after barrier placement in the chick leg: further evidence on the action of the polarizing zone. *Journal of Embryology and experimental Morphology* **69**, 1–6.

Rubin, L. and Saunders, J. W. (1972). Ectodermal–mesodermal interactions in the growth of limb buds in the chick embryo. Constancy and temporal limits of ectodermal induction. *Developmental Biology* **28**, 94–112.

Russell, E. S. Lawson, F. and Schbtach, G. (1957). Evidence for a new allele at the W locus of the mouse. *Journal of Heredity* **48**, 119–23.

Saunders, J. W. (1948). The proximo-distal sequence of origin of the parts of the chick wing and the role of the ectoderm. *Journal of experimental Zoology* **108**, 363–404.

—— (1972). Developmental control of three dimensional polarity in the avian limb. *Annals of the New York Academy of Science* **193**, 29–42.

—— (1977). The experimental analysis of chick limb development. In *Vertebrate limb and somite morphogenesis* (ed. D. A. Ede, J. R. Hinchliffe, and M. Balls), pp. 1–24. Cambridge University Press.

—— (1982). *Developmental biology*. Macmillan, New York.

—— and Gasseling, M. T. (1968). Ectodermal–mesenchymal interactions in the origin of limb asymmetry. In *Epithelial–mesenchymal interactions* (ed. R. Fleischmajor and R. F. Billingham), pp. 78–97. Williams & Wilkins, Baltimore.

—— —— (1982). New insights into the problem of pattern regulation in the limb bud of the chick embryo. In *Limb development and regeneration* (ed. J. F. Fallon and A. I. Caplan), pp. 67–76. Alan R. Liss, New York.

Sawyer, L. M. (1982). Fine structural analysis of limb development in the wingless mutant chick embryo. *Journal of Embryology and experimental Morphology* **68**, 69–86.

Scott, J. P. (1937). The embryology of the guinea pig. III. The development of the polydactylous monster. A case of growth accelerated at a particular period by a semi-dominant lethal gene. *Journal of experimental Zoology* **77**, 123–57.

—— (1938). The embryology of the guinea pig. II. The polydactylous monster. A new teras produced by the genes Px/Px. *Journal of Morphology* **62**, 299–321

Searle, A. G. (1953). Hereditary 'split hand' in the domestic cat. *Annals of Eugenics* **17**, 279–82.

—— (1964). The genetics and morphology of two luxoid mutants in the house mouse. *Genetical Research* **5**, 171–97.

Searls, R. L. and Janners, M. T. (1969). The stabilisation of cartilage properties in the cartilage forming mesenchyme of the embryonic chick limb bud. *Journal of Embryology and experimental Morphology* **41**, 269–77.

Shellswell, G. B. (1977). The formation of discrete muscles from the chick wing dorsal and ventral muscle masses in the absence of nerves. *Journal of Embryology and experimental Morphology* **41**, 269–71.

—— and Wolpert, L. (1977). The pattern of muscle and tendon development in the chick wing. In *Vertebrate limb and somite morphogenesis* (ed. D. A. Ede, J. R. Hinchliffe, and M. Balls), pp. 71–86. Cambridge University Press.

Slack, J. M. W. (1977). Control of anteroposterior pattern in the axolotl forelimb by a smoothly graded signal. *Journal of Embryology and experimental Morphology* **39**, 169–82.

Smith, J. C. (1979). Evidence for a positional memory in the development of the chick wing bud. *Journal of Embryology and experimental Morphology* **52**, 105–13.

—— and Wolpert, L. (1981). Pattern formation along the antero-posterior axis of the chick wing: the increase in width following a polarising region graft and the effect of X-irradiation. *Journal of Embryology and experimental Morphology* **63**, 127–44.

Staats, J. (1976). Standardised nomenclature for inbred strains of mice: sixth listing. *Cancer Research* **36**, 4353–77.

Stark, R. J. and Searls, R. L. (1973). A description of chick wing bud development and a model of limb morphogenesis. *Developmental Biology* **33**, 138–53.

Stewart, A. D. and Stewart, I. (1969). Studies of syndrome of diabetes insipidus associated with oligosyndactylism in mice. *American Journal of Physiology* **217**, 1191–8.

Stirling, R. V. (1976). Functional innervation of abnormally constructed limbs in chicken embryos. *Neuroscience Letters* **3**, 110.

—— and Summerbell, D. (1977). The development of functional innervation in the chick wing bud following truncations and deletions of the proximo-distal axis. *Journal of Embryology and experimental Morphology* **41**, 189–207.

Straznicky, K. (1967). The development and innervation of the musculature of wings innervated by thoracic nerves. *Acta biologica hungarica* **18**, 437–48.

Strong, L. C. (1961). The Springville mouse, further observations on a new 'luxoid' mouse. *Journal of Heredity* **52**, 122–4.

—— and Hardy, L. B. (1956). A new 'luxoid' mutant in mice. *Journal of Heredity* **47**, 277–84.

Summerbell, D. (1977). Regulation of deficiencies along the proximo-distal axis of the chick wing bud: a quantitative analysis. *Journal of Embryology and experimental Morphology* **41**, 137–59.

—— (1979). The zone of polarising activity: evidence for a role in normal chick limb morphogenesis. *Journal of Embryology and experimental Morphology* **50**, 217–33.

—— and Honig, L. S. (1982). The control of pattern across the antero-posterior axis of the chick limb bud by a unique signalling region. *American Zoologist* **22**, 105–16.

—— and Lewis, J. H. (1975). Time, place and positional value in the chick limb bud. *Journal of Embryology and experimental Morphology* **33**, 621–43.

—— —— and Wolpert, L. (1973). Positional information in chick limb morphogenesis. *Nature* **244**, 492–6.

Sweet, H. O. and Lane, P. W. (1980). X linked polydactyly (Xpl), a new mutation in the mouse. *Journal of Heredity* **71**, 207–9.

Taylor, L. W. (1972). Further studies on diplopodia. V. Diplopodia[3]. *Canadian Journal of Genetics and Cytology* **14**, 417–22.

—— Abbott, U. K., and Gunns, C. A. (1959). Further studies on diplopodia. I. Modifications of phenotypic segregation ratios by selection. *Journal of Genetics* **56**, 161–78.

—— and Gunns, C. A. (1947). Diplopodia, a lethal form of polydactyly in the chicken. *Journal of Heredity* **38**, 66–76.

Tickle, C. (1980). The polarising activity and limb development. In *Development in mammals* 4 (ed. M. H. Johnson), pp. 101–36. Elsevier, North Holland, Amsterdam.

—— Summerbell, D., and Wolpert, L. (1975). Positional signalling and specification of digits in chicken limb morphogenesis. *Nature* **254**, 199–202.

Truslove, G. M. (1956). The anatomy and development of the fidgit mouse. *Journal of Genetics* **54**, 64–86.

Van Valen, P. (1966). Oligosyndactylism, an early embryonic lethal in the mouse. *Journal of Embryology and experimental Morphology* **15**, 119–24.

Warkany, J. (1971) *Congenital malformations*. Year Book Medical Publishers Inc, Chicago.

Warren, D.C. (1941). Inheritance of polydactylism in the fowl. *Genetics* **29**, 217–31.

—— (1944). A new type of polydactyly in the fowl. *Journal of Heredity* **32**, 3–5.

Waters, N. F. and Bywaters, J. F. (1943). A lethal embryonic wing mutation in the domestic fowl. *Journal of Heredity* **34**, 213–17.

Wilby, O. K. and Ede, D. A. (1975). A model generating the pattern of cartilage skeletal elements in the embryonic chick limb. *Journal of Theoretical Biology* **52**, 199–207.

—— —— (1976). Computer simulation of vertebrate limb development. In *Automata, languages, development* (ed. A. Lindenmayer and G. Rozenberg), pp. 15–24. North Holland, Amsterdam.

Wolpert, L. S. (1969). Positional information and the spatial pattern of cellular differentiation. *Journal of Theoretical Biology* **25**, 1–47.

—— (1978). Pattern formation and the development of the chick limb. *Birth Defects* **14**, 547–59.

—— (1981). Positional information and pattern formation. *Philosophical Transactions of the Royal Society* **B295**, 441–50.

Wright, S. (1934). Polydactylous guinea pigs. Two types respectively heterozygous and homozygous for the same mutant gene. *Journal of Heredity* **25**, 359–62.

—— (1935). A mutation of the guinea pig tending to restore the pentadactyl foot when heterozygous, producing a monstrosity when homozygous. *Genetics* **20**, 84–107.

Yallup, B. L. and Hinchliffe, J. R. (1983). Regulation along the anteroposterior axis of the chick wing bud. In *Limb morphogenesis and regeneration* (ed. J. F. Fallon). Alan R. Liss, New York.

Zwilling, E. (1949). Role of the epithelial components in origin of wingless syndrome of chick embryos. *Journal of experimental Zoology* **11**, 175–87.

—— (1956). Interaction between limb bud ectoderm and mesoderm in the chick embryo. IV. Experiments with a wingless mutant. *Journal of experimental Zoology* **132**, 241–53.

—— (1961). Limb morphogenesis. *Advances in Morphology* **1**, 301–30.

—— (1964). Controlled degeneration during development. In *Cellular injury* (ed. A. V. S. de Reuck and J. Knight), pp. 352–68. Ciba Foundation Symposium, Churchill, London.

—— (1966). Cartilage formation from so-called myogenic tissue of chick embryo limb buds. *Annales medicinae experimentalis et biologiae Fenniae* **44**, 134–9.

—— and Ames, J. F. (1958). Polydactyly, retarded effects and axial shifts, a critique. *American Naturalist* **92**, 257–66.

—— and Hansborough, L. A. (1956). Interaction between limb bud ectoderm and mesoderm in the chick embryo. III. Experiments with polydactylous limbs. *Journal of experimental Zoology* **132**, 219–39.

9. Genes and teratogens

This is obviously not an appropriate place to attempt a review of the field of teratology, the effect of physical or chemical factors on the developing fetus and the subsequent production of monsters. However two points of intersection between teratology and genetics are of importance here. The first is the question of phenocopies, the second that of interaction between genes and teratogens. These two entities are, in fact, different aspects of the same process.

The term phenocopy was coined by Goldschmidt to describe non-hereditary varieties of *Drosphila* imagos which he had produced by treating various larval stages with heat shocks; these conditions closely resembled known mutants of the fly. Similar conditions were described by Jollos and by Friesen after exposure of larvae to X-irradiation (for references see Landauer 1959). The resemblance between a whole series of these conditions and known hereditary variants was too great to be due to chance. Goldschmidt thought that the resemblance was due to the 'definite tracks' to which genetic and phenocopic agent were both limited. Landauer (1959) traced the subsequent history of the concept, and the idea of the 'true' phenocopy—produced when the stage of development at which a particular modification may be produced is equivalent to the time at which the development of the homologous mutant deviates from the normal course of development. This is not a useful distinction since we still know little of the real point of inception of mutant action. Consider also a mutation producing insufficient of an essential compound at a specific developmental stage. A deficiency of the same compound at a later stage may cause regression of already formed parts in normal embryos and, by identical physiological and morphological steps produce a phenocopy closely resembling the mutant. Furthermore the stage at which a given organ system is most vulnerable to teratogenesis probably reflects only this vulnerability; the same system is likely to be vulnerable to genetically mediated change at the same stage. An externally applied agent may also achieve mimicry by routes quite other than those of a corresponding mutation, or produce homologous ends by interference with quite dissimilar primary events.

Whenever an external agent interferes with the development of an organism it acts upon factors under genetic control, such as differential growth processes or enzyme-catalysed steps in differentiation. Evidence that agent and mutant often act on the same pathway may be summarized under five headings, but the caveats mentioned above must be borne in mind. The evidence does no more than indicate that conditions suitable for the expression of a mutant gene are also favourable to teratogens producing similar effects.

1. *Additive activity.* Mutant genes and teratogens which have similar phenotypic effects may be additive in effect.

2. *Shift of the heterozygous phenotype.* Animals heterozygous for a given mutant phenotype may be made to resemble the homozygous condition more closely by treatment with a teratogen. For example Landauer (1960) showed that the treatment of chick embryos with nicotine produces a shortening of the neck, due to the production of abnormal cervical vertebrae and that higher doses also effect the upper beak and lead to hypoplasia of skeletal muscle and oedema. This phenotype is similar to a recessive 'crooked neck dwarf'; heterozygotes for this gene have a much higher response to nicotine than normals. Embryos of the brown Leghorn chicken, where such anomalies are found spontaneously (presumably due to the presence of genes with low penetrance) also show a high susceptibility to nicotine.

3. *Increase of penetrance by teratogens.* The brown Leghorn experiment of Landauer (1960) mentioned above also serves as an example of this phenomenon. If a gene shows a low penetrance on a particular genetic background then the penetrance can often be increased by exposure to a suitable teratogen. Landauer (1957) was also able to increase the penetrance of genetically determined rumplessness in chick by the administration of insulin.

4. *Genetic modifiers of teratogenic effects.* The accumulation of genetic modifiers which will reduce the effect of a particular gene, by lowering its penetrance, will also reduce the effect of a phenocopying teratogen.

5. *The maternal environment as a modifier.* McLaren and Michie (1958) established the following facts. First, the inbred mouse strains C3H and C57BL differ in the number of lumbar vertebrae, C3H having a mode of five, C57BL six. Reciprocal hybrids resemble the maternal strain—the trait is therefore not sex-linked. Second the transfer of F_1 zygotes from reciprocal crosses into uteri of either inbred strain produced offspring with vertebral counts typical of the maternal stock. The strains clearly differ from each other by genetic factors determining, eventually, the number of lumbar vertebrae. The uterine environment overrides these genes. A teratogen, as part of the uterine environment is theoretically capable of having the same effect.

Phenocopy thus may be a useful concept, but is by no means a universal one. It is clear that the rubella embryopathy is a phenocopy of nothing; there is no known gene-directed process leading to the same phenotype. The destruction by surgery or X-irradiation of specific areas is also not a phenocopy, nor is the sex reversal of female fowls (Nachsteim 1957), produced when an ovary is destroyed by tubular infection or malignancy.

The important idea which emerges is this: the results produced by a teratogen depend on the genetic make-up of an individual. In our artificial breeding populations this will be closely controlled and uniform. In the real world it is uncontrolled and very variable. There is thus a possibility that a few per cent, or a fraction of one per cent, of individuals in a population has the appropriate genetic make-up to produce an abnormal child when exposed to a substance extensively tested for teratogenicity and passed as harmless. This is unfortunate, but not something that can be eliminated without enormous difficulty.

Teratogens in general show a dosage relationship so that a very low dose produces little effect, a slightly higher dose begins to produce abnormalities, and a higher dose still increases the severity and frequency of the defect until we reach a point where the applied agent acts as an embryonic, or indeed maternal lethal agent. The effect produced, for any given teratogen, also depends on the stage at which it is administered; a single teratogen may have two or more quite disparate effects at different stages of development. The wide view is that systems or areas developing most rapidly, or perhaps passing through a critical stage at the time of application are most affected. It is a corollary of this that teratogens are usually given within a closely defined period in order to limit their effects to a given system: we therefore rarely see a teratogen which affects all mesodermal condensations, all cartilage, all bone, although the potential to do so may be present. It follows that teratogens allied in some way to known genes will reflect areas of effect rather than tissues: the conditions described below are accordingly grouped according to effects in the same way as Chapters 5–8 are arranged: axial skeleton, face, teeth, limbs.

AXIAL SKELETON

Extra-toes (Xt, Chr 13) and trypan blue

Johnson (1970) was able to demonstrate interaction between trypan blue and the extra-toes locus in the mouse. This combination was chosen for two reasons. First the Xt/Xt phenotype included many abnormalities in common with trypan blue teratogenesis and, second, trypan blue has little effect on the limbs at the stage at which it is effective as a teratogen. The genotype of a particular individual could thus be easily ascertained from limb structure in segregating litters.

He found that trypan blue administered on day 8 of pregnancy reduced the number of surviving young from segregating matings, but did not affect segregation ratios. The teratogen increased the frequency of exencephaly and microphthalmia in homozygotes and induced them in heterozygotes and normals. Oedema was increased in frequency in homozygotes and in frequency and severity in heterozygotes. Some treated homozygotes developed a kinky neural tube, an extreme form of the wavy neural tube usually seen in this genotype.

The arrangement of mystacial vibrissae was unaffected by trypan blue, as was the development of the limbs; the upset of the former is thought to be due to the overgrowth of the pharyngeal arches. It is interesting to note that both pharyngeal arches and limb buds, although visible and abnormal at stages when the teratogen is known to be active are unaffected by its presence.

Curly-tail (ct, Chr ?), vitamin A, and cytotoxics

Seller *et al.* (1983) looked at the effect of vitamin A on mice carrying the curly-tail gene. Normally 60 per cent of homozygotes for curly tail have neural tube defects. Hybrids between curly tail on its normal background and Strong A mice have no neural tube defects. However, they can be produced in these +/ct mice by the maternal administration of vitamin A on the eighth day of pregnancy.

Seller *et al.* used reciprocal crosses between Strong A and ct, and also noted the sex of affected fetuses. They found significantly more fetuses had neural tube defects when the ct mouse was used as the mother, thus establishing an interaction between teratogen and maternal genotype (the genotype of the fetus is, of course, identical in each case except for sex chromosomes). They were also able to demonstrate that significantly more of the female +/ct fetuses had neural tube defects when the mother was ct/ct, but that the proportion in the reciprocal cross was unaffected. This shows, they maintained, a contribution from the fetal genotype to the liability to produce neural tube defects. Seller (1983) and Seller and Perkins (1983) treated a homozygous stock of ct mice with hydroxyurea, mitomycin C or 5-fluorouracil on day 9 of pregnancy. In untreated mice 60 per cent of offspring from ct/ct × ct/ct matings had exencephaly, spina bifida, or a curly tail, classified by Seller as a neural tube defect (NTD). Hydroxyurea in various doses *decreased* the amount of NTD without inducing other major malformations or increasing embryotoxicity. Mitomycin C and 5-fluorouracil had similar effects, although the highest dose of 5 FU used (0.8 mg/kg) was lethal. All of these chemicals are cytotoxic and it is surprising that they show a curative effect on ct. A similar effect was also shown by maternal administration of vitamin A (Seller *et al.* 1979). However, vitamin A, besides decreasing the neural tube defects in ct mice, also produces other wide-ranging embryological abnormalities and may act via an entirely different mechanism. Hydroxyurea, mitomycin C, and 5FU however are all inhibitors of DNA synthesis and secondarily of cell division. Seller supposed that curly tail mice have a dissynchrony of DNA synthesis and that the NTD is due to a specific deficiency at one step in the

DNA pathway; inhibition of DNA synthesis, short term at about the right stage of development, might allow a build-up of this supposed deficient metabolite to an acceptable level.

In man (Seller and Nevin 1984) periconceptional vitamin supplementation reduces the incidence of recurrence of neural tube defects in mothers who have already given birth to one affected child. In a large study based largely on London and Belfast (areas with low and high incidence of neural tube defects) they found that the recurrence rate was lessened by a factor of approximately 2 and 3.6, respectively. The recurrences of neural tube defect in pregnancies with vitamin supplementation were heavily biased towards males, i.e. the female fetus is more susceptible to environmental manipulation—exactly the effect noted in ct mice.

For a further discussion of neural tube defects in man the reader is referred to Dobbing (1983).

Crooked tail (Cd, Chr 6), rib fusions (Rf, Chr ?) and insulin

Cole and Trasler (1980) looked at the heterozygotes of both crooked and rib-fusions treated with teratogenic doses of insulin. The presence of either gene was found to predispose fetuses to exencephaly. The heterozygotes from insulin-treated litters also had abnormally shortened bodies and were considered to be phenocopies of Cd/Cd and Rf/Rf. Since insulin delays the closure of the neural tube, it was postulated that extensive delays in turning could provide a mechanical basis for the failure of the neural tube to close.

The T-locus, cyclohexamide, trypan blue, and actinomycin D

Lary *et al.* (1982) looked at the interaction of cyclohexamide and T-locus alleles. They found that the offspring of CD-1 × T/+ crosses and CD-1 × t^{w18} crosses treated with cyclohexamide (30 mg/kg on day 9) showed polydactyly, oligodactyly, and non-vertebral skeletal and visceral abnormalities. In all cases these exceeded the frequency of similar malformations seen in CD-1 × CD-1 crosses and the incidence of prenatal death was also increased. Lary concluded that the presence of either T ot t^{w18} predisposes fetuses to the effects of this protein synthesis inhibitor.

Hamburgh *et al.* (1970) found that low doses of trypan blue induced taillessness in T/+ mice without affecting normal littermates. Winfield and Bennett (1971) found that T potentiated the teratogenic effect of actinomycin D.

Splotch (Sp, Chr 1) and retinoic acid

Dempsey and Trasler (1983) treated mice carrying litters segregating for Splotch and control litters on day 8.5 with 50 mg/kg of retinoic acid. Their data clearly showed an increase in susceptibility to exencephaly and spina bifida in litters segregating for Splotch. Other non-Splotch-related defects (micrognathia, exophthalmos, anophthalmia) were more frequent in litters not segregating for Splotch.

CLEFT PALATE

Inbred strains H-2, and cortisone

Baxter and Fraser (1950) observed that cortisone was able to induce cleft palate in mice. It was subsequently noted (Fraser and Faintstat 1951; Walker and Fraser 1957) that certain strains (A/J, DB1) were susceptible and others (C57BL, CBA) resistant to this teratogen, and that C3H was intermediate (Kalter 1965). Bonner and Slavkin (1975) associated this susceptibility with the presence of certain H-2 genes. $H-2^a$ mice are susceptible but mice coisogenic except for $H-2^b$ are resistant (Table 9.1). Biddle and Fraser (1977) supported these findings. However non-H-2 genes are also involved. Kalter (1965) and Goldman *et al.* (1977) showed that CBA/J and C3H/HEJ (Table 9.1) differ in susceptibility although both carry the $H-2^k$ locus. Biddle and Fraser (1977) concluded that at least two or three genes govern susceptibility.

TABLE 9.1. *Frequency of cortisone-induced cleft palate in inbred strains of mice*

Strain	H-2	Susceptibility (per cent)
A/J	$H-2^a$	100
DBA/1J	$H-2^q$	92
C3H/HEJ	$H-2^k$	68
C57BL/6	$H-2^b$	20
B10	$H-2^b$	22
CBA/J	$H-2^k$	12
Congenic		
B10.A	$H-2^a$	81

After Slavkin (1978).

H-2 and phenytoin

The strain dependence of phenytoin-induced cleft palate (Massey 1966; Gibson and Becker 1968; Johnston *et al.* 1978) is similar to that of cortisone, A being susceptible and C57BL resistant. Phenytoin is chemically dissimilar to cortisone but Goldman *et al* (1968) suggested (Table 9.2) that at least part of the phenytoin effect resides in the H-2 locus.

TABLE 9.2. *Frequency of cleft palate due to phenytoin in various mouse strains*

Strain	H-2	Susceptibility (per cent)
A/J	H-2^a	50.0
DBA/lJ	H-2^q	1.5
C3H/HEJ	H-2^k	0.0
B.10	H-2^b	1.6
CBA/J	H-2^k	0.0
Congenic		
B10.A	H-2^a	22.2
B10.D2	H-2^d	0.0
B10.BR	H-2^k	0.0

After Goldman and Katsuama (1978).

Significance of H-2 involvement

Pla *et al.* (1976) showed that the proportion of cortisone-sensitive lymphoid cells is determined by H-2 linked genes. These genes thus control the number of cytoplasmic corticoid receptors. Goldman *et al.* (1977) showed that the binding protein of fetal palatal cytosols was correlated with susceptibility to cleft palate in mouse strains isogenic except at the H-2 locus. The synthetic ^{3}H-exametasone has technical advantages over ^{3}H cortisone, and Goldman and Katsumata (1978) showed that the binding of this substance is significantly different in strains of C57BL.B10A and A/J from that seen in C57BL.B10 itself.

Slavkin (1978) pointed out that the A genotype is susceptible to very many teratogens causing clefting (he lists 20) and that H-2^a, H-2^q, and H-2^k alleles seem to confer susceptibility to strains carrying them. The haplotype of the mother, rather than that of the fetus seems to be the key to susceptibility. Slavkin suggested that the maternal H-2 haplotype, which governs her ability to metabolize lyophilic compounds, may determine whether or not teratogenic compounds are passed to the fetus. He suggested in particular that inducible mixed-function oxidases, ironically important at an early stage in drug detoxification, may be the basis of this transformation.

On the other hand Nebert and Shum (1978) considered that the Ah cluster in mice (concerned with the regulation of polycyclic aromatic compound inducible P-450 mediated mono-oxygenase activities), which is correlated with aryl hydrocarbon toxicity *in utero*, may act either in the mother, as described by Slavkin, or, in a non-responsive mother, on fetal tissue. They considered that the genetic predisposition of a fetus towards metabolism of a specific compound is important in the aetiology of some birth defects; this could explain why a birth defect is found in one specific individual in a litter rather than its sibs.

The relationship between human cleft palate and corticoids is controversial (Bongiovanni and McPaden 1960; Rolf 1966; Popert 1962; Fraser 1962) but Teraski *et al.* (1970) showed that women with cytotoxic HLA antibodies had a significantly higher incidence of children with cleft palate than those without.

HYPERVITAMINOSIS A AND THE TEETH

Knudsen (1966) summarized the results on the dentition of the administration of 4–15 000 IU of vitamin A to AK mice on days 7–9 of pregnancy. This treatment produces exencephaly. Exencephalic embryos have abnormalities of the teeth. Lower incisors may be fused, totally or partially, or absent (Fig. 9.1). Molar germs were occasionally missing (first and second molars only were included in this study). It was difficult to tell if the missing m2 germs were upper or lower, as the m2 germ which was present

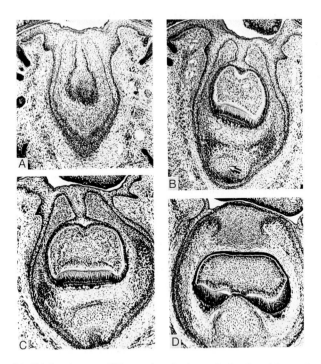

Fig. 9.1. (A–D) Sections at different levels through the fused lower incisors of an exencephalic mouse produced by exposure to hypervitaminosis A. (From Knudsen 1966.)

was often situated between the two jaws and directed laterally towards the cheek (Fig. 9.2). Knudsen thought that such teeth were the product of fusion between upper and lower molar germs. Abnormal molars were usually accompanied by abnormal incisors in the upper jaw. Knudsen was unable to decide whether the abnormalities were due to a direct effect of vitamin A, or were secondary to the cranial anomalies caused by the teratogen.

LIMBS

Polydactyly (py, Chr 1), TEM, and cyclophosphamide

Kocher and Kocher-Becker (1980) looked at the effect of TEM and cyclophosphamide (*Endoxan*) on the incidence of polydactyly in polydactyly (py) heterozygotes and their normal littermates. In parental strains C57BL and CBA they found the incidence of polydactyly was 16/4135 (0.39 per cent) and 0/3530 respectively. In py heterozygotes it was 36/1457 (4 per cent).

They found that TEM and cyclophosphamide administered at 10 days induced a py-like polydactyly in C57BL embryos not carrying py. In py heterozygotes polydactyly was induced by a much lower dose. The teratogens had no effect on CBA embryos, even at the embryological LD100—i.e. a lethal dose.

These results are classical. CBA will not respond even to a huge dose of teratogen since it has no genes for polydactyly, or their threshold is far too low for them to break through and no spontaneous polydactyly is therefore seen. C57BL has a low spontaneous level of polydactyly, indicating the presence of polydactyly genes acting near their threshold, and responds to the teratogen. +/py has an even greater susceptibility to polydactyly and responds to lower doses of the teratogen.

Polydactyly Nagoya (Pdn, Chr ?) 5FU and Ara-C

Naruse and Kameyama (1983) recently published most interesting findings on polydactyly Nagoya (Pdn). This semidominant gene (see Chapter 8) produces a single extra-preaxial digit in the hindlimb and a rudimentary extra phalanx of the first digit of the forelimb when heterozygous. In the homozygote there is polydactyly of all limbs, with duplicated or triplicated metacarpals/metatarsals.

Digital malformations including preaxial polydactyly can be produced in high frequency by 5-fluorouracil (5FU, Dagg 1963) and cytosine arabinose (Ara-C, Chaube *et al.* 1968) in mice and rats. Naruse and Kameyama combined these effects. Treatment of segregating litters on day 10 of pregnancy with a subteratogenic dose of either teratogen increased the

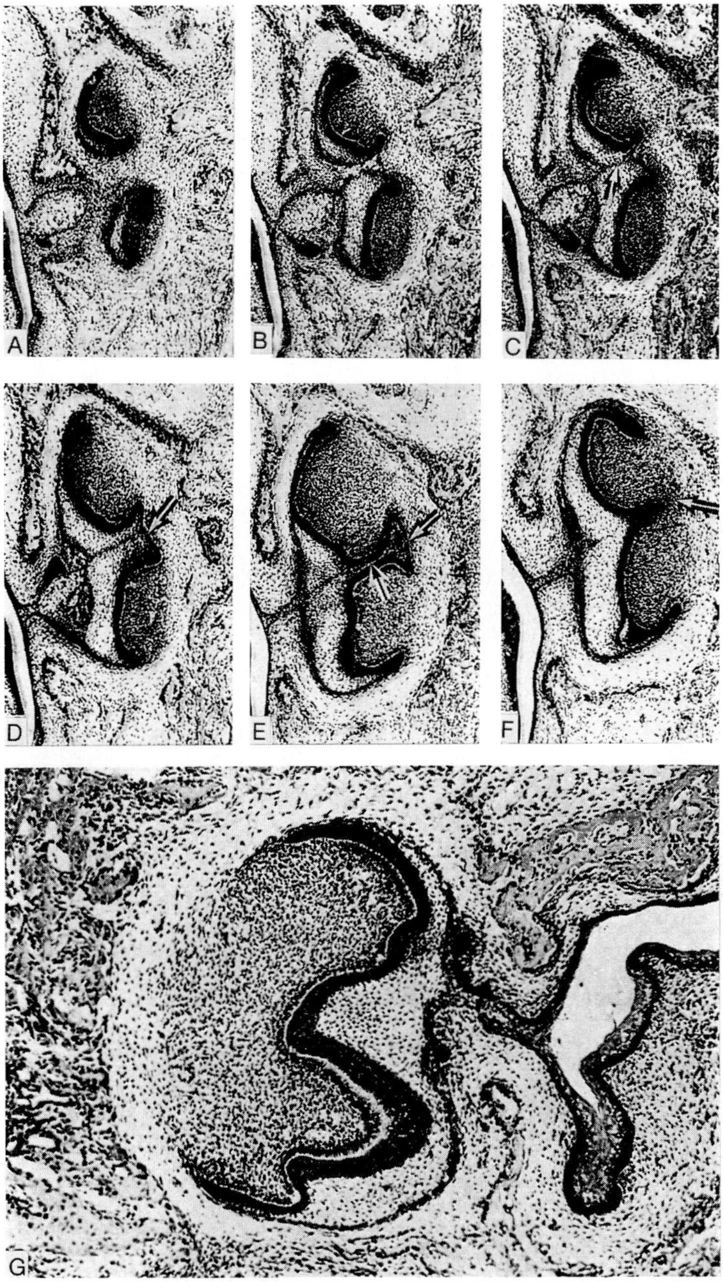

Fig. 9.2. (A–F) Sections at different levels through the partially fused upper and lower first molar germs of an exencephalic mouse produced by exposure to hypervitaminosis A. (G) In another individual upper and lower molar germs were completely fused. (From Knudsen 1966.)

expressivity of Pdn/+ (Fig. 9.3); +/+ embryos were not, of course, affected. A teratogenic dose on the same day produced polydactyly resembling that seen in Pdn/Pdn. Clearly the effect is as expected; the expressivity is increased by either teratogen.

Naruse and Kameyama (1984) extended these findings. In a further set of experiments they found that a subteratogenic dose of Ara-C also decreased the incidence of preaxial polydactyly in Pdn heterozygotes.

In Pdn/Pdn, however, the degree of polydactyly was reduced by a non-teratogenic dose of either teratogen. A teratogenic dose had the effect of

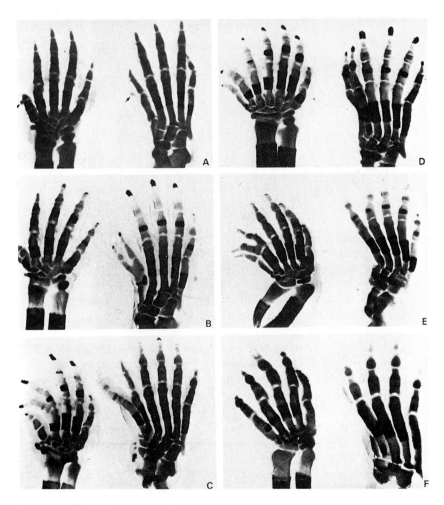

Fig. 9.3. (A,D) Non-treated Pdn/+ has its phenotype modified by doses of (B) 10 mg/kg; (C,E) 20 mg/kg of 5FU; or (F) 5 mg/kg Ara-C. (From Naruse and Kameyama 1983.)

reducing the number of extra digits and producing tibial hememelia. In fetuses treated with Ara-C, polydactyly disappeared in some cases and ectrodactylous limbs with involvement of the femur, fibula, and radius were produced.

This is interesting, as it demonstrates that a gene known to us as a simple mirror-image polydactyly is in fact a low expression of a luxate series gene. We may suppose that other simple polydactylys showing mirror-image symmetry will also turn out to be luxoids with low expressivities.

These results also fit neatly with the explanation of the hemimelic series proposed in Chapter 8. The most extreme effect of the gene/teratogen producing the syndrome is to inhibit the development of mesoderm producing ectrodactyly; a lesser effect is to delay the development so that the radius/tibia is involved. The mesoderm when produced is used to produce extra digits. The lowest level of the defect produces only a little extra mesoderm, perhaps enough to produce one extra digit.

Luxoid, (lu) luxate (lx), Strong's luxoid (lst) and 5FU

Dagg (1965, 1967) treated mice carrying luxoid (lu) and luxate (lx) with 5-fluorouracil (5FU) at 10 days of pregnancy. He found that in both strains the penetrance of the gene (0.6 for lx, 0.85 for lu on C57BL, Forsthoefel 1958) was increased by the treatment. The degree of abnormality was also increased by the teratogen, many heterozygotes showing tibial hemimelia, a characteristic of the homozygote, or of normals treated with much higher doses of 5FU.

Dagg (1972) found that 5FU did not interact with lu and Strong's luxoid (lst) if given on day 11 or 12 of pregnancy at the same dose causing interaction (recognized by the induction of hemimelia in heterozygotes) on day 10. He concluded that if the chemical had the same effect on days 10, 11, and 12, when it is known to be teratogenic then the effect of lx, lst on the limb was complete by day 11.

Forsthoefel (1972) also defined the critical period of treatment with 5FU as 10 days, finding little or no effect at 8 or 9 days, and only a reduction in digital number on day 11 with a high dose of the teratogen. Increasing the dose in general brought about more extreme manifestations on day 10, and allowed them to be recognized on day 9 and 11. Modifying genes for lst also responded to the teratogens; in the normal × normal cross, where lst was absent, the strain carrying + modifiers for lst was more affected by the teratogen that that carrying − modifiers.

Forsthoefel and Williams (1975) also looked at C57BL mice carrying Strong's luxoid (lst) and selected for plus and minus modifiers of the gene. They subjected these lines to 10–20 mg/kg 5FU and/or 5FUDR on day 10 of gestation. In a rather complicated set of experiments they demonstrated that the co-treatment of mice, receiving 5FU with equimolar amounts of thymidine, thymine, or uracil, greatly increased the teratogenic effects of the former whereas thymidine protected against 5FUDR and thymine and uracil had little effect. In the main, a modifier increasing the expression of lst also increased its sensitivity to teratogens and vice versa. Again the effect of the teratogen is to increase the expression of the gene, with heterozygotes showing forefoot polydactyly and hemimelia normally seen in homozygotes, and the hemimelia of homozygotes becoming exaggerated.

Brachydactyly (br) in the rabbit, hypoxia, phenylhydrazine and folic acid

Brachydactyly in the rabbit (Greene and Saxton 1939; Jost *et al.* 1969) is an autosomal recessive characterized by congenital limb amputations resulting from haemorrhagic blebs which develop on the limbs between 16 and 21 days of development (Inman 1941).

Petter *et al.* (1971) were able to prevent these spontaneous amputations by subjecting the mother of br/br offspring to hyperoxia, suggesting that hypoxia might be involved in the genesis of the abnormality. However Petter *et al.* (1973) were also able to demonstrate a haematological defect in br/br fetuses; they had an excessive number of large red corpuscles ($> 12\,\mu$m diameter). The injection of penylhydrazine into pregnant females completely suppressed these large blood cells and protected about half the br fetuses against amputations. Petter *et al.* (1977) investigated the defect in br/br and found that thrombosis in limb vessels was the probable cause of ischaemia and the local hypoxia, leading to oedema, haemorrhage, and necrosis of the extremities. Polycyclic blebs in the liver were also considered to be due to hypoxia. In the fetal liver, erythropoiesis was also affected, but the cause-and-effect relationship between this and hypoxia was unclear. Petter *et al.* speculated that the large number of large rbcs present in br/br may be due to intense haemopoietic activity by the yolk sac, whose products are normally larger in diameter than those of the liver. A similar process was described (Borghese 1952; Russel *et al.* 1957) in the W/W^v mouse. The large cells appear to obstruct small vessels, perhaps as a consequence of their size, number, poor deformability, or clumping properties.

Folic acid on days 9–14 of pregnancy reduced macrocytosis, but fetuses still had haemorrhagic blebs. Vitamin B_{12} had no effect, but a combination of folic acid and vitamin B_{12} reduced the number of large rbcs by 65 per cent at 5 mg+500 μg and by almost 100 per cent at twice this dose. The treated fetuses were without amputations, but had slight haemorrhages or clear blebs at the tips of their fingers. Folic acid had a similar, though lesser action.

The implication of this work is clear; the defect in br could be one of folate metabolism which is cured by vitamin supplementation. The blebs and haemorrhages are clearly consequent upon the abnormality of the blood.

REFERENCES

Baxter, H. and Fraser, F. C. (1950). Production of congenital defects in the offspring of female mice treated with cortisone. *McGill Medical Journal* **19**, 245–9.

Biddle, F. G. and Fraser, F. C. (1977). Cortisone induced cleft palate in the mouse. A search for the genetic control of the embryonic response. *Genetics* **85**, 298–302.

Bongiovanni, A. M. and McPaden, A. J. (1960). Steroids during pregnancy and possible fetal consequences. *Fertility and Sterility* **1**, 181–6.

Bonner, J. J. and Slavkin, H. C. (1975). Cleft palate susceptibility linked to histocompatibility-2 (H-2) in the mouse. *Immunogenetics* **2**, 213–18.

Borghese, E. (1952). Richerche embryologiche ed istologische sui mutanti del locus W nei topi. *Genetica* **3**, 86–118.

Chaube, S., Kreis, W., Uchida, K., and Murphy, L. M. (1968). The teratogenic effect of 1-beta-D-arabinofurasylcytosine. *Biochemical Pharmacology* **17**, 1213–16.

Cole, W. A. and Trasler, D. G. (1980). Gene-teratogen in insulin-induced mouse exencephaly. *Teratology* **22**, 125–39.

Dagg, C. P. (1963). The interaction of environmental stimuli and inherited susceptibility to congenital deformity. *American Zoologist* **3**, 223–33.

—— (1965). Effects of fluorouracil on the penetrance of two skeletal mutants in mice. *Anatomical Record* **151**, 341.

—— (1967). Combined action of fluorouracil and two mutant genes on limb development in the mouse. *Journal of experimental Zoology* **164**, 479–90.

—— (1972). Independent effects of fluorouracil and two mutant genes in mouse embryos. *Teratology* **5**, 377–82.

Dempsey, E. E. and Trasler, D. G. (1983). Early morphological abnormalities in Splotch mouse embryos and predisposition to gene and retinoic acid induced neural tube defect. *Teratology* **28**, 461–72.

Dobbing, J. (ed.) (1983). *Prevention of spina bifida and other neural tube defects.* Academic Press, London.

Forsthoefel, P. F. (1958). The skeletal effects of the luxoid gene in the mouse including its interaction with the luxate gene. *Journal of Morphology* **102**, 247–88.

—— (1972). The effects on mouse development of interactions of 5-fluorouracil with Strong's luxoid gene and its plus and minus modifiers. *Teratology* **6**, 5–18.

—— and Williams, M. L. (1975). The effects of 5-fluorouracil and 5-fluoro-deoxyuridine used alone and in combination on development of mice in lines selected for low and high expression of Strong's luxoid gene. *Teratology* **11**, 1–20.

Fraser, F. C. (1962). Pregnancy and adrenocortical hormones. *British Medical Journal* **2**, 479.

—— and Faintstat, T. D. (1951). Production of congenital defects in the offspring of pregnant mice treated with cortisone. *Pediatrics* **8**, 527–33.

Gibson, J. E. and Becker, B. A. (1968). Teratogenic effects of diphenylhydantoin in Swiss Webster and A/J mice. *Proceedings of the Society for experimental Biology and Medicine* **128**, 905–9.

—— —— and Gasser, D. L. (1968). Fetal hydantoin syndrome: susceptibility to phenytoin-induced cleft (lip) and palate linked to H-2 locus in mice. *Pediatrics Research* **12**, 517.

Goldman, A. S. and Katsumata, M. (1978). The genetics of clefting in the mouse: a critique. In *Etiology of cleft lip and palate* (ed. M. Melnick, D. Bixler, and E. D. Shields), pp. 91–120. Alan R. Liss, New York.

—— —— Yaffe, S. J., and Gasser, D. L. (1977). Palatal cytosol cortisol-binding protein associated with cleft palate susceptibility and H-2 genotype. *Nature* **265**, 643–5.

Greene, H. S. N. and Saxton, J. A. (1939). Hereditary brachydactylia and allied anomalies in the rabbit. *Journal of Embryology and experimental Morphology* **69**, 301–14.

Hamburgh, M., Herz, R., and Landa, G. (1970). The effect of trypan blue on expressivity of the brachyury gene T in mice. *Teratology* **3**, 111–18.

Inman, O. R. (1941). Embryology of hereditary brachydactyly in the rabbit. *Anatomical Record* **79**, 483–505.

Johnson, D. R. (1970). Trypan blue and the extra-toes locus in the mouse. *Teratology* **3**, 105–10.

Johnston, M. C., Sulik, K. K., and Dudley, K. H. (1978). Phenytoin (diphenyl-hydantoin, dilantin)-induced cleft lip and palate in mice. *Teratology* **17**, 29A.

Jost, A., Roffi, J., and Courtat, M. (1969). Congenital amputation determined by the br gene and those induced by adrenalin injection in the rabbit fetus. In *Limb development and deformity* (ed. C. A. Swingard), pp. 187–99. Thomas, Springfield, Illinois.

Kalter, H. (1965). In *Teratology, principles and techniques* (ed. J. G. Wilson and J. Warkany, pp. 57–95. University of Chicago Press.

Knudsen, P. A. (1966). Congenital malformations of the lower incisors and molars in exencephalic mouse embryos induced by hypervitaminosis A. *Acta Odontologica Scandanavica* **24**, 55–71.

Kocher, W. and Kocher-Becker, U. (1980). Bilaterally symmetric realisation of limb malformations. In *Teratology of the limbs* (ed. H-J. Merker, H. Nau, and D. Neubert), pp.259–72. de Gruyter, Berlin.

Landauer, W. (1957). Phenocopies and genotype, with special reference to sporadically occurring developmental variants. *American Naturalist* **91**, 71–90.

—— (1959). The phenocopy concept; illusion or reality? *Experientia* **15**, 409–12.

—— (1960). Nicotine induced malformations of chicken embryos and their bearing on the phenocopy problem. *Journal of experimental Zoology* **143**, 107–22.

Lary, J. M., Hood, R. D., and Lindahl, R. (1982). Interactions between cyclohexamide and T locus alleles during mouse embryogenesis. *Teratology* **25**, 345–9.

Massey, K. M. (1966). Teratogenic effects of diphenylhydantoin sodium. *Journal of Oral and Theoretical Pharmacology* **2**, 380–5.

McLaren, A. and Michie, D. (1958). Factors affecting vertebral variation in mice. 4. Experimental proof of the uterine basis of a maternal effect. *Journal of Embryology and experimental Morphology* **6**, 645–59.

Nachsteim, H. (1957). Mutation and Phanokopie bei Saugetier und Mensch. *Experientia* **13**, 57–68.

Naruse, I. and Kameyama, T. (1983). Effects of 5-fluorouracil and cytosine arabinoside on the manifestation of digital malformations in Pdn mice. *Congenital Anomalies* **23**, 211–21.

—— —— (1984). Prevention of genetic expression of polydactyly in heterozygotes of polydactyly Nagoya (Pdn) mice by Cytosine arabinoside. *Environmental Medicine* **28**, 89–92.

Nebert, D. and Shum, S. (1978). The murine Ah locus: genetic differences in birth defects among individuals in the same uterus. In *Etiology of cleft lip and cleft palate* (ed. M. Melnick, D. Bixler, and E. D. Shields), pp. 173–96. Alan R. Liss, New York.

Petter, C., Bourbon, I., Maltier, J. P. and Jost, A. (1971). Prévention des amputations congénitales héréditaires du lapin par une hyperoxie maternelle.

Comptes rendus de l'Académie des Sciences **273**, 2639–42.

—— —— —— —— (1973). Hématies primordiales et amputations congénitales chez les fetuses de lapin porteurs du gène br. *Comptes rendus de 'Académie des Sciences* **277**, 801–3.

—— —— —— —— (1977). Simultaneous prevention of blood abnormalities and hereditary congenital amputations in a brachydactylous rabbit stock. *Teratology* **15**, 149–58.

Pla, M., Zakany, J. and Fachet, J. (1976). H-2 influence on corticosteroid effects on thymus cells. *Folia Biologica, Praha* **22**, 49–50.

Popert, A. J. (1962). Pregnancy and adrenocortical hormones. *British Medical Journal* **1**, 967–71.

Rolf, B. B. (1966). Corticosteroids and pregnancy. *American Journal of Obstetrics and Gynaecology* **95**, 339–44.

Russel, E. S., Lawson, F., and Schbtach, G. (1957). Evidence for a new allele at the W-locus of the mouse. *Journal of Heredity* **48**, 119–23.

Seller, M. J. (1983). The cause of neural tube defects—some experiments and a hypothesis. *Journal of Medical Genetics* **20**, 164–8.

—— and Nevin, C. (1984). Periconceptional vitamin supplementation and the prevention of neural tube defects in south east England and Northern Ireland. *Journal of Medical Genetics* **21**, 325–30.

—— and Perkins, K. J. (1983). Effect of hydroxyurea on neural tube defects in the curly tail mouse. *Journal of Craniofacial Genetics and Developmental Biology* **3**, 11–17.

—— Embury, S., Adinoff, M., and Polani, P. E. (1979). Neural tube defects in curly tail mice. II. Effects of maternal administration of vitamin A. *Proceedings of the Royal Society, London* **B206**, 95–107.

—— —— —— and Adinoff, M. (1983). Differential response of heterozygous curly tail mouse embryos to vitamin A teratogenesis depending on maternal genotype. *Teratology* **28**, 123–9.

Slavkin, H. C. (1978). Major histocompatability complex (H-2)-linked genes affecting teratogen induced congenital craniofacial modifications in mice. In *Etiology of cleft lip and palate* (ed. M. Melnick, D. Bixler, and E. D. Shields), pp. 121–47. Alan R. Liss, New York.

Teraski, P. I., Mickey, M. R., Yamazaki, J. N., and Vredevoe, D. (1970). Maternal–fetal incompatability. I. Incidence of HL-A antibodies and possible association with congenital anomalies. *Transplantation* **9**, 538–43.

Walker, B. E. and Fraser, F. C. (1957). The embryology of cortisone-induced cleft palate. *Journal of Embryology and experimental Morphology* **5**, 201–9.

Winfield, J. B. and Bennett, D. (1971). Gene–teratogen interaction: potentiation of actinomycin D teratogenesis in the house mouse by the lethal gene brachyury. *Teratology* **4**, 157–69.

10. Conclusion: The validity of animal models

When we look at a mutation in a laboratory animal we do so with various purposes in mind. We may consider that the condition is of interest in its own right, or that the malformation's novel arrangement will tell us something about normal development. Alternatively, we may change our viewpoint a little and suppose that the mutation will give us a good experimental system which will enable us to perform experiments not feasible in man and hence give an insight into a disease of man rather than one of mice. The first viewpoint is obviously valid, although we cannot guarantee the information gained is worth the effort expended; all science is something of a lottery. The second viewpoint embraces the idea that a disease of mouse, or rat, or chick is homologous with a disease of man. Is this a reasonable assumption?

The best animal models certainly fulfil this criterion. They are also all of a clearly distinguishable type. In an increasing number of inherited animal models of disease a defective enzyme has been identified and found to be chemically identical to the enzyme affected by a disease of man. Migaki (1982) produced a compendium of disorders with 'inborn errors of metabolism' usually arising from a specific enzyme defect, or a disorder of transport or synthesis of a particular metabolite. The list is extensive and includes, for example, diabetes mellitus from mouse, cat, rat, hamster, guinea pig, *Mastomys*, monkey, carp, and dog, and glycogenesis II from cat, sheep, dog, cow, and quail.

But when we look at conditions, not simply or exclusively linked with a known biochemical defect, such as the chondrodystrophies, the osteo-petroses, or even the polydactylys, we see that at least one new link of organization has been added to the chain. The relationship defective enzyme $\rightarrow$ inherited disease is the shortest possible one. If we are dealing with a disease of development rather than one of simple biochemistry the chain is lengthened and we see that $? \rightarrow$ symptoms. The symptoms may be similar or identical but the ? may represent a finite, or as far as we know, infinite series of processes. Think of the number of known defects which can produce chondrodystrophy, as listed in Chapter 3. These include: collagen defect, link protein defect, mucopolysaccharide backbone defect, sulphation defect, cell division defect, oxidative phosphorylation defect, cell death defect. The list is, of course, incomplete. All these produce chondrodystrophy because they all inhibit cartilage growth. No mouse chondrodystrophy is necessarily a good model for any known human chondrodystrophy, any or all of which may have causes not yet described in mouse or chick. Until we can biochemically define the chondrodystrophy in a single affected person we cannot ascribe him or her to a specific disease modelled by a specific condition in a laboratory animal. Once we can do so, of course, the need for the model partially disappears.

This general problem is particularly acute in the case of skeletal disorders. Skeletal structures are, of their very nature, complex. Grüneberg (1975) was disenchanted when he realized that the 'skeletal gene' for which he had sought for 40 years does not, in fact, exist. Genes affecting the skeleton do so secondarily. He quoted as examples vestigial tail, tail-short, and congenital hydrocephalus and ascribes their defects to the primitive streak, yolk sac haemopoiesis, and 'some generalized defect of the mesenchyme', respectively. Although I would not agree with his interpretation of Ts the examples serve. At this point Grüneberg stated that although these are all skeletal genes they tell us nothing of the development of the skeleton. I cannot agree with this statement. It seems extremely unlikely that there *are* skeletal genes *per se*, any more than there are liver genes, hand genes, or nose genes, yet we know that the development of all these organs is under genetic control. Genes are acting at, or have perceived activities at, various levels. We know of genes which affect a single enzyme, because its absence often leads to failure of a vital metabolic step. We know of genes which change the surface properties of cells, and can only imagine that they do so by disturbing the macro-molecules normally resident there. If this interference is simply physical, although with a chemical basis, and cell adhesion is modified, then the defects observed at a gross level, as in talpid, brachypod, or Ts, tell us that a close control of cell adhesion is a necessary prerequisite in some places and at some times in the developing embryo. The effect on the skeleton may be secondary, but is far from being therefore trivial. It leads us on, it adds another letter to the crossword puzzle, another tile to the Scrabble board. Some clues to the problem of morphogenesis must reside at the cell surface.

The central problem of animal models has changed. In 1947 Grüneberg saw it as one of communication, of making sure that clinicians knew what was available and how to deal with it. Now we see that the concept of a small number of diseases and congenital defects represented by a small number of genetic models was over simplistic. Close study of the diseases has shown that the same syndrome, previously classified as a disease, is in fact a portmanteau entity covering many possible first causes. The more we understand these conditions the more are we likely to be able to dissect them into separate entities. Exactly the same process governs research on mutants; the more we look at them the more disparate will a group of conditions become. None of the mutants will necessarily correspond to any of the human conditions. We must also ask ourselves the converse question. If a given disease, with a common cause, were to arise in several species would the final facies be identical? Probably not. The identical defect, in mucopolysaccharide link protein in mouse and chick, gave phenotypes which were not even particularly similar, let alone identical. In biochemical defects we can apply the acid test and prove homology. In defects of unknown aetiogy we classify by appearance, and not necessarily in a logical manner. Syndactyly in mouse and cow may well be due to similar or identical processes; the resulting condition, although recognizable as syndactyly has important differences imposed by foot structure.

The use of animal models which we cannot classify as first-order biochemical defects are thus not necessarily of much use in medicine. What of their use in science? Here they are much more at home. We need not worry if a particular mouse osteopetrosis corresponds to a particular human condition; instead we can say 'how interesting, this osteopetrosis is based a defect in. . .' and is perhaps quite different from its fellows. From this approach we shall learn more.

REFERENCES

Grüneberg, H. (1975). How do genes affect the skeleton? In *New approaches to the evolution of abnormal embryonic development* (ed. D. Neubert and H-J. Merker), pp. 354–9. George Thieme, Stuttgart.

Migaki, G. (1982). Compendium of inherited metabolic diseases in animals. In *Animal models of inherited metabolic diseases*, pp. 473–50. Alan R. Liss, New York.

Appendix 1. Stockholders of genetic material

UNITED KINGDOM

Queries should be addressed to:

Medical Research Council Laboratories, Woodmansterne Rd, Carshalton, Surrey SM5 4EF, England.

or

Medical Research Council Radiobiology Unit, Harwell, Didcot, Oxon OX11 ORD, England.

UNITED STATES

Queries should be addressed to:

Jackson Laboratories, Bar Harbor, Maine 04609, USA.

SOURCES OF INFORMATION

Mouse News Letter. Distributed twice yearly by the Jackson Laboratory, Bar Harbor, Maine 04609, USA, and MRC Laboratories, Woodmansterne Rd, Carshalton, Surrey SM5 4EF, England.
  Inbred strains of mice—distributed with *Mouse News Letter*.
  Standardized nomenclature for inbred strains of mice. Appears each fourth year in *Cancer Research*.
  List of mutations and mutant stocks of the mouse. Distributed by the Animal Resources Department, Jackson Laboratory.

Appendix 2. Comparative times of development for mouse, rat, and chick

All times are approximate and tend to vary according to strain.

Developmental stages

	Age in days		
	Mouse	Rat	Chick
Primitive streak	7	> 8.5	0.5
Neural plate	7.5	—	0.5
Heart primordium	7.5–8.3	8.5	1.25
Anterior neuropore closes	9–10	10.5–11	1.5–2.0
Posterior neuropore closes	9.5–10.0	11.75	—
Anterior limb bud arises	9.0–9.5	11	—
Posterior limb bud arises	10.0–10.5	11.75–12.0	< 3
First trace of forepaw digits	12.0	> 14.0	—
Digits fully separated	15.0	16.75	—

Somites

Number of somites	Age in days		
	Mouse	Rat	Chick
1	8.0	9.0	1.0
5	8.5	9.5	1.0+
10	8.5+	9.0–10.0	1.25–1.5
15	9.0	—	1.5
20	9.0–10.0	10.0–11.0	1.75–2.0
25	9.5–10.0	11.0–12.0	2.0+
30	10.0–10.5	11.0–11.5	2.25–2.5
35	10.5	12.0	—
40	11.0	12.0–12.5	3.0–3.25
45	11.5–12.0	13.5	3.0–4.5
50	12.0–12.5	—	4.0+
60	13.0+	14.75	—
65	15.0	—	—

REFERENCES

Edwards, J. A. (1968). The external development of the rabbit and rat embryo. In *Advances in teratology* (ed. D. H. M. Woolam), pp. 239–63. Academic Press, New York.

Gruneberg, H. (1943). The development of some external features in mouse embryos. *Journal of Heredity* **34**, 88–92.

Hamburger, V. and Hamilton, H. L. (1951). A series of normal stages in the development of the chick embryo. *Journal of Morphology* **88**, 49–92.

Ogawa, T. (1967). Comparative study on development in the stage of organo-genesis in the mouse and rat. *Congenital Anomalies* **7**, 27–31.

Otis, E. M. and Brent, R. (1954). Equivalent ages in mouse and human embryos. *Anatomical Record* **120**, 33–63.

Patten, B. M. (1951). *Early embryology of the chick*, 4th edn. McGraw-Hill, New York.

Schneider, B. F. and Norton, S. (1979). Equivalent ages in rat, mouse and chick embryos. *Teratology* **19**, 273–8.

Theiler, K. (1972). *The house mouse* (Development and normal stages from fertilization to four weeks of age). Springer-Verlag, Berlin.

Witschi, E. (1956). *Development of vertebrates*. W. B. Saunders, Philadelphia.

Index

Abbreviations

cAMP cyclic adenosine monophosphate